INDUSTRIAL LOCOMOTIVES

including preserved and
minor railway locomotives

HANDBOOK

18EL

INDUSTRIAL RAILWAY SOCIETY
A Charitable Incorporated Organisation
Registered Charity (Engalnd & Wales) No. 1177413
www.irsociety.co.uk

Published by the INDUSTRIAL RAILWAY SOCIETY
at 24 Dulverton Road, Melton Mowbray, Leicestershire, LE13 0SF

ISBN 978 1 901556 99 5 (hardbound)

ISBN 978 1 912995 00 4 (softbound)

British Library Cataloguing-in-Publication Data
A catalogue record for this book is available from the British Library

Visit the Society at www.irsociety.co.uk

The content of this book is based on records held by the Industrial Railway Society, originally created by the late Eric Hackett. These records have now been continuously updated over the years by Alex Betteney, Ian Bendall, John Beechey, Colin Billinghurst, Adrian Booth, Bob Darvill, Ted Knotwell, Mick Morgan, George Morton, and Andrew Waldron.

The Society obtains update information from a wide variety of sources, but primarily from reports of visits and other observations submitted by our readership, both members and non-members, plus some information kindly supplied by locomotive manufacturers, owners and operators. The publishers are extremely grateful for all this essential co-operation, without which production of our Handbooks would be impossible. All readers are encouraged to submit data, correctional, additional, or confirmatory, to the most seemingly appropriate Assistant Records Officer (see list on page 5) or by email to : hist.records@irsociety.co.uk

Production co-ordinated by Alex Betteney

Distributed by IRS Publications, 24 Dulverton Road, Melton Mowbray, Leicestershire, LE13 0SF

Produced for the IRS by Print Rite, Witney, Oxon. 01993 881662.

CONTENTS

Cover Photographs :

Front Cover – 'SACCHARINE' JF 13355 of 1912 pauses outside the locomotive shed at the Statfold Barn Railway. [Photo : Statfold Barn Railway]

Rear Cover – 'RYAM SUGAR CO 1' Davenport 1650 of 1918 calls at Oak Tree Halt, Statfold Barn Railway on 18th April 2015 during an IRS AGM meeting. [Photo : Statfold Barn Railway]

The text of this book incorporates all amendments notified before 1st December 2018

18EL – 50 YEARS

Fifty years of publishing "Industrial Locomotives"

With the end of steam locomotives on the main line in 1968, attention and cameras were turned towards the industrial steam locomotives that were still at work with UK industry. At the time, very little information was available to those outside the Society who were interested in industrial railways and locomotives so it was decided to publish details of the existing industrial locomotives in the UK and make it available to a wider audience.

The very first "British Industrial Locomotives" published in January 1969 was in two parts, stapled and a total of 248 pages. It was edited by Roy Etherington, Peter Excell and Eric Tonks. Clearly emblazoned on the front cover was the proud proclamation "First Edition". Did they know then that we would continue to publish a new edition every three years and celebrate 50 years with this 18th edition in 2019?

From 2 EL the title changed to "Industrial, Preserved, and Minor Railway Locomotives in Great Britain". 3EL "Industrial Locomotives 1973 was sub-titled "Including Preserved and Minor Railway Locomotives" and included contractors locomotives for the first time. Editions appeared in 1976, 1979, and 1982. From 7EL the year was dropped from the title and industry began to sponsor the cover. From 10EL hardback as well as softback was offered.

By 11 EL computers and word-processors were assisting the production team but not until 13EL did county and company titles appear in bold type. The word-processor was a god-send for the production team, if a company changed its name – just lift it out of the list and move it to the appropriate new position in the text. Whereas before, when the text was produced on a type-writer, changes and amendments resulted in vast sections being retyped manually.

Now expanded to 400 pages plus, this increase was a result of the amount of information included. Colour photographs are now the norm.

The first locomotive listed in 1EL (page 2) is RH 321727 which I am pleased to say is still with us and appears in 18EL. But how industrial railways have changed in these 50 years! The steel industry was nationalised and, later, what remained was privatised. NCB underground locomotives were not included until 12EL but sadly the last deep mines are recorded in 17EL and now the entire coal mining industry has all but gone. The end of wagon load freight on the main line saw the demise of many private sidings and hence the industrial locomotives that worked there.

Over the 50 years the biggest change has been in preservation and the number of locomotives saved. Dedicated individuals and groups raising funds to save a locomotive from scrap, saving rolling stock and indeed entire railways. Heritage sites and railways now form an integral part of the leisure and tourist market.

The gathering of all the information contained within our publications would not be possible without all those who volunteer as records officers to collate and maintain the various categories of information, those who help prepare the amendments, select the photographs, check the final draft etc. But this would not be possible without the support of our members and others who sent in their reports of visits, observations and changes to the industrial railways. Thank you.

Ian R Bendall

Chairman.

Industrial Railway Society

FOREWORD to the 18th Edition

This is the 18th in a line of EL (Existing Locomotive information as opposed to historical) Handbooks, which commenced publication in 1968. We are therefore extremely proud to celebrate 50 years of continuous production which would not have been possible without our members visiting the sites and submitting their observations to enable us to keep our records as up to date and comprehensive as possible. The lists include all known existing locomotives of 1ft 3in gauge and above, but excluding the Capital Stock of London Transport and the Train Operating Companies (ex British Railways). In addition, several significant monorail locomotives are also listed.

The lists are in accordance with our records for December 1st, 2018. We wish to stress that the majority of the information derives from the observations and researches of both members and other readers. We shall always rely on similar enthusiastic support to keep all the records up to date, so please send your observations (and your queries also) to whoever you judge to be the most appropriate Hon. Assistant Records Officer selected from the list below.

Or email our sorting office at : hist.records@irsociety.co.uk

All relevant information received is subsequently distributed to members of the Industrial Railway Society by means of regular Bulletins, thus enabling the reader to keep this book up to date.

Current Observations
I.R. Bendall (Asst. Hon. Records Officer)
25 Byron Way
Melton Mowbray
Leicestershire
LE13 1NY

All historical matters
R. Waywell (Hon. Records Officer)
29 Caldbeck Close
Gunthorpe
Peterborough
PE4 7NE.

Preservation (Current & Historical)
M. Morgan (Asst. Hon. Records Officer)
29 Groby Road
Ratby
Leicestershire
LE6 0LJ

Ireland
A.J. Waldron (Asst. Hon. Records Officer)
Riverside
1 Hazel Pear Close
Fairlawns
Horwich
Bolton
BL6 5GS

The Society also has Assistant Records Officers who collate information which is appropriate to the various industrial railway subjects such as MOD, rolling stock etc. Information supplied to any of the above officers will be circulated to the appropriate officer. All information supplied is ultimately deposited in our archive.

INTRODUCTION

Within the main Country Sections of this book, information is presented in a similar format to that used in previous editions, with locomotives listed under owners' titles arranged alphabetically, within sub-groups of Industrial and Preservation Sites. Within these sub-groups, each location is sequenced alphabetically, using its main title, or surname if a person's name is included.

LOCATION HEADINGS include both National Grid References and postcodes where known; details of missing grid references and/or postcodes will be most welcome.

Postcodes and Grid References are particularly useful for readers with Internet access, as either code can be requested at one of the free mapping sites (such as www.streetmap.co.uk or http://www.multimap.com) to obtain a printable map of the precise location of the railway site.

In general, we try to fix the Grid Reference/s to indicate the locomotive shed or stabling point, particularly at the larger sites.

DIVISION OF COUNTRIES. In recent years there have been a number of reorganisations of the Administrative Areas in England, Scotland and Wales, in some cases the traditional "geographic counties" have been re-mapped and divided to create small independent areas now administered by Unitary Authorities of one sort or another. In order to avoid fragmenting our listing data, in this edition we have amalgamated some listings under headings of their "most recent" larger-landmass titles (many of which are still in current use for certain administrative purposes).

In the case of Wales and Northern Ireland, which have in fact relatively few locomotive locations, we have re-listed the entries on an Land Area basis, supported by maps indicating the geographical position of the locations and by extending the location titles to include the name of the actual Authority in which the site is now located. The boundaries of the Land Areas have been selected to coincide with the coverage of books in our companion "Historic Handbook" series. We hope that readers will find this arrangement more convenient to use and, if so, it may be extended to some other areas in the next edition.

Locations are indicated on such maps by numbers, which correspond to the serial numbers that appear immediately to the right of the location titles.

LOCOMOTIVES AT SCRAPYARDS purely for scrapping are not generally listed, unless they have been 'resident' for a year or more. Those at dealers' yards are, however, listed in order that readers can keep a full picture of their movements.

CONTRACTORS' LOCOMOTIVES can be found on sites in all parts of the country, working on sewer and tunnel schemes, etc. The locomotives of such firms are shown, in the usual way, in the list of the County in which the firm's plant depot (or main plant depot if more than one is in use) is situated. A list of such Contractors and Plant Hire specialists and their relevant base counties appears later in this Introduction, following the table listing Locomotive Builders.

LOCOMOTIVE OWNERSHIP. The information within this publication does not confirm ownership of any of the locomotives listed, only their physical presence at a location. When a locomotive is sold, the transaction may have taken place sometime before the date that we record the locomotive moving to another location.

> **Industrial/Commercial location:** Some locomotives may be hired or leased from third parties. Where the hirer or leasing company is known, this information is shown in a footnote.
> **Preservation Sites:** Locomotives may be owned by individuals, groups, trusts, etc and may not be the property of the host site under which they are listed.

HIRE FLEET OPERATORS. A small number of firms maintain a fleet of locomotives which are hired out to both Industrial Sites and Main-line Train Operating (TOC) Companies. The principal firms involved are listed on page 22. Within the listings of these fleet-owners, we detail their full fleet (in so far as we know it) under an entry for their operational base. Such locomotives are duplicated in other Industrial or Preservation Site lists, to record their known current locations.

LOCATIONS THAT ARE CLOSED, but where locomotives are still present, are indicated either by "(Closed)" appearing beside their title or subtitle, or else in a line of text immediately beneath that title. When a location is still in operation but no longer uses rail traffic, although the locomotives remain on site, this is indicated by "R.T.C." after the gauge concerned. The letters RTC are an abbreviation of the phrase "Rail Traffic Ceased".

TRACK GAUGE. The gauge(s) of the railway system(s) are given at the head of the locomotive list(s). At preservation sites and museums, etc, where several gauges differ by only a small fraction they are usually all listed under the nominal gauge of the majority.

WEBSITE. The official website relating to a location is listed. Unofficial websites or affiliated websites are not normally listed. Websites are subject to change from time to time. Details of websites were correct at the time of going to press.

POSTCODE. These are shown in bold type above the Grid Reference. Postcodes shown in brackets, are the nearest known, but not necessarily officially used locations. Some postcodes relate to a street or village and the respective location may be some distance away. All postcodes were checked and correct at the time of going to press.

SEQUENCE OF LOCOMOTIVES WITHIN THE LISTS is generally in accordance with our basic formula of "list steam first, followed by non-steam" and, in each of these groups, "list ex-mainline locomotives first in main-line running number order, followed by non-mainline locomotives in builder/works number order". Permanent-way motorised trolleys and other miscellaneous units are usually relegated to the end of the lists. However, in some cases, the nature or details of the actual fleet dictates a different arrangement of listing.

LOCOMOTIVE NUMBER and NAME. The title of the locomotive - number, name or both - is given in the first two columns. A name unofficially bestowed and used by staff but not carried on the locomotive is indicated by quotes (inverted commas " "). Locomotives under renovation at preservation sites, etc, may not currently carry their intended title; nevertheless these are shown unless it is definitely known that the name will not be retained.

Ex British Railways / Train Operating Company / Leasing Company (ROSCO) locomotives are further identified by the inclusion of formerly carried numbers in brackets even if not carried now.

TYPE The type of locomotive is given in column three.

The Whyte system of wheel classification is used in the main, but when driving wheels are connected not by outside rods but by chains or other means (as in various 'Sentinel' steam locomotives and diesel locomotives) they are shown as 4w (four-wheeled), 6w (six-wheeled), or if only one axle is motorised it is shown as 2w-2.

Trapped Rail System locomotives are shown using wheel type "ad" (meaning "axles driven") – thus "3ad" indicates a locomotive which may have 3 or more axles, but only 3 are driven.

For ex British Railways / Train Operating Company / Leasing Company (ROSCO) diesel and electric locomotives, the usual development of the Continental notation is used.

The following abbreviations are used :-

CT	Crane Tank - a T type locomotive fitted with load lifting apparatus
F	Fireless steam locomotive
IST	Inverted Saddle Tank
PT	Pannier Tank - side tanks not fastened to the frame
ST	Saddle Tank
STT	Saddle Tank with Tender
T	Side Tank or similar - a tank positioned externally and fastened to the frame
VB	Vertical Boilered locomotive
WT	Well Tank - a tank located between the frames under the boiler
BE	Battery powered Electric locomotive
BH	Battery powered electric locomotive - Hydraulic transmission
CA	Compressed Air powered locomotive
CE	Conduit powered Electric locomotive
D	Diesel locomotive - unknown transmission
DC	Diesel locomotive - Compressed air transmission
DE	Diesel locomotive - Electrical transmission
DH	Diesel locomotive - Hydraulic transmission

DM	Diesel locomotive - Mechanical transmission
F	(as a suffix, for example BEF, DMF) – Flameproof (see following paragraph)
FE	Flywheel Electric locomotive
GTE	Gas Turbine Electric locomotive
P	Petrol or Paraffin locomotive - unknown transmission
PE	Petrol or Paraffin locomotive - Electrical transmission
PH	Petrol or Paraffin locomotive - Hydraulic transmission
PM	Petrol or Paraffin locomotive - Mechanical transmission
R	Railcar - a vehicle primarily designed to carry passengers
RE	Third Rail powered Electric locomotive
WE	Overhead Wire powered Electric locomotive

FLAMEPROOF locomotives, usually battery but sometimes diesel powered, are denoted by the addition of the letter **F** to the wheel arrangement in column three.

CYLINDER POSITION is shown in column four for steam locomotives.
In each case, a prefix numeral (**3, 4, etc**) denotes more than the usual two cylinders.

IC	Inside cylinders
OC	Outside cylinders
VC	Vertical cylinders
G	Geared transmission - suffixed to IC, OC or VC

RACK DRIVE. Certain locomotives are fitted with rack-drive equipment to enable them to climb steep inclines and were mainly used in the UK underground in coal mines. The pattern of rack used was the "Abt" design. Locomotives that were built with (and are thought to still retain) this equipment are indicated by the word **RACK** in column four. However, due to contraction of the coal industry, such locomotives have often been transferred to sites where the rack equipment is no longer in use.

ROAD/RAIL. Certain locomotives are capable of working on either rail or road, many being of either Unilok or Unimog manufacture. These are indicated by **R/R** in column four.

STEAM OUTLINE. Non-steam locomotives, with a steam locomotive appearance added, are shown by **s/o** in column four.

MAKERS. The builder of the locomotive is shown in column five; the abbreviations used are listed on pages 10 to 22.

MAKERS NUMBER and DATE. The sixth column shows the works number, the next column the date which appears on the plate, or the date the locomotive was built if none appears on the plate.

DOUBTFUL INFORMATION. Information known to be doubtful is denoted as such by the wording, or printed in brackets with a question mark. Thus, Wkm (7573 ?) 1956 denotes that the locomotive is a Wickham built in 1956, and possibly of the works number 7573.

MISCELLANEOUS NOTES.

c	denotes circa, i.e. about the time of the date quoted
Dsm	indicates a "long-term" dismantled locomotive (as an aid to photographers, etc) Note: battery locomotives not carrying a battery are not regarded as Dsm
DsmT	used for motorised trolleys which have been converted to engineless trailers
Pvd	denotes Preserved on site
reb	denotes the locomotive was rebuilt (by and when, as per list) Note: "rebuilt" infers significant alteration to transmission or appearance; not a routine overhaul no matter how thorough that may have been
rep	indicates a major repair, and is ONLY detailed in the listings when a "repair plate" is (or was) affixed to the locomotive (and thus an aid to identification).
OOU	"Out Of Use" - indicates a locomotive which has, apparently permanently or at least for a significant period of time, ceased work

There are cases where all locomotives at a site are OOU but the railway, siding etc, is still in use, utilising a road tractor, cable-haulage or man-power for example.

Electric locomotives at steelworks are usually to be found working at Coke Ovens, the grid reference for the Ovens being shown in the usual place, when known.

HINTS ON RECORDING OBSERVATIONS

As already explained, the Society is largely reliant on the observations of enthusiasts to keep the records up to date and we are always pleased to receive reports of visits, which should be sent to the Hon. Records Officer or his assistants. The following will, we hope, be of assistance to those sending in reports.

Report anything you see. With regard to locomotives this means reporting locomotives present, even if there is no change to the published list. Someone else may visit in the next six months and find a change, the date of which can then be narrowed down to this six months. In addition to locomotives, details of rolling stock, track layouts and items of historical information gleaned are all welcome and will be added to our “Historic Handbook” files. It is surprising how often these items are required later by other enthusiasts. There is nothing to beat a note made on the spot; a note now is far better than trying to get the information from the recollection of employees in five years time.

Care should always be taken to distinguish actual observation from personal inference. If you see a 4wDH with worksplate say S 10019 and numbered 19, all well and good; but if it does not carry a worksplate, that fact should be stated. You may infer the loco is S 10019 from the running number, or by reference to this Handbook. If you say 'number on bonnet', 'number on cabside' or 'locomotive assumed to be S 10019' as the case may be, we will know the position precisely, which is important when bonnets are exchanged or cabs rebuilt, for example. If a locomotive carries no identification, you will probably be able to guess its make from design features and a note of livery may help. Also with diesel locomotives, a note should be made of the make of its engine, its type or model and its serial number, if these can be obtained. “Anonymous” locomotives can often be positively identified from our records given such details.

A thorough search of all premises is worthwhile, as locomotives (particularly if OOU) are frequently hidden away - and how often have surprises turned up in this way. Further, if you search diligently and a locomotive is missing, it may be presumed to have gone and enquiries can be made, as to where it has gone or if it has been scrapped. Please try to ascertain such information from the staff. Similarly, in the case of new arrivals, exact dates of arrival should always be obtained if possible.

Firms' titles change from time to time, so please check from the office or board at the entrance. Subsidiary companies frequently display the title of a parent company, but we always use the name under which the company trades, i.e. the subsidiary where such exists.

Preservation locations often take locomotives on short term hire, or hire locomotives out to other railways. Such movements are generally not noted within the EL series of books, unless it is known that the locomotive is intended to become a resident of a railway for at least a year or more, however, any movements should be recorded with dates, as often a temporary movement can become a permanent one.

It goes without saying that it is essential to obtain permission to enter premises to see locomotives and this applies with equal force to 'preserved' locations other than public parks and museums. Established systems which operate trains or have 'Open Days' will usually permit inspection of locomotives by arrangement.

Locomotives shown as at private homes or farms are very often in storage or in the course of restoration and, as a rule, our members and readers are strongly advised to NOT, in any way possible, embarrass the owners by writing for permission to view, but rather to wait until the owners announce that their machines are available for inspection.

You may note that whilst we list details of all known existing locomotives, some appear under incomplete or imprecise heading titles. This is regrettable, but is entirely a direct consequence of a few individuals having ignored the advice given in the paragraph above.

Finally, please try to establish friendly relations with the firms visited, as we are only allowed access by their courtesy. Do not be a nuisance or hold up production, nor expose yourself to danger in any way, but show a healthy interest in the processes being carried out. In this way, not only you, but other enthusiasts, will also be welcome there at a later date.

BUILDERS and REBUILDERS of LOCOMOTIVES

A

A&O	Alldays & Onions Ltd, Birmingham
AB	Andrew Barclay, Sons & Co Ltd, Caledonia Works, Kilmarnock, Ayrshire
ACAB	Adam Charles Albert Barber, Tan-Y-Dool Works, Llangollen
ACCars	A.C.Cars Ltd, Thames Ditton, Surrey
ACI	ACI Élévation, Isles sur Suippe, France
Acton	Acton Works, Acton, Greater London London Transport Executive / London Underground Ltd (from /1987)
ADC	Associated Daimler Co Ltd, Coventry
AE	Avonside Engine Co Ltd, Fishponds, Bristol
AEC	Associated Equipment Co Ltd, Southall, Middlesex
AEG	Allgemeine Elektrizitäts Gesellschaft, Berlin-Hennigsdorf, Germany
AEI	Associated Electrical Industries Ltd, Trafford Park, Greater Manchester
AFB	Société Anglo Franco-Belge des Ateliers de la Croyère, Seneffe et Godarville, Belgium
Afd	Ashford Works, Kent South Eastern & Chatham Railway / Southern Railway / British Railways
AH	A. Horlock & Co, North Fleet Iron Works, Kent
AK	Alan Keef Ltd, Cote Farm, Cote, near Bampton, Oxfordshire, later at Lea Line, Ross-on-Wye, Herefordshire (successor to SMH)
Albion	Albion Motors Ltd, Scotstoun, Glasgow W4
Alco	American Locomotive Co, USA; and/or Montreal, Canada
Alco(C)	American Locomotive Co, Cooke Works, Paterson, New Jersey, USA
AllenJ	Jason Allen, Grimoldby, Lincolnshire
ALR	Abbey Light Railway, P.N.Lowe & Sons, Bridge Road, Kirkstall, Leeds
Alstom	Alstom Rail Ltd, Washwood Heath, Birmingham
AP	Aveling & Porter Ltd, Invicta Works, Rochester, Kent
APEL	Adrian Phillips Engineering Ltd, Pontypool, Torfaen
Aquarius	Aquarius Railroad Technologies, Mickley, North Yorkshire
ARC	Amalgamated Roadstone Corporation Ltd, Stanton Harcourt Depot, near Witney, Oxfordshire
Artisair	Artisair Ltd, Moorwell Road, Yaddlethorpe, Scunthorpe, Lincolnshire.
AS&W	Allied Steel & Wire Ltd, Castle Works, Cardiff
Ashbury	Ashbury Railway Carriage & Iron Co Ltd, Openshaw, Manchester, Lancashire
AshbyI	I. Ashby, Valley Nurseries, Evesham, Worcestershire
Ashford	Ashford Plant Depot, Newtown Road, Ashford, Kent (B.R./Balfour Beatty/Network Rail)
Atlas	Atlas Loco & Mfg Co Ltd, Cleveland, Ohio, U.S.A.
AtW	Atkinson Walker Wagons Ltd, Frenchwood Works, Preston, Lancashire
AVB	A.V. Access Ltd, Bridgnorth, Shropshire
AW	Sir W.G.Armstrong, Whitworth & Co (Engineers) Ltd, Newcastle-upon-Tyne
Ayle	Ayle Colliery Co Ltd, Alston, Cumbria

B

B&S	Bellis & Seekings Ltd, Birmingham
BabyDeltic	Baby Deltic Project, Barrow Hill, Derbyshire
Bala	Bala Lake Railway, Llanuwchllyn, Gwynedd
Balfour Beatty	Balfour Beatty Rail Services, Raynesway, Derby
Bance	R. Bance & Co.Ltd, Cockrow Hill House, St. Mary's Road, Surbiton, Surrey
Barlow	H.N. Barlow, Southport, Lancashire
Barnes	A. Barnes & Co, Albion Works, Rhyl, North Wales (a subsidiary of Rhyl Amusements Ltd)
BateJ	John L. H. Bate, Reigate, West Sussex
Battison	Samuel Battison, c/o Tathams Ltd, Nottingham Road, Ilkeston, Derbyshire
BBC(S)	AG Brown Boveri & Cie, Baden, Aargau, Switzerland
BBM	Officine Mecciniche BBM S.p.A. Rossano Veneto, Via Mottinello, Italy

BBT	Brush Bagnall Traction Ltd, Loughborough, Leicestershire; and Stafford
B(C)	Peter Brotherhood, Engineers, Chippenham, Wiltshire
BCM	Big Country Motioneering Ltd, Kingswinford, West Midlands
BD	Baguley-Drewry Ltd, Burton-on-Trent, Staffordshire
BE	Brush Electrical Engineering Co Ltd, Loughborough, Leicestershire
Beamish	Beamish Museum, Stanley, Co. Durham
BEMO	BemoRail, Debbemeerweg 59, 1749 DK Alkmaar-Warmenhuizen, Netherlands
Benford	Benford (Terex) Ltd, Warwick, Warwickshire
Berwyn	Berwyn Engineering, Thickwood, Chippenham, Wiltshire
BES	Beech Engineering Services, Unit 4, Wetmore Industrial Estate, Wharf Road, Burton-on-Trent, Staffordshire
BEV	British Electric Vehicles Ltd, Churchtown, Southport, Lancashire ("BEV" branded locomotives were later built by Wingrove & Rogers – see "WR")
Bg	E.E. Baguley Ltd, Burton-on-Trent, Staffordshire
BgC	Baguley Cars Ltd, Shobnall Road Works, Burton-on-Trent, Staffordshire
Bg/DC	built by Bg for DC; makers numbers identical
BGB	Becorit (Mining) Ltd, Grove Street, Mansfield Woodhouse, Nottingham, then Hallam Fields Road, Ilkeston, Derbyshire from /1984
BH	Black, Hawthorn & Co Ltd, Gateshead
BHSC	Barrow Haematite Steel Co Ltd, Barrow-in-Furness, Lancashire
BIS	War Department, Arncott Workshops, Bicester Depot, Oxfordshire and including Ministry of Defence, Bicester Depot Workshops
BL	W.J. Bassett Lowke Ltd, Northampton
Bluebell	Bluebell Railway Co Ltd, Sheffield Park, East Sussex
BLW	Baldwin Locomotive Works, Philadelphia, Pennsylvania, USA
BM	Brown, Marshalls & Co Ltd, Adderley Park, Birmingham (in 1902 became part of what later became Metropolitan-Cammell Ltd)
BMR	Brecon Mountain Railway Co Ltd, Pant, Mid Glamorgan
BnM	Bord na Móna (Irish Turf Board) : Various of the larger sites (e.g. Bl, Bo, Dg & M) have built their own locomotives & railcars – see Section 4 (Ireland)
Bonnymount	Mr Taylor, Bonnymount Farm, Siston Common, Bristol
BoothR	R. Booth, Isle of Man Railways, Douglas, Isle of Man
BoothWKelly	Booth W Kelly, Boatbuilders, Ramsey, Isle of Man
Borsig	A. Borsig GmbH, Berlin-Tegel, Germany
Bow	Bow Locomotive Works, North London Railway
Bowman	N.Bowman, Launceston Steam Railway, Cornwall
BP	Beyer, Peacock & Co Ltd, Gorton, Manchester
BPH	Beyer, Peacock (Hymek) Ltd, Gorton, Manchester
Bradshaw	Bradshaw Electric Vehicles, Stibbington, Peterborough, Cambridgeshire
Braithwaite	Bevan Braithwaite / Bressingham Steam Preservation Company, Bressingham, near Diss
BRCW	Birmingham Railway Carriage & Wagon Co Ltd, Smethwick, West Midlands
BRE	British Rail Engineering Ltd
BRE(D)	British Rail Engineering Ltd, Derby Locomotive Works, Derby
BRE(S)	British Rail Engineering Ltd, Shildon Works, Co.Durham
Bredbury	Bredbury & Romiley Urban District Council, Cheshire [Greater Manchester]
Brennan	Louis Brennan, Gillingham, Kent
BriddonA	Andrew Briddon, AFRPS, Scunthorpe, Lincolnshire
BrightonTC	Brighton Corporation Tramways Dept, Brighton, Sussex
Brookside	Brookside Engine Co Ltd, Whaley Bridge, Derbyshire
BrownGM	G.M.Brown Ltd, Stanhopeburn, Stanhope, Co.Durham
BrownJ	Joe Brown, Farington Moss, Preston, Lancashire
Bruff	Bruff Rail Ltd, Suckley, Worcestershire
Brunning	H. Brunning (? of Crewe) – see Rhiw Valley Railway, Mid Wales
Bryce	B. Bryce, Corrofin, Co. Clare, Ireland
BT	Brush Electrical Machines Ltd, Traction Division, Falcon Works, Loughborough, Leicestershire
BTH	British Thomson-Houston Co Ltd, Rugby, Warwickshire
Bton	Brighton Works, Sussex London, Brighton and South Coast Railway / Southern Railway / British Railways
Bulrush(B)	Bulrush Peat Co Ltd, Bellaghy Works, Magherafelt, Co. Derry, Northern Ireland

BuryC&K	Bury, Curtis & Kennedy, Clarence Foundry, Love Lane, Liverpool
Bush Mill Rly	Bush Mill Railway, Port Arthur, Tasmania, Australia
BV	Brook Victor Electric Vehicles Ltd, Burscough Bridge, Ormskirk, Lancashire
BVR	Bure Valley Railway Ltd, Aylsham, Norfolk
Byers	R.S. Byers Ltd, Houghton, Carlisle, Cumbria
Byworth	Byworth Engineering Ltd, Keighley, West Yorkshire

C

CAF	Construcciones y Auxiliar de Ferrocarriles, Beasain, Spain
Campagne	E. Campagne & Cie, Paris, France
Cannon	Cannon, Dudley, West Midlands
Carland	Carland Engineering Ltd, Harold Wood, Essex
Carter	D.Carter, Tucking Mill Tramway, Midford, Somerset
Case	Case IH Agriculture, Steyrer Strasse 32, St. Valentin, USA
CastleGKN	Castle Works (GKN Sankey Ltd), Hadley, Telford, Shropshire
CCLR	Cleethorpes Coast Light Railway, Cleethorpes, Lincolnshire
Cdf	Cardiff West Yard Locomotive Works (Taff Vale Railway), Glamorgan
CE	Clayton Equipment Ltd, Burton-on-Trent, Staffordshire previously at Hatton, Derbyshire, where also traded as NEI Mining Equipment Ltd, and originally Clarke Chapman Ltd
Chance	Chance Manufacturing Co Inc, Wichita, Kansas, USA
Chaplin	Alexander Chaplin & Co Ltd, Cranstonhill Works, Glasgow
Chrz	Fabryka Lokomotywim "Feliksa Dzierzynskiego" Chrzanów, Poland
CityTrain	City Train GmbH, t/a Tschu-Tschu, Neumarkt, Germany
Civil	T.D.A. Civil, near Uttoxeter, Staffordshire
Clarkson	H. Clarkson & Son, York
ClayCross	Clay Cross Co Ltd, Spun Pipe Plant, Clay Cross Iron Works, Derbyshire
CM	Century Millwrights, Platts Eyot's Works, Sunbury-on-Thames, Surrey
CNES	Corus Northern Engineering Services, Scunthorpe, Lincolnshire
Coalbrookdale	Coalbrookdale Co, Coalbrookdale Ironworks, Shropshire
Cockerill	Société pour L'Exploitation des Etablissements John Cockerill, Seraing, Belgium
Coferna	Coferna, Constructions Ferroviaires et Navales de l'Ouest, Sables d'Olonne, Vendée, France
ColdChon	Cold Chon Ltd, Oranmore, Co Galway
ColebySim	Coleby-Simkins Engineering, Melton Mowbray, Leicestershire
CollinsD	Dennis Collins, Kanturk, Co. Cork, Ireland
Cometi	Cometi S.r.l., Italy
Consett	Consett Iron Co Ltd, Consett Works, Co.Durham
Consillia	Consillia Ltd, Shardlow, Derbyshire
Corpet	L. Corpet, Avenue Philippe-Auguste, Paris, France
Couillet	Société Anonyme des Usines Métallurgiques du Hainaut, Couillet, Marcinelle, near Charleroi, Belgium
Cowlairs	Cowlairs Works, Glasgow North British Railway / London & North Eastern Railway / British Railways
CPM	CPM S.r.L., Marmaloda, Milano, Italy
Cranmore	East Somerset Railway Co Ltd, Cranmore, Somerset
CravenEA	E.A. Craven, North Shields, Northumberland
CravenJ	J. Craven, Walesby, Nottinghamshire
Cravens	Cravens Ltd, Darnall, Sheffield
Crewe	Crewe Works, Cheshire London and North Western Railway / London, Midland & Scottish Railway / British Railways
Crome/Loxley	R. Crome & R. Loxley, Doncaster
CurleA	A. Curle, Skipsea, East Yorkshire
CurwenD	A. Curwen, All Cannings, near Devizes, Wiltshire
CW	Cowlishaw Walker Engineering Co Ltd, Biddulph, Staffordshire

D

D	Dübs & Co, Glasgow Locomotive Works, Glasgow
Dar	Darlington Works, Co.Durham North Eastern Railway / London & North Eastern Railway / British Railways
Darlington	A1 Locomotive Trust, Hopetown Carriage Works, Darlington, Co. Durham
Dav	Davenport Locomotive Works, Davenport, Iowa, USA
DB	Sir Arthur P. Heywood, Duffield Bank Works, Derbyshire
DC	Drewry Car Co Ltd, London (supply agents only)
ĐĐak	Đuro Đakovic, Industrije Lokomotive, Strojeva I Mostova, Slavonski Brod, Yugoslavia
Decauville	Société Nouvelle des Etablissements Decauville Aîne, Petit Bourg, Corbeil, Essonne, France (later Société Anonyme Decauville, Corbeil, Essonne, France)
DeDietrich	Société Lorraine des Anciens Établissements de Dietrich & Cie, Reichshoffen, Bas Rhin, France
Derby	Derby Locomotive Works Midland Railway / London, Midland & Scottish Railway / British Railways
DerbyC&W	Derby Carriage & Wagon Works, Litchurch Lane, Derby Midland Railway / London, Midland & Scottish Railway / British Railways
DeW	DeWinton & Co, Union Works, Caernarfon, North Wales
Dieci	Deici S.r.L., Majorana, Montecchio Emilia, Italy
Diema	Diepholzer Maschinenfabrik Fritz Schöttler GmbH., Diepholz, Niedersachsen, Germany
DK	Dick Kerr & Co Ltd, Preston, Lancashire (Britannia Works, Kilmarnock, Ayrshire until 1919)
DL	Dorman Long (Steel) Ltd, Middlesbrough, North Yorkshire
DLR	Difflin Lake Railway, Raphoe, Co. Donegal, Ireland
DM	Davies & Metcalfe Ltd, Romiley, Stockport, Greater Manchester
Dodman	Alfred Dodman & Co, Highgate Works, Kings Lynn, Norfolk
Don	Doncaster Works, South Yorkshire Great Northern Railway / London & North Eastern Railway / British Railways
Donelli	F.L. Donelli SpA, Reggio Emilia, Italy
Donelon	J.F. Donelon Ltd, Horwich, Greater Manchester
DonM	British Rail, Marshgate Permanent Way Depot, Doncaster, South Yorkshire
Dorothea	Dorothea Restorations, Mevril Spring Works, New Road, Whaley Bridge, Derbyshire
Dotto	Dotto Trains, Castelfranco, Italy
DP	Davey, Paxman & Co Ltd, Colchester, Essex
Dtz	Motorenfabrik Deutz AG / Humboldt-Deutz-Motoren AG / Klöckner-Humboldt-Deutz AG, Köln (Cologne), Nordrhein-Westfalen, Germany
Dundalk	Dundalk Works, Great Northern Railway of Ireland, Co. Louth, Republic of Ireland
DunEW	Dundalk Engineering Works, Dundalk, Co. Louth, Republic of Ireland Locomotives built from parts supplied by Gleismac

E

EAGIT	East Anglian Group for Industrial Training, Norwich, Norfolk
EARM	East Anglian Railway Museum, Chappel & Wakes Colne Station, near Colchester, Essex
EB	E. Borrows & Sons, St.Helens, Lancashire
EBW	E.B. Wilson & Co, Railway Foundry, Leeds
Eclipse	Eclipse Peat Co Ltd, Ashcott, Somerset
Eddy	M. Eddy
EdwardsE&J	E & J Edwards
EE	English Electric Co Ltd, London
EEDK	English Electric Co Ltd, Dick Kerr Works, Preston, Lancashire
EES	English Electric Co Ltd, Stephenson Works, Darlington (successors to RSHD)
EEV	English Electric Co Ltd, Vulcan Works, Newton-le-Willows, Lancashire (successors to VF)
EEV-AEI	English Electric Co Ltd & Associated Electrical Industries Ltd, Vulcan Foundry, Newton-le-Willows, Lancashire (predecessors to GECT)
EIMCO	Eastern Iron & Metal Corporation, Salt Lake City, Utah, USA
Eiv de Brive	L'Etablissement Industriel Équipement SNCF de Brive, Brive-de-Gaillarde, Corrèze, France
EK	Märstaverken, Eksjö, Sweden

ELC	East Lancashire Coachbuilders
Electro	Uzinele Electroputere, Craiova, Romania
Elh	Eastleigh Works, Hampshire London and South Western Railway / Southern Railway / British Railways
ENG/GEM	ENG/GEM, Wisbech, Cambridgeshire
ESCA	ESCA Engineering Ltd, 6 Wetheral Close, Hindley Industrial Estate, off Swan Lane, Hindley Green, Wigan, Greater Manchester
ESR	Exmoor Steam Railway, Devon
EvansJ	J. Evans, Northampton
EV	Ebbw Vale Steel, Iron & Coal Co Ltd, Ebbw Vale Works, Gwent
EVM	Elham Valley Museum, Peene Yard, Peene, Folkestone, Kent
EVRA	Ecclesbourne Valley Railway Association, Wirksworth, Derbyshire
Express	Express Service OOD, Ruse, Bulgaria

F

Fairbourne	Fairbourne Railway Co, Fairbourne, Gwynedd, North Wales
Fairmont	Fairmont Railway Motors Ltd, Toronto, Ontario, Canada
FE	Falcon Engine Works, Loughborough, Leicestershire
Ferndale	K. Watson & K. Tingle, Ferndale Engineering, Canning Vale, Western Australia
FFP	Fire Fly Project, Great Western Preservations Ltd, Bristol (1987) and Didcot, Oxfordshire (from 1989)
FH	F.C. Hibberd & Co Ltd, Park Royal, London ("Planet" locomotives) later at Butterley Works, Ripley, Derbyshire
Fiat	Fabrica Italiana Automobili Torino, Italy
Fisons	Fisons Ltd, British Moss Works, Swinefleet, near Goole, East Yorkshire constructed from parts supplied by Diema
FJ	Fletcher Jennings & Co, Lowca Engine Works, Whitehaven, Cumbria
FlourMill	The Flour Mill, Bream, Forest of Dean, Gloucestershire
FMB	F.M.B. Engineering Co Ltd, Unit 10, Southlands, Latchford Lane, Oakhanger, near Bordon, Hampshire
Ford	Ford Motor Co Ltd, Dagenham, Essex
FordTTC	Ford Motor Co Ltd, Thameside Technical Training Centre, Dagenham, Essex
ForshawJJ	J.J. Forshaw, Clifton, Bedfordshire
FosterRastrick	Foster, Rastrick & Co, Stourbridge, Worcestershire
Foulds/Collins	Andrew Foulds & Robert Collins, c/o E.A. Foulds Ltd, Albert Works, Clifton Street, Colne, Lancashire
FoxA	A. Fox
FRCo	Festiniog Railway Company, Boston Lodge Works, Porthmadog, North Wales
Frenze	Frenze Engineering, (Diss area), Norfolk
Freud	Stahlbahnwerke Freudenstein & Co, Berlin-Tempelhof, Germany
FRgroup4	Festiniog Railway, Group 4, Birmingham
Frichs	A/S Frichs Maskinfabrik & Kedelsmedie, Århus, Denmark
FRSociety	Festiniog Railway Society, North Staffs Group
FRT	Furness Railway Trust, Marconi Marine / Vickers Shipbuilding & Engineering Ltd, Barrow-in-Furness
Funkey	C H Funkey & Co. (Pty) Ltd., Alrode, Transvaal, Republic of South Africa [now in Gauteng]; formerly at Alberton, Transvaal; later a part of Dorbyl Transport Products, Rolling Stock Division, Boksburg, Transvaal, Republic of South Africa
FW	Fox, Walker & Co, Atlas Engine Works, Bristol

G

G&S	G.& S. Light Engineering Co Ltd, Stourbridge, Worcestershire
Gartell	Alan Gartell, Common Lane, Yenston, near Templecombe, Somerset
GB	Greenwood & Batley Ltd, Albion Ironworks, Armley, Leeds
GE	George England & Co Ltd, Hatcham Ironworks, London
GEC	General Electric Co Ltd, Witton, Birmingham
GEC-Alsthom	GEC-Alsthom, France
GECT	G.E.C.Traction Ltd, Vulcan Works, Newton-le-Willows, Lancashire
Geevor	Geevor Tin Mines Ltd, Pendeen, near St. Just, Cornwall

Geismar	Geismar (UK) Ltd, Salthouse Road, Brackmills Industrial Estate, Northampton and 68006 Colmar, Alsace, France
GGR	Groudle Glen Railway Co Ltd, Onchan, Isle of Man
GH	Gibb & Hogg, Victoria Engine Works, Airdrie, North Lanarkshire
Ghd	Gateshead Works, Co.Durham North Eastern Railway / London & North Eastern Railway / British Railways
GIA	GIA Industria AB, Grãngesberg, Sweden
GibbonsCL	C.L. Gibbons, c/o Bury Transport Museum, Bury Depot, Greater Manchester
Gleismac	Gleismac Italiana SpA, Viale Delia Stazione 3, 46030 Bigarello, Mantova, Italy
GM	General Motors Ltd, Electro-Motive Division, La Grange, Illinois, U.S.A.
GMC	General Motors Ltd, Electro-Motive Division, London, Ontario, Canada
Gmd	Gmeinder & Co GmbH, Mosbach, Germany
GMT	Gyro Mining Transport Ltd, Victoria Road, Barnsley, South Yorkshire, then Bramley Way, Hellaby Industrial Estate, Hellaby, near Rotherham, South Yorkshire from c/1987
GNS	Great Northern Steam Co Ltd, Unit 3, Forge Way, Cleveland Industrial Estate, Darlington, Co.Durham
Goold	J.R. Goold Engineering Ltd, Camerton, near Radstock, Somerset
Gorton	Gorton Works, Manchester Great Central Railway / London & North Eastern Railway / British Railways
Govan	Govan Workshops, Broomloan Road, Glasgow Glasgow Corporation Transport Department, Underground Railway
GR	Grant, Ritchie & Co, Townholme Engine Works, Kilmarnock, Ayrshire
GRC&W	Gloucester Railway Carriage & Wagon Co Ltd, Gloucester
Greaves	J.W. Greaves & Sons Ltd, Llechwedd Quarry, Blaenau Ffestiniog, North Wales
Greensburg	Greensburg Machine Co, Greensburg, Pennsylvania, USA
Group Eng	Group Engineering
GS	George Stephenson
GS(H)	George Stephenson, Hetton, Co.Durham
GS(K)	George Stephenson, West Moor Workshops, Killingworth, Northumberland
Guest	Guest Engineering & Maintenance Co Ltd, Stourbridge, Worcestershire
GuinnessNL	Nigel L. Guinness, Cobham, Surrey
Gullivers	Gullivers Land, Milton Keynes, Bedfordshire
GWS	Great Western Society, Didcot, Oxfordshire

H

H	James & Fredk. Howard Ltd, Britannia Ironworks, Bedford
HAB	Hunslet-Barclay Ltd, Caledonia Works, Kilmarnock, Ayrshire
Hackworth	Timothy Hackworth, Soho Works, Shildon, Co.Durham
Hako	Hako GmbH, Bad Oldesloe, Germany
HallamP	Peter Hallam, St. Austell, Cornwall
HallT	T.Hall, North Ings Farm Museum, Dorrington, near Ruskington, Lincolnshire
Hano	Hannoversche Maschinenbau-AG, vormals Georg Egestorff, Hannover-Linden, Niedersachsen, Germany
Harbin	Harbin Forest Machinery Factory, Harbin, Heilongjiang Province, China
HardyK	K. Hardy, Brookhouse, Badgeworth, near Cheltenham, Gloucestershire
Harmill	Harmill Systems Ltd, Leighton Buzzard, Bedfordshire
Harsco	Harsco Track Technologies, Unit 1, Chewton Street, Eastwood, Nottinghamshire. (successors to Permaquip).
Hart	Sächsische Maschinenfabrik, vormals Richard Hartmann AG, Chemnitz, Germany
HartlepoolBSC	Hartlepool Works (British Steel Corporation), North Teesside
Haydock	Haydock Foundry Co Ltd, Haydock, Lancashire
HaylockJ	J. Haylock, Moors Valley Railway, Ringwood, Dorset
Hayne	N. Hayne, Sheppards Tea Rooms & Boat House, near Saltford, Somerset, later Blaise Castle, Henbury, Bristol [Gloucestershire]
Haytor	M.P. Haytor & Son, Frensham, Surrey
HB	Hudswell Badger Ltd, Railway Foundry, Hunslet, Leeds
HC	Hudswell, Clarke & Co Ltd, Railway Foundry, Hunslet, Leeds

HE	Hunslet Engine Co Ltd, Hunslet, Leeds
	Graycar Industrial Estate, Barton-under-Needwood, Staffordshire. (A Division of LH Group Services Ltd)
	Statfold, Tamworth, Staffordshire
Heath	Robert Heath & Sons Ltd, Norton Ironworks, Stoke-on-Trent
Heatherslaw	P. Smith, Heatherslaw Light Railway Co, Cornhill-on-Tweed, Northumberland
Hedley	William Hedley, Wylam Colliery, Northumberland
Hegenscheidt	Hegenscheidt-MFD GmbH & Co. KG, Hegenscheidt Platz, Erkelenz, Germany (supply agents only)
Hen	Henschel & Sohn GmbH, Kassel, Germany later traded as Thyssen Henschel and Adtranz
Herschell	Allan Herschell, North Tonawanda, New York, USA
H(L)	Hawthorns & Co, Leith Engine Works, Edinburgh
HL	R.& W.Hawthorn, Leslie & Co Ltd, Forth Bank Works, Newcastle-upon-Tyne
HLH	Hunslet Locomotive Hire Ltd, Station Road, Killamarsh, Derbyshire
HLT	Hughes Locomotive & Tramway Engine Works Ltd, Loughborough, Leicestershire
HN(R)	Harry Needle Railroad Company, Barrow Hill, Derbyshire
HopleyCP	C. P. Hopley
Hor	Horwich Works, Lancashire — Lancashire and Yorkshire Railway / London, Midland & Scottish Railway / British Railways
House	B. House
HPET	H.P.E.Tredegar Ltd, Tafarnaubach Industrial Estate, near Tredegar, Gwent
HSE	Harry Steer Engineering, Breaston, near Derby
HT	Hunslet Taylor Consolidated (Pty.) Ltd, Germiston, Transvaal, South Africa
HU	Robert Hudson Ltd, Leeds
HuntTG	T.G. Hunt, Oldbury, West Midlands
Hutchings	R. Hutchings
HW	Head, Wrightson & Co, Teesdale Ironworks, Thornaby-on-Tees, North Yorkshire
Hy-rail	This is a trademark of Harsco Track Technologies / Permaquip (see- Harsco / Perm)

I

IFA	Industrieverband Farzeugbau Association, VEB IFA-Automobilwerke, Ludwigsfeldt, Germany
Inchicore	Inchicore Works, Dublin — Great Southern & Western Railway / Great Southern Railways / Córas Iompair Éireann / Iarnród Éireann
Iso	Iso Speedic Co Ltd, Fabrications & Electric Vehicles, Charles Street, Warwick

J

Jaco	Jaco Engineering Co Ltd, Edwards Road, Birmingham
Jaywick	Jaywick Light Railway, near Clacton, Essex
Jesty	Bedford & Jesty Ltd, Doddings Farm, Bere Regis, Dorset
Jenbach	Jenbachwerke A G, Jenbach, Austria.
JF	John Fowler & Co (Leeds) Ltd, Hunslet, Leeds
JMR	J.M.R.(Sales) Ltd, 173 Liverpool Road South, Birkdale, Southport, Lancashire
Jung	Arnold Jung Lokomotivfabrik GmbH, Jungenthal bei Kirchen-an-der-Sieg, Germany

K

K	Kitson & Co, Airedale Foundry, Leeds
Kambarka	Kambarka Machine, Kambarka, Republic of Udmurtia, Russia
Kawasaki	Kawasaki Heavy Industries Ltd, Motorcycle & Engine Company, Japan
KC	Kent Construction & Engineering Co Ltd, Ashford, Kent
Kearsley	Kearsley Power Station (Central Electricity Generating Board), Radcliffe, Gtr. Manchester
Keltec	Keltec Engineering Ltd, Co. Limerick, Ireland
Kennan	Thos Kennan & Son, Dublin, Ireland
Kershaw	Kershaw Manufacturing Co. Inc, Montgomery, Alabama, USA
Kew	Kew Bridge Steam Museum, Green Dragon Lane, Brentford, London
Kierstead	Kierstead Systems & Controls Ltd, Ketley Bank Hall, Telford, Shropshire
Kilmarnock	Kilmarnock Regional Civil Engineers Workshops (British Rail, Scottish Region) Kilmarnock, Ayrshire (conversions only)
King	King Rail, Market Harborough, Leicestershire (UK agent for Zagro)

Kitching	A.Kitching, Hope Town Foundry, Darlington, Co.Durham
Knowell	M. Knowell
Krauss	Lokomotivfabrik Krauss & Co
KraussL	Lokomotivfabrik Krauss & Co, Linz, Austria
KraussM	Lokomotivfabrik Krauss & Co, Marsfeld Works, München (Munich), Germany
KraussS	Lokomotivfabrik Krauss & Co, Sendeling Works, München (Munich), Germany
Krupp	Friedrich Krupp, Maschinenfabriken Essen, Abt. Lokomotivbau, Essen, Nordrhein-Westfalen, Germany
KS	Kerr, Stuart & Co Ltd, California Works, Stoke-on-Trent, Staffordshire

L

L	R.A.Lister & Co Ltd, Dursley, Gloucestershire
LaMeuse	Société Anonyme des Ateliers de Construction de la Meuse, Sclessin, Liège, Belgium
Lake&Elliot	Lake & Elliot Ltd, Braintree, Essex
Lancing	Lancing Carriage Works, Sussex Southern Railway / British Railways
LancTan	Lancashire Tanning Co Ltd, Littleborough, Lancashire
Landore	British Rail, Landore Depot, Swansea
Landrover	Landrover Ltd, Solihull, West Midlands
Lawson	C. Lawson, 11 Okeley Lane, Highfield Estate, Tring, Hertfordshire
LB	Lister Blackstone Traction Ltd, Dursley, Gloucestershire (successor to L)
LBNGRS	Leighton Buzzard Narrow Gauge Railway Society, Stonehenge Workshops, Leighton Buzzard, Bedfordshire
Leake	J. Leake, Lytchett Matravers, Dorset
LemonB	J.Lemon-Burton, Paynesfield, Albourne Green, West Sussex, and Shelmerdine & Mulley Ltd, Edgeware Road, Cricklewood, London NW2
Lesmac	Lesmac (Fasteners) Ltd, Dykehead Street, Glasgow
Lewin	Stephen Lewin, Dorset Foundry, Poole, Dorset
Leyland	Leyland Vehicles Ltd, Workington, Cumbria
LHGroup	L.H.Group Services Ltd, Barton-under-Needwood, Staffordshire
Lima	Lima Locomotive Works Inc., Lima, Ohio, U.S.A.
Lind	James Lind & Sons
LJ	Lester Jones, Hobart, Tasmania.
Llangollen	Llangollen Railway Engineering Services, Llangollen, Denbighshire, Wales
LlanwernBSC	Llanwern Works (British Steel Corporation), Newport, Gwent
LMM	Logan Mining & Machinery Co Ltd, Dundee
LO	Lokomo Oy, Tampere, Finland
LocoEnt	Locomotion Enterprises (1975) Ltd, Bowes Railway, Springwell, Gateshead, Co.Durham
Locospoor	NV Locospoor International, Den Haag, Netherlands; formerly CV Locospoor
Longhedge	Longhedge Works, London — South Eastern & Chatham Railway
Longleat	Longleat Light Railway, Longleat, Warminster, Wiltshire
Lumb	James Lumb & Son, Elland, Leeds, West Yorkshire

M

M&P	Mather & Platt Ltd, Park Works, Manchester
Mace	C.Mace, The Woodland Railway, Kent
Maffei	J.A. Maffei AG, Locomotiv & Maschinenfabrik, München (Munich), Bavaria
MaK	Maschinenbau Kiel GmbH, Kiel-Friedrichsort, Germany
MaLoWa	MaLoWa Bahnwerkstatt GmbH, Klostermansfeld, Germany
MalyanG	Garry Malyan, Lappa Valley Railway, St. Newlyn East, near Newquay, Cornwall
MarshallJ	John Marshall, Spring Lane, Hockley Heath, Warwickshire
Massey	G.D. Massey, 57 Silver Street, Thorverton, Exeter, Devon
Matisa	Matisa Material Industriel SA, Arc-En-Ciel 2, Crissier, Lausanne, Switzerland
MatisaSPA	Matisa SpA, S.Palomba, Rome, Italy
MaxEng	Max Engineering, Station Road, Epworth, Doncaster, South Yorkshire
Maxitrack	Maxitrack, "Rothiemay", Offham Road, West Malling, Kent
McCulloch	W & D McCulloch, Ballantrae, Ayrshire
McDowall	Wallace McDowall Ltd, Monkton, Ayrshire

McGarigle	P.McGarigle, Niagara Falls, near Buffalo, New York, USA ("Cagney" locomotives)
Mechan	Mechan Ltd, Davy Industrial Park, Sheffield, South Yorkshire
MER	Manx Electric Railway, Derby Castle Works, Douglas, Isle of Man
Mercedes	Mercedes-Benz AG, Stuttgart, Germany
Mercia	Mercia Fabrications Ltd, Steel Fabrications, Units K1 & K3, Dudley Central Trading Estate, Shaw Road, Dudley, West Midlands
Mercury	Mercury Truck & Tractor Co Ltd, Gloucester
Meridian	Meridian (Motioneering) Ltd, Bradley Way, Hellaby Industrial Estate, Hellaby, near Rotherham, South Yorkshire
Metalair	Metalair Ltd, Wokingham, Berkshire
MetallbauE	Metallbau Emmeln GmbH, Haren (Ems), Germany
MetAmal	Metropolitan Amalgamated Railway Carriage & Wagon Co Ltd (until 6/1912)
MetC&W	Metropolitan Carriage, Wagon & Finance Co Ltd (6/1912 to 12/1928)
MetCam	Metropolitan-Cammell Carriage, Wagon & Finance Co Ltd (1/1929 to 10/1934) Metropolitan-Cammell Carriage & Wagon Co Ltd (1/1935 to 12/1964) Metropolitan-Cammell Ltd (from 1/1965 to 5/1989) — all located in Birmingham
MH	Muir-Hill (Engineers) Ltd, Trafford Park, Manchester
Middleton	Middleton Railway Trust, Hunslet, Leeds, West Yorkshire
Minilok	allrad-Rangiertecknik GmbH, D-5628 Heiligenhaus Bez, Dusseldorf, Nordrhein-Westfalen, Germany *[Note that the firm spells their name with a small 'a']*
Minirail	Minirail Ltd, Frampton Cotterell, Bristol
Mitsubishi	Mitsubishi Group, Japan
Mkm	Markham & Co Ltd, Chesterfield, Derbyshire
Moës	S.A. Moteurs Moës, Waremme, Belgium
Montania	Gerlach and König, Nordhausen, Germany (later became OK, but continued using the Montania name until 1945)
MoorsValley	Moors Valley Railway, Moors Valley Country Park, Horton Road, Ashley Heath, Ringwood, Hampshire
MorrisRP	R.P. Morris, 193 Main Road, Longfield, Kent
Morse	R.H. Morse, Potter Heigham, Norfolk
Mortimer	Mortimer Manufacturing Ltd, Cottesmore, Rutland
MossAJ	A.J. Moss, 97 Martin Lane Ends, Scarisbrick, Lancashire
MossDW	Derek W Moss, Burscough, Lancashire
Motala	AB Motala Verkstad, Motala, Sweden
Moyse	Locotracteurs Gaston Moyse, La Courneuve, Seine St.Denis, France Société Anonyme Moyse, La Courneuve, Seine St.Denis (after 1965)
MPES	Motive Power & Equipment Solutions, Greenville, South Carolina, USA
MR	Motor Rail & Tramcar Co Ltd / Motor Rail Ltd, Simplex Works, Bedford
MRWRS	MRW Railways Ltd, Sheffield, South Yorkshire
MSI	The Museum of Science & Industry in Manchester, Liverpool Road, Castlefield, Manchester
Multicar	VEB Fahrzeugwerk Waltershausen, Germany (until 1991) Multicar - Hako-Werke GmbH, Waltershausen, Germany (see also Hako - from 1996)
MV	Metropolitan-Vickers Electrical Co Ltd, Trafford Park, Manchester
MW	Manning, Wardle & Co Ltd, Boyne Engine Works, Hunslet, Leeds

N

N	Neilson & Co, Hyde Park Works, Springburn, Glasgow
NB	North British Locomotive Co Ltd, Glasgow
NBH	North British Locomotive Co Ltd, Hyde Park Works, Glasgow
NBQ	North British Locomotive Co Ltd, Queens Park Works, Glasgow
NBRES	NBR Engineering Services Ltd, Scarborough, North Yorkshire then Darlington, Co. Durham (from 2017)
NC	Northern Counties Coach Builders, Wigan, Lancashire
NCC	Northern Counties Committee (of London, Midland & Scottish Rly etc) York Road Works, Belfast, Northern Ireland
NDLW	North Dorset Locomotive Works, Motcombe, Dorset
Neasden	Neasden Works, London Metropolitan Railway

NemethJ	Joe Nemeth Engineering Ltd, Washingpool Farm, Main Road, Easter Compton, Bristol
Newag	Newag, Ripshorster Strasse, Oberhausen, Nordrhein-Westfalen, Germany
New Holland	New Holland Agricultural
Nissan	Nissan Motor Co Ltd, Japan
Niteq	Niteq BV, Overspoor 21, 1688 J.G. Nibbexwould, Netherlands
NMW	National Museum of Wales, Industrial & Maritime Museum, Butetown, Cardiff, South Glamorgan
NNM	Noord Nederlandsche Machinefabriek BV, Winschoten, Netherlands
Nohab	Nydquist & Holm AB, Trollhättan, Sweden
NR	Neilson Reid & Co, Glasgow
NW	Nasmyth, Wilson & Co Ltd, Bridgewater Foundry, Patricroft, Manchester

O

OAE	Olympic Aquatic Engineers, Norwich, Norfolk
Oerlikon	Maschinenfabrik Oerlikon, Zurich, Switzerland
OIM	OIM Ltd, Munster Railway Works, Kealkill, Co. Cork, Ireland
OK	Orenstein & Koppel AG, Berlin-Drewitz and Abt.Montania, Nordhausen, Germany then, from 1945 : – Orenstein-Koppel und Lübecker Maschinenbau AG, Dortmund-Dorstfeld, Germany
OldburyC&W	Oldbury Carriage & Wagon Co Ltd, Birmingham

P

P	Peckett & Sons Ltd, Atlas Locomotive Works, St.George, Bristol
P&T	Plasser & Theurer GmbH, Austria
ParmenterC	C. Parmenter, Launceston, Cornwall
PBR	Pleasure Beach Railway, Blackpool, Lancashire
Pendre	Pendre Works (Talyllyn Railway Co), Tywyn, Gwynedd, North Wales
Perm	The Permanent Way Equipment Co Ltd, Pweco Works, Lillington Road North, Bulwell, Nottingham — later at 1 Giltway, Giltbrook, Nottingham
Plasser	Plasser Railway Machinery (GB) Ltd, Drayton Green Road, West Ealing, London
Plymouth	Plymouth Locomotive Works, Plymouth, Ohio, USA
Porter	H. K. Porter Co, Inc, Pittsburgh, Pennsylvania, USA
PortTalbotBSC	Port Talbot Works (British Steel), Port Talbot, West Glamorgan
Potter	D.C. Potter, Yaxham Park, Yaxham, near Dereham, Norfolk
Powell	Alan Powell, Norwich, Norfolk
PPM	Parry People Movers Ltd, Corngreaves Trading Estate, Overend Road, Cradley Heath, West Midlands
Prestige	Prestige Engineering, Abbotskerwell, Newton Abbot, Devon
PRoyal	Park Royal Vehicles, Park Royal, London
Pritchard	William Pritchard, c/o Manchester, Bury, Rochdale & Oldham Steam Tramway, Oldham, Lancashire
PSteel	Pressed Steel Ltd, Linwood, Paisley, Renfrewshire
PVRA	Plym Valley Railway Association, Marsh Mills, Plympton, Devon
PWR	Pikrose & Co Ltd, Wingrove & Rogers Division, Delta Road, Audenshaw, Greater Manchester (successors to WR)

R

R&R	Ransomes & Rapier Ltd, Riverside Works, Ipswich, Suffolk
RADev	RA Developments, Scunthorpe, Lincolnshire
Ravenglass	Ravenglass & Eskdale Railway Co Ltd, Ravenglass, Cumbria
Red(F)	Redland Bricks Ltd, Funton Works, near Sittingbourne, Kent
Red(T)	Redland Bricks Ltd, Baltic Road, Tonbridge, Kent
Redstone	Mr Redstone, Penmaenmawr, North Wales
RegentSt	Regent Street Polytechnic, London
Renault	Régie Nationale des Usines Renault, Division Matériel Ferroviaire, Choisy-le-Roi, near Paris, France (from 1945) earlier:- SA de Usines Renault, Boulogne-Billancourt, Hauts de Seine, near Paris
Resco	Resco (Railways) Ltd, Manor Road Industrial Estate, Erith, Greater London
Resita	Uzinele de Fier si Domeniile din Resita Societate Anonima Resita, Resita, Romania (later: Combinatul Metalurgic Resita)

RFSD	R.F.S. Engineering Ltd, Doncaster Works, Hexthorpe Road, Doncaster
RFSK	R.F.S. Engineering Ltd, Kilnhurst Works, Hooton Road, Kilnhurst, South Yorkshire (successors to TH)
RH	Ruston & Hornsby Ltd, Lincoln
RHDR	Romney Hythe & Dymchurch Railway, New Romney, Kent
Rhiw	Rhiw Valley Light Railway (J.Woodruffe), Lower House, Manafon, near Welshpool, Powys, Mid Wales
Rhiwbach	Rhiwbach Quarries Ltd, Rhiwbach Slate Quarry, North Wales
Richard	Establishments B Richard, Saint-Denis-de l' Hôtel, Loire, France
Riley	Ian V.Riley Engineering, Arbour Locomotive Works, Kirkby, Merseyside later at Bury Car Sheds, Bury, Greater Manchester later at Premier Locomotive Works, Sefton Street, Heywood, Greater Manchester
Riordan	Riordan Engineering Ltd, Surbiton, Surrey
RM	Road Machines (Drayton) Ltd, West Drayton, Middlesex later at Iver, Buckinghamshire
RMS	RMS Locotec Ltd, West Yorkshire
Roanoke	Roanoke Engineering, Grange Hill Industrial Estate, Bratton Fleming, Devon
Robel	Robel & Co, Maschinenfabrik, München, Bayern, Germany
Rosewall	K. Rosewall, Cross Elms Nursery, Bristol
RP	Ruston, Proctor & Co Ltd, Lincoln
RPSI(W)	Railway Preservation Society of Ireland, Whitehead, Co.Antrim, Northern Ireland
RR	Rolls Royce Ltd, Sentinel Works, Harlescott, Shrewsbury, Shropshire (successors to Sentinel)
RRS	Rapido Rail Systems, Dudley, West Midlands
RS	Robert Stephenson & Co Ltd, Forth Street, Newcastle-upon-Tyne and Darlington, Co.Durham
RSH	Robert Stephenson & Hawthorns Ltd
RSHD	Robert Stephenson & Hawthorns Ltd, Darlington Works, Co.Durham
RSHD/WB	built by RSHD but ordered by WB
RSHN	Robert Stephenson & Hawthorns Ltd, Newcastle-upon-Tyne Works (successors to HL)
RSM	Royal Scottish Museum, Chambers Street, Edinburgh
Ruhrthaler	Ruhrthaler Maschinenfabrik Schwarz & Dyckerhoff AG, Mülheim/Ruhr, Germany
RVEL	RVEL Ltd, London Road, Derby, Derbyshire
RVR	Rother Valley Railway Ltd, Robertsbridge, East Sussex
RWH	R.& W.Hawthorn & Co, Forth Bank Works, Newcastle-upon-Tyne (later HL)

S

S	Sentinel (Shrewsbury) Ltd, Harlescott, Shrewsbury, Shropshire. (Diesel locomotives numbered between 10001 and 10183 were actually designed and built by Rolls Royce Ltd, but were fitted with Sentinel worksplates)
Sabero	Hulleras de Sabero y Anexas SA, Sabero, Spain
Sanline	Sanline Systems Ltd, Lucan, Dublin, Ireland
Sara	Sara & Burgess, Penryn, Cornwall
Sartori	Sartori Rides Srl, Montagnana, Italy
Saxby	Frank Saxby & Co, Guildford Works, Surrey
Saxton	C. Saxton, Cheadle Hulme, Greater Manchester
Scarrott	D.J. Scarrott, Kingsteignton, Newton Abbot, Devon
Schalke	Gewerkschaft Schalke Eisenhütte, Gelsenkirchen-Schalke, Nordrhein-Westfalen, Germany
Schichau	Maschinenbau F.Schichau, Maschinen-und Lokomotivfabrik, Elbing, Germany later at Elbtag, Poland
Schöma	Christoph Schöttler Maschinenfabrik GmbH, Diepholz, Niedersachsen, Germany
SchörlingB	Schörling-Brock GmbH, Gehrden, Germany
Schw	L. Schwartzkopff, Berlin, Germany (later became Berliner Maschinenbau AG — see BMAG)
ScienceMus	Science Museum, South Kensington, London.
ScottP	Peter Scott, Hillsborough, Northern Ireland
Scul	Sculfort, Zac des fonds St Jaques, Feignies, France
SCW	I.C.I. Ltd, South Central Workshops, Tunstead, Derbyshire

Sdn	Swindon Works, Wiltshire — Great Western Railway / British Railways
SdnCol	Swindon College, Department of Engineering, North Star Avenue, Swindon
SDSI(S)	South Durham Steel & Iron Co Ltd, Stockton Works, Co.Durham
Selhurst	Selhurst Maintenance Depot, Greater London — British Rail, Southern Region
SET	Stored Energy Technology, Litchurch Lane, Derby
SGLR	Steeple Grange Light Railway, Steeplehouse Junction, Wirksworth, Derbyshire
S&H	Strachan & Henshaw Ltd, Ashton, Bristol
Shackerstone	Shackerstone Station (The Battlefield Line), Leicestershire
SharmanJ	J. Sharman, Stone, Stoke-on-Trent, Staffordshire
Sharon	Sharon Engineering Ltd, Leek, Staffordshire
—	Shelmerdine & Mulley Ltd, Edgeware Road, Cricklewood, London NW2 (see code "LemonB")
ShepherdFG	F.G. Shepherd, Flow Edge Colliery, Middle Fell, Alston, Cumbria
SherwoodF	Sherwood Forest Railway, Edwinstowe, Mansfield, Nottinghamshire
ShooterA	A. Shooter, Oxfordshire
Siemens	Siemens Bros Ltd, London – possibly agents for :-
S&H(B)	Siemens & Halske, Berlin, Germany (until 1903)
SSW	Siemens-Schuckert-Werke, Berlin, Germany (from 1903)
SkinnerD	D. Skinner, 660 Streetsbrook Road, Solihull, West Midlands
SL	Severn-Lamb UK Ltd, Western Road, Stratford-Upon-Avon, Warwickshire then Alcester, Warwickshire.
SLM	Schweizerische Lokomotiv- und Maschinenfabrik, Winterthur, Switzerland
SM	Southern Motors Ltd, Ringsend, Dublin, Ireland
SMH	Simplex Mechanical Handling Ltd, Elstow Road, Bedford (successor to MR)
SmithEL	E.L. Smith, Garsington, Oxfordshire
SmithN	N. Smith, Heatherslaw Light Railway Co, Heatherslaw Mill, Northumberland
SmithP	Smith, Pengam, South Glamorgan
SMR	Saltburn Miniature Railway, Valley Gardens, Saltburn, North Yorkshire
SolHütte	Federal-Mogul Sollinger Hütte GmbH, Uslar, Germany
SouthCrofty	South Crofty Ltd, Pool, near Camborne, Cornwall
SPA	Specialist Plant Associates Ltd, 23 Podington Airfield, Hinwick, Bedfordshire
Spence	Wm.Spence, Cork Street Foundry, Dublin
SPL	Science Projects Ltd, (Constructors), Hammersmith, London
Spondon	Spondon Power Station, Derbyshire Derbyshire & Nottinghamshire Electric Power Co Ltd
Spoor	NV Spoorijzer, Delft, Netherlands
SRS	Swedish Rail Systems Euroc, Solna, Sweden
SS	Sharp, Stewart & Co Ltd, Atlas Works, Manchester (until 1888) later at Atlas Works, Glasgow
StanhopeT	T.Stanhope, Arthington Station, near Leeds, West Yorkshire
Statfold	Statfold Barn Railway, Tamworth, Staffordshire
StationRoad	Station Road Steam Ltd, Metheringham, Lincolnshire
Steamtown	Steamtown Railway Museum, Warton Road, Carnforth, Lancashire
StewartWP	W.P.Stewart, Washington Sheet Metal Works, Industrial Road, Hertburn Industrial Estate, Washington, Co.Durham
Stoke	Stoke Works, Stoke-on-Trent North Staffordshire Railway
StokesMJ	M.J. Stokes, Little West Garden Railway, Southerndown, Mid Glamorgan
Str	Stratford Works, London Great Eastern Railway / London & North Eastern Railway / British Railways
StrawberryHill	Strawberry Hill Depot, South London (British Rail)
StRollox	St.Rollox Works, Glasgow Caledonian Railway / London, Midland & Scottish Railway / British Railways
STRPS	South Tynedale Railway Preservation Society, Alston Station, Cumbria
Strüver	Ad.Strüver AG, Hamburg, Germany
SUSTRACO	Sustainable Transport Co Ltd, Bristol
SVI	SVI S.p.A., Perugia, Italy
SVR	Severn Valley Railway, Bridgnorth, Shropshire
Swanhaven	Swanhaven, Hull, East Yorkshire

Sweet	Brendon Sweet, Polgooth, Cornwall
Syl	Sylvester Steel Co, Lindsay, Ontario, Canada

T

Tambling	N.J. Tambling, Lappa Valley Railway, St. Newlyn East, near Newquay, Cornwall
TargettR	R.C. & J. Targett, Shrewsbury, Shropshire
TaylorB	B.Taylor, 7 Abbey Road, Shepley, Huddersfield, West Yorkshire
TaylorJ	J. Taylor, The Ford, Woolhope, Herefordshire
TeasdaleS	S. Teasdale, Sleights, near Whitby, North Yorkshire
TEE	The Engineering Emporium, Bramcote Works, Bramcote, Warwickshire
Terberg	Terberg Benschop B.V., Benschop, Netherlands
TG	T. Green & Son Ltd, Leeds
TH	Thomas Hill (Rotherham) Ltd, Vanguard Works, Kilnhurst, South Yorkshire
Thakeham	Thakeham Tiles Ltd, Thakeham, Sussex
Thursley	The Thursley Railway, Hampshire
ThurstonS	T.S. Thurston
Thwaites	Thwaites Ltd, Leamington Spa, Warwickshire
TK	Oy Tampella Ab, Tampere, Finland
TK & L	Todd, Kitson & Laird, Leeds
TMA	TMA Automation Ltd, Feeds Automated Systems, Jubilee Works, Erdington, Birmingham
Trackmobile	Trackmobile Ltd, La Grange, Illinois, USA (originally part of the Whiting Corporation [WhC])
TRRF	The Railroad Factory Ltd, Cork, Ireland
TS&S	Track Supplies & Services Ltd, Old Wolverton, Milton Keynes, Buckinghamshire
TU	Task Undertakings Ltd, Birmingham
Tunn	Tunnequip Ltd, Nowhurst Lane, Broadbridge Heath, Horsham, West Sussex
TurnerT	T. Turner, Long Eaton, Derbyshire
TyseleyLW	Tyseley Locomotive Works (Standard Gauge Steam Trust), Tyseley, Birmingham

U

U23A	Uzinele 23 August, Bucuresti, Romania
UCA	UCA, Antwerp, Belgium
UK Loco	UK Loco Ltd, Unit 1, The Heath Works, Main Road, Cropthorne, Worcestershire
Unilok	Unilok locomotives, but whether "(G)" or "(H)" as yet unrecorded.
Unilok(G)	Unilokomotive Ltd, International Division, Mervue Industrial Estate, Galway, Co.Galway, Republic of Ireland
Unilok(H)	Hugo Aeckerle & Co, Hamburg, Germany [some early locomotives had frames constructed by Jung]
Unimog	Unimog road/rail locomotives — Mercedes Benz AG, Stuttgart, Germany
UphillJ	John Uphill

V

Vanstone	D. Vanstone, Pixieland Mini-Zoo, Kilkhampton, near Bude, Cornwall
VE	Victor Electrics Ltd, Burscough Bridge, Lancashire
VER	Volks Electric Railway (Magnus Volk), Madeira Drive, Brighton, Sussex
VF	Vulcan Foundry Ltd, Newton-le-Willows, Lancashire
VIW	Vulcan Iron Works, Wilkes-Barre, Pennsylvania, USA
VL	Vickers Ltd, Barrow-in-Furness, Cumbria
Vollert	Hermann Vollert GmbH & Co KG, Maschinenfabrik, Weinsberg/Wurtt, Germany
Volvo	AB Volvo, Gothenburg, Sweden

W

WalkerG	G. Walker, Lakeside Miniature Railway, Marine Lake, Southport, Merseyside
WalkerS	Samuel Walker
Waterfield	James Waterfield, Boston, Lincolnshire
Watson&Haig	Watson & Haig Ltd, Andover, Hampshire
WB	W.G.Bagnall Ltd, Castle Engine Works, Stafford
Wbton	Wilbrighton Wagon Works, Shropshire
WCI	Wigan Coal & Iron Co Ltd, Kirkless, Wigan, Lancashire
WeaverP	P. Weaver, New Farm, Lacock, near Corsham, Wiltshire

Werk	Werkspoor NV, Utrecht, Holland
WhC	Whiting Corporation, Harvey, Illinois, U.S.A.
WHR(GF)	Welsh Highland Light Railway (1964) Ltd, Gelert's Farm Works, Porthmadog, North Wales
WilliamsWJ	W.J. Williams, Blaenau Ffestiniog, North Wales
Wilmott	Wilmott Bros (Plant Services) Ltd, Ilkeston, Derbyshire
WilsonAJ	A.J. Wilson, 6 Trentdale Road, Carlton, Nottingham
Wilton(ICI)	Wilton Works (Imperial Chemical Industries Ltd), Middlesbrough, North Yorkshire
Windhoff	Rheiner Maschinenfabrik Windhoff AG, Rheine, Germany later WINDHOFF Bahn-und Anlagentechnik GmbH, Rheine, Germany
Winson	Winson Engineering, Porthmadog (later Penrhyndeudraeth), North Wales and at Daventry, Northamptonshire
WkB	Walker Bros (Wigan) Ltd, Wigan, Lancashire
Wkm	D.Wickham & Co Ltd, Ware, Hertfordshire
WkmR	Wickham Rail, Bush Bank, Suckley, Worcestershire (successors to Bruff & Wkm)
WLLR	West Lancashire Light Railway, Hesketh Bank, near Preston, Lancashire
W&LLR	Welshpool & Llanfair Light Railway Preservation Co Ltd, Llanfair Caereinion, Powys, Mid Wales
WMD	Waggon & Maschinenbau GmbH, Donauwörth, Germany
Wolverton	Wolverton Works, Buckinghamshire — British Rail Engineering Ltd
Woodings	Woodings Railcar Ltd, Alexandria, Ontario, Canada
Woolwich	Woolwich Arsenal, London
WR	Wingrove & Rogers Ltd, Kirkby, Liverpool (successors to BEV)
WSR	West Somerset Railway, Williton, Somerset
WVanHeiden	W. Van der Heiden, Rotterdam, Netherlands

Y

YE	Yorkshire Engine Co Ltd, Meadow Hall Works, Sheffield
YEC	Yorkshire Engine Company Ltd, Unit 7, Meadow Bank Industrial Estate, Harrison Street, Rotherham, South Yorkshire; *and later at* Unit A3, Templeborough Enterprise Park, Bowbridge Close, Rotherham, South Yorkshire
York	York Works, York, North Yorkshire North Eastern Railway/ London & North Eastern Railway/ British Railways
York(BRE)	York Works (British Rail Engineering Ltd), North Yorkshire
Young&Co	J. Young & Co, Leeds

Z

Zagro	ZAGRO Bahn-und Baumaschinen GmbH, Bad Rappenau-Grombach, Germany
Zephir	Zephir S.p.a., via Salvador Allende 85, 41100 Modena, Italy
Zetor	The Zetor Tractor Factory Co, Brno, Czech Republic
Zweiweg	Zweiweg International GmbH & Co KG, Leichlingen, Germany (also Sehnde, Germany)
Zwiehoff	G. Zwiehoff GmbH, Tegernseestrabe 15, 83022 Rosenheim, Germany

9

9E	Nine Elms Works, London — London & South Western Railway

CONTRACTORS

Listed below are the Civil Engineering Contractors / Plant Hire specialists who own locomotives for use on tunnelling and sewer contracts, etc. The locomotives are to be found in all parts of the Country but the details of the locomotive fleets are listed under the firm's main depot in the County shown below.

TITLE OF FIRM	COUNTY
Amey plc / Byzac Contractors Ltd	Lancashire
Costain Group plc	Warwickshire
D.C.T. Civil Engineering	Greater Manchester
V.J. Doneghan (Plant) Ltd	Greater Manchester
Fineturret Ltd	Greater Manchester
Magnor Plant	Staffordshire
McNicholas Construction Co Ltd	Hertfordshire
J.Murphy & Sons Ltd	Greater London
Edmund Nuttall Ltd	Greater London
South Western Mining & Tunnelling Ltd	Cornwall
Specialist Plant Associates Ltd	Bedfordshire
A.E.Yates Ltd	Greater Manchester

HIRE FLEET OPERATORS

A number of companies are particularly active in the locomotive hire business, maintaining a fleet of locomotives for hire or sale. Such locomotives are hired out to industrial users and main-line "TOC" (Train Operating Companies). As such the locomotives can be seen all over the country, frequently on short-term loan. Hire Fleet Operators are listed in this book in the county of their principal or headquarters address, together with known details of their current fleet. In several cases, however, few or no locomotives are stabled at the addresses given.

Alan Keef Ltd (narrow gauge locomotives)	Herefordshire
British American Railway Services Ltd	Co. Durham
L H Group Services Ltd,	Staffordshire
Ed Murray & Sons Ltd	Teesside
Harry Needle Railroad Co Ltd	Derbyshire
Northumbria Rail Ltd	Northumberland
E.G. Steele & Co Ltd	North Lanarkshire
Railway Supply Services Ltd	Warwickshire
Traditional Traction	Essex
Wabtec Rail Ltd	South Yorkshire

RAILWAY CONTRACTORS

Listed below are the Railway Engineering Contractors / Plant Hire specialists who own locomotives for use on railway contracts etc. The locomotives are to be found in all parts of the Country but the details of the locomotive fleets are listed under the firm's main depot in the County shown below.

Avondale Environmental Services Ltd	Kent
Contracked Lands Ltd	Kent
Marsh Trackwork	Cheshire
W & D McCulloch	Ayrshire
QTS Rail Ltd	Ayrshire
Stobart Rail Ltd	Cumbria
Story Rail Ltd	Cumbria
TRAC International Ltd	Lanarkshire
TXM Plant Ltd	Greater Manchester

SECTION 1 — ENGLAND

‡ "Tyne & Wear" has been replaced by a number of Unitary Authorities. Rather than fragment our listings, we have amalgamated locations South of the River Tyne into the County Durham listing and locations North of the river into the Northumberland listing.

BEDFORDSHIRE

INDUSTRIAL SITES

CRANFIELD SAFETY & ACCIDENT INVESTIGATION CENTRE, SCHOOL OF ENGINEERING, ACADEMIC OPERATIONS UNIT, CRANFIELD UNIVERSITY, CRANFIELD — MK43 0AL

Gauge 4ft 8½in www.cranfield.ac.uk

69933	(390033)	4w-4wWER	Alstom		2002

GOVIA THAMESLINK RAILWAY LTD, BEDFORD CAULDWELL LIGHT MAINTENANCE DEPOT, CAULDWELL WALK, BEDFORD — MK42 9DT

(subsidiary of Govia Ltd, a Go-Ahead Group and Keolis Joint Venture)

Gauge 4ft 8½in www.thameslinkrailway.com TL 044486

See section 7 for details

HARMILL SYSTEMS LTD, UNIT B1, CHERRYCOURT WAY, LEIGHTON BUZZARD — LU7 4UH

www.harmill.co.uk **SP 940249**

New Harmill locomotives under construction or repair occasionally present

SPECIALIST PLANT LTD, PODINGTON

Locomotives present in yard between contracts and in store for third parties.

Gauge 2ft 6in www.specialistplant.co.uk **SP 948608**

03-418	(SP 418)	4wDH		CE	B1563Q	1978
	SP 419	4wDH		CE		
SP 543		4wBE		CE	B4056B	1995
SP 1122		4wBE		CE	B4057B	1994

Gauge 2ft 0in / 1ft 6in

SP 83		4wBE		CE	5942A	1972
	RR51-069	4wBE		CE	5943	1972
	RR51-071	4wBE		CE	5949D	1972
SP 88		4wBE		CE	B0402A	1974
SP 90		4wBE		CE	B0402C	1974
SP 86		4wBE		CE		1970
			reb	SPA		1995
	RR51 7002	4wBE		CE	B3686D	1990
			reb	CE	B4075.2	1995

Gauge 2ft 0in

SP 1014		4wBE		CE		1968
SP 1015		4wBE		CE		1968
(SP 972)		4wBE		CE	B3686A	1990
(SP 973)	72292/2	4wBE		CE	B3686B	1990

RR51 7001	4wBE		CE	B3686C	1990	
		reb	CE	B4075.1	1995	
4	4wBE		CE	B4010A	1994	
3	4wBE		CE	B4010B	1994	

Gauge 1ft 6in

SP 84	4wBE		CE	5942B	1972	
(SP 1273)	4wBE		CE	B0111C	1973	OOU

PRESERVATION SITES

BEDFORDSHIRE STEAM ENGINE PRESERVATION SOCIETY, CLIFTON

Gauge 2ft 0in www.bseps.org.uk

THE IRON & STEEL RAIL CO LTD No.4	4wVBT	VCG	ForshawJJ		c1998
1 VERACRUZ	0-6-0WT	OC	OK	11009	1925

GREAT WOBURN RAILWAY, PETER SCOTT WOBURN SAFARI PARK, WOBURN, near MILTON KEYNES **MK17 9QN**

Gauge 1ft 8in www.woburnsafari.co.uk **SP 962343**

LADY ALEXANDRA	0-4-0DH	s/o	AK	70	2004

LEIGHTON BUZZARD NARROW GAUGE RAILWAY SOCIETY

Locomotives are kept at :- Pages Park Shed LU7 4TG SP 929242
Stonehenge Works SP 941275

Gauge 2ft 6in www.buzzrail.co.uk

5	LBC 1 NUTTY	4wVBT	ICG	S	7701	1929
	WD 767139	2w-2PMR		Wkm	3282	1943

Gauge 2ft 0in

4	DOLL	0-6-0T	OC	AB	1641	1919
	SEZELA No.4	0-4-0T	OC	AE	1738	1915
	ELIDIR	0-4-0T	OC	AE	2071	1933
No.3	RISHRA	0-4-0T	OC	BgC	2007	1921
778		4-6-0T	OC	BLW	44656	1917
1	CHALONER	0-4-0VBT	VC	DeW		1877
	PENLEE	0-4-0WT	OC	Freud	73	1901
			reb	ARC		c1983
(HF 2023) (5)		0-8-0T	OC	KraussM	7455	1918
	PETER PAN	0-4-0ST	OC	KS	4256	1922
2	PIXIE	0-4-0ST	OC	KS	4260	1922
No.11	P.C.ALLEN	0-4-0WT	OC	OK	5834	1912
	PEDEMOURA	0-6-0WT	OC	OK	10808	1924
"5"	ELF	0-6-0WT	OC	OK	12740	1936
–		0-4-0WT	OC	OK	2544	1907
NG 46		4wDH		BD	3698	1973
NG 23		4wBE		BD	3702	1973
			reb	AB		1987

81	PETER WOOD	4wDH		HE	9347	1994	
No.5	ISABEL	4wDM		MR	5608	1931	a
No.6	1	4wDM		MR	5875	1935	b
"7"	8986 FALCON	4wDM		OK	8986	1938	
No.8		2w-2DMR		Bg	3539	1959	
			reb	AK		1988	c
A.M.W. No.165	FIRE "TRUMPTON"	4wDM		RH	194784	1939	
8	"GOLLUM"	4wDM		RH	217999	1942	d
"9"	"MADGE"	4wDM		OK	7600	1938	
10	21 HAYDN TAYLOR	4wDM		MR	7956	1945	
12	CARBON	4wPM		MR	6012	1930	
13	ARKLE	4wDM		MR	7108	1936	
"14"		4wDM		HE	3646	1946	d
"15"		4wDM		FH	2514	1941	
"16"		4wDM		L	11221	1939	
No.17	DAMREDUB	4wDM		MR	7036	1936	
"18"	FËANOR	4wDM		MR	11003	1956	
"19"		4wDM		MR	11298	1965	Dsm
20	60S317 "VILYA"	4wDM		MR	60S317	1966	
No.21	FESTOON	4wPM		MR	4570	1929	
"22"	"FINGOLFIN"	4wDM		LBNGRS	1	1989	e
23		4wDM		RH	164346	1932	
24		4wDM		MR	11297	1965	
"25"		4wDM		MR	7214	1938	
"26"	YIMKIN	4wDM		RH	203026	1942	f
27	POPPY	4wDH		RH	408430	1957	
28	R.A.F. STANBRIDGE	4wDM		RH	200516	1940	g
29	CREEPY YARD No. P 19774	4wDM		HE	6008	1963	
No.30	MANIFOLD MAGGIE	4wDM		MR	8695	1941	
31	No.37658 THORIN OAKENSHIELD	4wPM		L	4228	1931	
"32"		4wDM		RH	172892	1934	
No.34	RED RUM	4wDM		MR	7105	1936	
"35"	9303/507	0-4-0DM		HE	6619	1966	
"36"	"CARAVAN"	4wDM		MR	7129	1936	
No.38	HARRY BARNETT	4wDM		L	37170	1951	
LM 39	T.W.LEWIS	4wDM		RH		1954	d h
"40"	TRENT	4wDM		RH	283507	1949	
41	LOD/758054 SOMME	4wDM		HE	2536	1941	i
"42"	"SARAH"	4wDM		RH	223692	1943	
43	TROTTER	4wDM		MR	10409	1954	
44		4wDM		MR	7933	1941	
"45"		4wDM		MR	21615	1957	Dsm
"46"		4wDM		RH	209430	1942	
"48"	MCNAMARA	4wDM		HE	4351	1952	
80	BEAUDESERT	4wDH		SMH	101T018	1979	
			reb	AK	59R	1999	
No.131		4wDM		MR	5613	1931	Dsm
No.1568		4wPM		FH	1568	1927	
WD2182	No. LR 10227	4wPM		MR	461	1917	
R8		4wDM		MR	5612	1931	Dsm
LOD	758009	4wDM		MR	8641	1941	

26	ANNA	4wDM	MR	8720	1941	
LOD	758220	4wDM	MR	8745	1942	
	BLUEBELL	4wDM	FH		1943	j
	"REDLANDS"	4wDM	MR	5603	1931	Dsm
–		4wDM	MR	8731	1941	Dsm
–		4wDM	MR			Dsm
–		4wDM	RH	187105	1937	
–		4wDM	RH	218016	1943	Dsm
–		4wDM	RH	229656	1944	
RTT/767182		2w-2PMR	Wkm	2522	1938	d

a converted into mobile compressor for air braking
b converted into a brake van
c rebuilt from 4ft 8½in gauge 2w-2DMR, now a bogie unpowered coach
d stored off site
e built from parts of RH 425798/1958 and RH 444207/1961
f nameplate is in Arabic
g carries plate RH 200513
h either RH 375315 or RH 375316
i on loan to the Chemin de Fer Froissy-Cappy-Dompierre, Somme, France
j either FH 2631 or FH 2834

Chamberlain's Branch Junction, Vandyke Road, Leighton Buzzard

Gauge 2ft 0in **SP 931262**

24	4wDM	MR	4805	1934	

THE LIGHT RAILWAY ASSOCIATION, STEVINGTON & TURVEY LIGHT RAILWAY, WOBURN

Gauge 2ft 0in

ROOBARB // JOY DIVISION		4wDM	Diema	1600	1953	
–		4wPM	FH	1767	1931	Dsm
No.15	OLDE	4wDM	HE	2176	1940	
No.1	PAUL COOPER	4wDM	MR	9655	1951	
–		4wDM	OK	3685	1929	
11	NEEDHAM	4wDM	OK	6504	1936	
–		4wDM	OK	7728	1937	Dsm
14	No.21	4wDM	RH	373359	1954	

Gauge 1ft 6in

1514	0-4-0BE	WR			a

a currently under renovation elsewhere

Mrs. MACKINNON

Gauge 1ft 8in

–	4wDM	OK	6703	1936	Pvd

Locomotive believed to be preserved by a member of the family at an unknown location.

RAY MASLEN & FRIENDS, PRIVATE RAILWAY, ARLESEY

Gauge 2ft 0in

CLARABEL	4wDMF		HE	4758	1954	
–	4wDM		L	37911	1952	
–	4wDM		RH	441951	1960	
RTT/767187	2w-2PM		Wkm	2559	1939	Dsm
RTT/767094	2w-2PM		Wkm	3033	1941	Dsm

WHIPSNADE WILD ANIMAL PARK LTD, 'JUMBO EXPRESS' GREAT WHIPSNADE RAILWAY, WHIPSNADE ZOO, DUNSTABLE **LU6 2LF**

Gauge 2ft 6in www.zsl.org/zsl-whipsnade-zoo **TL 004172**

No.2 EXCELSIOR	0-4-2ST	OC	KS	1049	1908
No.4 SUPERIOR	0-6-2T	OC	KS	4034	1920
3	4wDH		BD	3780	1983
No.8 VICTOR	0-6-0DM		JF	4160004	1951
No.9 HECTOR	0-6-0DM		JF	4160005	1951

BERKSHIRE

INDUSTRIAL SITES

FIRST GREATER WESTERN LTD, t/a GREAT WESTERN RAILWAY, READING TRAINCARE DEPOT, 101 COW LANE, READING **RG1 8FN**

(part of the First Group plc)

Gauge 4ft 8½in www.gwr.com SU 701741

See section 7 and Appendix 1 for details

PRESERVATION SITES

D. BUCK, near WINDSOR

Locomotives are kept at a private site.

Gauge 1524mm

1016 LADY PATRICIA	4-6-2	OC	TK	946	1955

Gauge 4ft 8½in

334\G102 SIR VINCENT	4wWT	G	AP	8800	1917
HORNPIPE	0-4-0ST	OC	P	1756	1928
–	0-4-0DM		AB	352	1941
ARMY 9112	4wDMR		Bg	3538	1959
"TRACKRAT"	2-2wDHR		Group Eng	RAT2001	c1998
–	2w-2PMR		Bance	062	1998

LEGOLAND WINDSOR PARK LTD, LEGOLAND WINDSOR, WINKFIELD ROAD, WINDSOR (part of the Merlin Entertainments Group) **SL4 4AY**
Gauge 2ft 0in www.legoland.co.uk **SU 939746**

	AMEY	4wDM		MR	7902	1939	OOU
–		4-4-0DH	s/o	SL	663	1995	

PETER SMITH, NEWBURY
Gauge 1ft 11½in

	(GELLI)	0-4-0VBT	VC	DeW		1893	Dsm
	SUZANNE	4wDM		LB	55730	1968	

NICK WILLIAMS, Private Railway, READING
Gauge 2ft 0in

9	JACK	0-4-0T	OC	AB	1871	1925	
2	SEZELA No.2	0-4-0T	OC	AE	1720	1915	
6	SEZELA No.6	0-4-0T	OC	AE	1928	1923	
–		4wDM		FH	2163	1938	
–		4wDM		LB	53225	1962	
	LR 2478	4wPM		MR	1757	1918	
–		4wDM		MR	7512	1938	
–		4wDM		MR	11264	1964	
–		4wDM		RH	277265	1949	
–		4wDM		RH	296091	1949	
–		0-4-0BE		WR	N7661	1974	
RTT/767186		2w-2PM		Wkm	2558	1939	Dsm

WINDSOR STATION LTD, WINDSOR & ETON CENTRAL STATION, THAMES STREET, WINDSOR **SL4 1PJ**
Gauge 4ft 8½in **SU 969773**

3041	THE QUEEN	4-2-2		Steamtown		1983	
	non-working replica of Sdn 1401 / 1894						

BRISTOL

INDUSTRIAL SITES

COUNTY OF AVON FIRE BRIGADE, AVONMOUTH FIRE STATION, ST. ANDREWS ROAD, AVONMOUTH **BS11 9HQ**
Gauge 4ft 8½in www.avonfire.gov.uk **ST 517784**

N842 HFB	99709 979098-9	4wDM	R/R	Renault		1995	

FIRST GREATER WESTERN LTD, t/a GREAT WESTERN RAILWAY,
ST. PHILIP'S MARSH T&RSMD, ALBERT ROAD, BRISTOL **BS2 0GW**
(part of the First Group plc)
Gauge 4ft 8½in www.gwr.com **ST 608721**

See Section 7 and Appendix 1 for details

R.& A. GUNN, BRISTOL
Gauge 1ft 10in — locomotive in storage.

–	4wBE	WR	2489	1943

HITACHI RAIL EUROPE LTD,
STOKE GIFFORD IEP DEPOT, LITTLE STOKE, BRISTOL **BS34 7QG**
Gauge 4ft 8½in www.hitachirail-eu.com **ST 612801**

–	4wBE	_(Express ¯(Hegenscheidt	2016 2016

LONDON & NORTH WESTERN RAILWAY CO LTD, t/a ARRIVA TRAINCARE,
BARTON HILL DEPOT, DAY'S ROAD, ST. PHILIPS, BRISTOL **BS2 0QS**
(Arriva - A DB company) Preserved locomotives between Mainline duties occasionally present.
Gauge 4ft 8½in www.arrivatc.com **ST 604728**

(D3678) 08516	0-6-0DE	Dar	1958

PRESERVATION SITES

M SHED, BRISTOL HARBOUR RAILWAY,
WAPPING ROAD, PRINCES WHARF, CITY DOCK, BRISTOL **BS1 4RN**
Gauge 4ft 8½in www.bristolmuseums.org.uk/m-shed **ST 585722**

I.W. & D. 34	0-6-0ST	OC	AE	1764	1917
3	0-6-0ST	OC	FW	242	1874
HENBURY	0-6-0ST	OC	P	1940	1937
–	0-4-0DM		RH	418792	1959

JOE NEMETH ENGINEERING LTD,
WASHINGPOOL FARM, MAIN ROAD, EASTER COMPTON, BRISTOL **BS35 5RE**
Gauge 2ft 0in www.joenemethengineeringltd.com **ST 570831**

DINORWIC	4wPM	NemethJ	AV0509	2009

Visitors welcome by appointment only.

BUCKINGHAMSHIRE

INDUSTRIAL SITES

CHILTERN RAILWAYS CO LTD, AYLESBURY DEPOT, LEACH ROAD, AYLESBURY

(A subsidiary of DB Regio, part of the Deutsche Bahn AG Group) **HP21 8LG**

Gauge 4ft 8½in www.chilterntrains.co.uk **SP 816134**

See Section 7 for details

GEMINI RAIL SERVICES UK LTD, WOLVERTON WORKS, STRATFORD ROAD, WOLVERTON (part of the Gemini Rail Group) **MK12 5NT**

Gauge 4ft 8½in www.geminirailgroup.co.uk **SP 812413**

(D3796)	08629 WOLVERTON	0-6-0DE	Derby		1959
(D3816)	08649 BRADWELL	0-6-0DE	Hor		1959
	(TITCHIE)	4wDM	SMH	103GA078	1978

a non-working replica

PRESERVATION SITES

Mrs. BRITT, c/o SUKAMI, 43 WEST STREET, STEEPLE CLAYDON **MK18 2NS**

Gauge 750mm **SP 693268**

–		0-6-0WT	OC	Chrz	2959	1951

THE FAWLEY HILL RAILWAY, FAWLEY HILL, FAWLEY GREEN, near HENLEY-on-THAMES

Gauge 4ft 8½in www.fawleyhill.co.uk **SU 755861**

SIR ROBERT McALPINE & SONS (LONDON) LIMITED No.31

		0-6-0ST	IC	HC	1026	1913
D2120	(03120)	0-6-0DM		Sdn		1959
–		4wDM		FH	3817	1956
	ERNIE	4wDM		FH	3894	1958
			reb	AB	6930	1988
	R436 XRA	4wDM	R/R	_(Landrover		1998
				(Perm		1998

GULLIVERS LAND, LIVINGSTON DRIVE, NEWLANDS, MILTON KEYNES **MK15 0DT**

Gauge 1ft 3in www.gulliversfun.co.uk **SP 872399**

–	4+6wDE	s/o	Gullivers	1999

LAVENDON NARROW GAUGE RAILWAY, HARROLD ROAD, LAVENDON MK46 4HU

Gauge 3ft 0in www.lavendonconnection.co/home/lngr/ **SP 922537**

–		4wBE		GB			

MILTON KEYNES MUSEUM, McCONNELL DRIVE, WOLVERTON, MILTON KEYNES MK12 5EL

Gauge 4ft 8½in www.mkmuseum.org.uk **SP 820404**

1009	WOLVERTON	2-2-2	IC	Mercia		1991	a

a non-working replica

QUAINTON RAILWAY SOCIETY LTD, BUCKINGHAMSHIRE RAILWAY CENTRE, QUAINTON ROAD STATION, STATION ROAD, QUAINTON, near AYLESBURY HP22 4BY

Gauge 4ft 8½in www.bucksrailcentre.org **SP 736189, 739190**

3020	CORNWALL	2-2-2	OC	Crewe		1858	
6989	WIGHTWICK HALL	4-6-0	OC	Sdn		1948	
7200		2-8-2T	OC	Sdn		1934	
(7715)	LT 99	0-6-0PT	IC	KS	4450	1930	
9466		0-6-0PT	IC	RSHN	7617	1952	
30585		2-4-0WT	OC	BP	1414	1874	
	SWANSCOMBE	0-4-0ST	OC	AB	699	1891	
–		0-4-0F	OC	AB	1477	1916	
	LAPORTE	0-4-0F	OC	AB	2243	1948	
–		4wWT	G	AP	807	1872	
	SYDENHAM	4wWT	G	AP	3567	1895	
No.1	SIR THOMAS	0-6-0T	OC	HC	1334	1918	
	MILLOM	0-4-0ST	OC	HC	1742	1946	
	ARTHUR	0-6-0ST	IC	HE	3782	1953	
–		0-6-0ST	IC	HE	3890	1964	
–		0-4-0ST	OC	HL	3717	1928	
No.4	"SWANSCOMBE"	0-4-0ST	OC	HL	3719	1928	a
	COVENTRY No.1	0-6-0T	IC	NBH	24564	1939	
	ANNIE	0-4-0ST	OC	P	1159	1908	
–		0-4-0T	OC	P	1900	1936	
	ROKEBY	0-4-0ST	OC	P	2105	1950	b
12		4wVBT	VCG	S	6515	1926	
11	CYNTHIA	4wVBT	VCG	S	9366	1945	
5208		2w-2-2-2w-4-4	12CGR	S	9418	1950	
	"SCOTT"	0-4-0ST	OC	WB	2469	1932	
	CHISLET	0-6-0ST	OC	YE	2498	1951	
D2298		0-6-0DM		_(RSHD (DC	8157 2679	1960 1960	
249	ESL 118A/ESL 118B	4w-4+4-4wRE	 reb	BRCW Acton		1932 1961	
T1		4wDM		FH	2102	1937	
	"WALRUS"	0-4-0DM		FH	3271	1949	
	TARMAC	4wDM		FH	3765	1955	
WD 849	ESSO	0-4-0DM		HE	2067	1940	

GWR No.1	OSRAM	0-4-0DM		JF	20067	1933	
	REDLAND	4wDM		KS	4428	1930	
–		0-4-0DE		RH	425477	1959	
			reb	Resco		1979	
(01585)		0-6-0DH		RH	459518	1961	
1139	HILSEA	4wDM		RH	463153	1961	
M51886		2-2w-2w-2DMR		DerbyC&W		1960	
M51899	AYLESBURY COLLEGE	2-2w-2w-2DMR		DerbyC&W		1960	
53028		2w-2-2-2wRER		BRCW		1938	
54233		2w-2-2-2wRER		GRC&W		1939/40	
			reb	Acton		1941	
ARMY 9040		2w-2PMR		Wkm	6963	1955	
RLC 009037		2w-2PMR		Wkm	8197	1958	Dsm
	TP 53P	2w-2PMR		Wkm	8263	1959	
–		2w-2PMR		Geismar	97/15	1997	

a includes parts from HL 3718
b actually built in 1948 but plates dated as shown

Gauge 3ft 6in

3405	JANICE	4-8-4	OC	NBH	27291	1953

Gauge 2ft 0in

803	2w-2-2-2wRE	EEDK	803	1931	a

a built 1931 but originally carried plates dated 1930

SIR JEREMY SULLIVAN, WOTTON LIGHT RAILWAY, near AYLESBURY

Gauge 1ft 3in

SANDY	0-6-0T	OC	ESR	301	1996
PAM	0-4-0DH		AK	52	1996
POMPEY	4w-4wDH		AK	64	2001

CAMBRIDGESHIRE

INDUSTRIAL SITES

ABELLIO EAST ANGLIA LTD, c/o ARRIVA TRAINCARE, CAMBRIDGE DEPOT, COLDHAMS LANE, CAMBRIDGE — CB1 3EW

(Arriva - a DB company)

Gauge 4ft 8½in www.greateranglia.co.uk / www.arrivatc.com **TL 468586**

(D3673)	08511	0-6-0DE	Dar	1958	a

a property of Railway Support Services Ltd, Wishaw, Warwickshire

JOHN BRADSHAW LTD, t/a BRADSHAW ELECTRIC VEHICLES, NEW LANE, STIBBINGTON, PETERBOROUGH — **PE8 6LW**

www.bradshawev.com — TL 083981

Bradshaw vehicles under construction or repair may be present

D.T. WAREHOUSING LTD, ROAD & RAIL DISTRIBUTION CENTRE, QUEEN ADELAIDE WAY, ELY — **CB7 4UB**

Gauge 4ft 8½in www.dtwarehousing.co.uk — TL 563810

–	4wDH	TH	184V	1967	a

a property of Railway Support Services Ltd, Wishaw, Warwickshire

EUROPEAN METAL RECYCLING LTD, MAYER PARRY RECYCLING, 111 FORDHAM ROAD, SNAILWELL, NEWMARKET — **CB8 7ND**

Gauge 4ft 8½in www.emrgroup.com — **TL 638678**

ARMY 410	0-4-0DH	NBQ	27645	1958
468048	0-6-0DH	RH	468048	1963

PRESERVATION SITES

C. CROSS, 4 PACIFIC CLOSE, MARCH

Gauge 4ft 8½in — **TF 415975**

RTU 7417	2w-2BER	Geismar 5E/0001	2003

IMPERIAL WAR MUSEUM, DUXFORD AERODROME — **CB22 4QR**

Gauge 2ft 0in www.iwm.org.uk — **TL 456456**

–	4wPM	MR	1364	1918
–	4wPM	MR	3849	1927

NENE VALLEY RAILWAY LTD

Locomotives are kept at :-

Wansford Depot and Yard PE8 6LR TL 091978
Wansford Tunnel TL 087976
Ferry Meadows Station PE2 5UU TL 151970

Gauge 4ft 8½in www.nvr.org.uk

34081	92 SQUADRON	4-6-2	3C	Bton		1948
73050	CITY OF PETERBOROUGH	4-6-0	OC	Derby		1954
	TOBY	0-4-0VBT	OC	Cockerill	1626	1890
Nr.656		0-6-0T	OC	Frichs	360	1949
	DEREK CROUCH	0-6-0ST	IC	HC	1539	1924
1	THOMAS	0-6-0T	OC	HC	1800	1947
	JACKS GREEN	0-6-0ST	IC	HE	1953	1939
75006		0-6-0ST	IC	HE	2855	1943
75008	SWIFTSURE	0-6-0ST	IC	HE	2857	1943
64.305		2-6-2T	OC	Krupp	1308	1934

1178		2-6-2T	OC	Motala	516	1914	
S.V.J.B	101A	4-6-0	OC	Nohab	2082	1944	
(D5801)	31271 STRATFORD 1840 – 2001	A1A-A1ADE		BT	302	1961	
D9520	45	0-6-0DH		Sdn		1964	
(D)9529		0-6-0DH		Sdn		1965	
801		Bo-BoDE		Alco	77120	1950	a
–		0-4-0DH		EEV	D1123	1966	
–		4wPM		FH	2895	1944	
–		4wDM		FH	2896	1944	OOU
323.674-2	SIMONSIDE / SPLUTTER	4wDH		Gmd	4991	1957	a
–		0-4-0DM		RH	304469	1951	
	BARABEL	0-4-0DH		RR	10202	1964	
DL83		0-6-0DH		RR	10271	1967	
DR 98500		4wDHR		Plasser	52788	1985	
1212	HELGA	4w-4DMR		EK		1958	a
–		2w-2PMR		Bance	023	1995	
MPP 10188		2w-2BER		Bance	098	2000	a
(960243)	(749) NVR 1612	2w-2PMR		Wkm	1642	1934	

a property of Northumbria Rail Ltd, Bedlington, Northumberland

The Night Mail Museum, Ferry Meadows, Nene Park, Peterborough **PE2 5UU**
Gauge 2ft 0in www.nightmail.org.uk **TL 151970**

807	[211 212]	2w-2-2-2wRE	EEDK	807	1931

S. OWEN, OLD NORTH ROAD STATION, LONGSTOWE, near BOURN **CB23 2TZ**

Gauge 4ft 8½in **TL 313546**

NL 1305	2-2w-2w-2RER	MetCam	1961

C.J. & A.M. PEARMAN, HUNTINGDON AREA

Gauge 1ft 11½in

–	4wPM	MR	2059	1920

loco stored at a private location

RAILWORLD, (MUSEUM OF WORLD RAILWAYS), OUNDLE ROAD, WOODSTON, PETERBOROUGH **PE2 9NR**

Gauge 4ft 8½in www.railworld.org.uk **TL 188982**

–	4-6-2	4C	Frichs	415	1950
804	Bo-BoDE		Alco	77778	1950

SHEPRETH WILDLIFE PARK, WILLERSMILL ROAD, SHEPRETH **SG8 6PZ**

Gauge 400mm www.sheprethwildlifepark.co.uk **TL 393481**

–	4-6wRE	s/o	UK Loco	2007

S. THOMASON, SPRINGFIELD AGRICULTURAL RAILWAY, PRIVATE LOCATION, HUNTINGDON

Gauge 2ft 0in www.ingr.co.uk

SP 203	4wBE	CE	B0176A	1974
–	4wBE	WR	3557	1946
12	4wDM	Moës		
112	4wDM	Spoor	112	1952

Gauge 600mm

2	4wDM	Diema	1553	1953

CHESHIRE

INDUSTRIAL SITES

ALSTOM TRANSPORT TECHNOLOGY CENTRE, LOVEL'S WAY, HALEBANK, WIDNES (part of Alstom Holdings S.A.) (WA8 8FZ)

Gauge 4ft 8½in www.alstom.com **SJ 484848**

(D3569) 08454	0-6-0DE	Derby		1958
(D3889) 08721	0-6-0DE	Crewe		1960

BOMBARDIER TRANSPORTATION UK LTD, CREWE WORKS, WEST STREET, CREWE CW1 3JB

Gauge 4ft 8½in www.bombardier.com **SJ 691561**

	4wDH	R/R	NNM	77501	1980	
1	4wDH	R/R	NNM	82503	1983	a
T2	4wDH	R/R	NNM	83501	1983	a
T5	4wDH	R/R	NNM	83504	1984	a

a one of these locomotives has been sold to E.G. Steele, Hamilton

CROGHAN PEAT INDUSTRIES LTD, WILMSLOW PEAT FARM, LINDOW MOSS, MOOR LANE, WILMSLOW SK9 6DN

Gauge 2ft 0in R.T.C. **SJ 823803**

–	4wDM	AK	No.4	1979	OOU
–	4wDH	LB	50888	1959	OOU
–	4wDM	LB	52528	1961	Dsm

LOCOMOTIVE STORAGE LTD, CREWE DIESEL DEPOT, off NANTWICH ROAD, CREWE CW2 6HR

Gauge 4ft 8½in www.iconsofsteam.com **SJ 711542**

5029 NUNNEY CASTLE	4-6-0	4C	Sdn	1936	
34046 BRAUNTON	4-6-2	3C	Bton	1946	a

(35022)	(HOLLAND-AMERICA LINE)	4-6-2	3C	Elh		1948	Dsm
(35027)	(PORT LINE)	4-6-2	3C	Elh		1948	
35028	CLAN LINE	4-6-2	3C	Elh		1948	
45231	THE SHERWOOD FORESTER	4-6-0	OC	AW	1286	1936	
(46100)	6100 ROYAL SCOT	4-6-0	3C reb	Derby Crewe		1930 1950	
60532	BLUE PETER	4-6-2	3C	Don	2023	1948	
70000	BRITANNIA	4-6-2	OC	Crewe		1951	
D213	(40013) ANDANIA	1Co-Co1DE		_(EE (VF	2669 D430	1959 1959	
(D3798)	08631	0-6-0DE		Derby		1959	
(D3905)	08737	0-6-0DE		Crewe		1960	
(D9000)	55018 BALLYMOSS (55022 ROYAL SCOTS GREY)	Co-CoDE		_(EE (VF	2905 D557	1960 1960	
D9016	(55016) GORDON HIGHLANDER	Co-CoDE		_(EE (VF	2921 D573	1961 1961	
W55034	121034	2-2w-2w-2DMR		PSteel		1960	
B30W	PWM 3956	2w-2PMR		Wkm	6941	1955	

a currently carries r/n 34052 and name LORD DOWDING

LONDON & NORTH WESTERN RAILWAY COMPANY LTD, t/a ARRIVA TRAINCARE, CREWE CARRIAGE WORKS, off WESTON ROAD, CREWE CW1 6NE

Mainline and preserved vehicles usually present.

Gauge 4ft 8½in www.arrivatc.com **SJ 715538**

(D3884	08717) 09204	0-6-0DE	Crewe	1960
(D4036)	08868	0-6-0DE	Dar	1960

MARSH TRACKWORKS - TRACK RECOVERY SERVICES, MANISTY WHARF, NORTH ROAD, ELLESMERE PORT

Gauge 4ft 8½in **SJ 393784**

5	N5 WGM MARSHY	4wDM	R/R	_(Landrover (Hy-rail 29932	1996 1996	
–		2w-2PMR		Geismar ST/99/37	1999	
–		2w-2PMR		Geismar ST/01/21	2001	
T111725	99709 975006-6	2-4wDMR	R/R	_(John Deere 029255 (Harsco DP270/005		
–		2-4wDMR	R/R	_(John Deere 039380 (Harsco DP270/039		OOU

STOBART PORTS, (part of the Stobart Group plc), MERSEYSIDE MULTIMODAL GATEWAY (3MG), WESTBANK DOCK ESTATE, DESOTO ROAD, WIDNES WA8 0PE

Gauge 4ft 8½in www.stobartgroup.co.uk **SJ 503846**

7189	GAFFER	0-6-0DH	HE	7189	1970	a

a property of Harry Needle Railway Co Ltd, Derbyshire

VIRGIN TRAINS TALENT ACADEMY, TATTON HOUSE, WESTMERE DRIVE, CREWE BUSINESS PARK, CREWE **CW1 6ZD**

Gauge 4ft 8½in www.virgintrains.co.uk **SJ 719548**

390033	69133	4w-4wWER	Alstom		2002

PRESERVATION SITES

CREWE HERITAGE TRUST LTD, CREWE HERITAGE CENTRE, VERNON WAY, CREWE **CW1 2DB**

Gauge 4ft 8½in www.crewehc.org **ST 708553**

(D1842)	47192	Co-CoDE	Crewe		1965
(D1948	47505) 47712 LADY DIANA SPENCER				
		Co-CoDE	BT	610	1966
(D2073)	03073	0-6-0DM	Don		1959
(D6808)	37108	Co-CoDE	(EE	3237	1962
			(VF	D762	1962
M49002		Bo-BoWE	Derby		1977
M49006		Bo-BoWE	Derby		1977
87035	ROBERT BURNS	Bo-BoWE	Crewe		1974
–		4wDH	S	10007	1959
–		0-4-0DH	TH	132C	1963
	a rebuild of	0-4-0DM	JF	22982	1942

DUKE OF WESTMINSTER, EATON HALL RAILWAY, EATON HALL, ECCLESTON, near CHESTER

Private location with no public access

Gauge 1ft 3in www.eatonestate.co.uk **SJ 386612**

KATIE	0-4-0T	OC	(Thursley		1994
			(FMB		1994
SIBELL	4wDH		HE	9331	1994

GULLIVERS WORLD, LOST WORLD RAILROAD, SHACKLETON CLOSE, WARRINGTON **WA5 9YZ**

Gauge 1ft 3in www.gulliversfun.co.uk **SJ 590900**

NEVILLE	0-6-0+6wDE	s/o	Meridian	1989	
INVICTA	4wPH		Maxitrack	1989	OOU

PAUL WALLEY, c/o CAMBI UK LTD, CONGLETON TECHNOLOGY PARK, RADNOR PARK INDUSTRIAL ESTATE, FIRST AVENUE, BACK LANE, CONGLETON **CW12 4XJ**

Gauge 4ft 8½in **SJ 846637**

2859	2-8-0	OC	Sdn	2765	1918

Gauge 2ft 0in

3	0-8-0T	OC	Hen	14928	1917
3010	0-6-0T	OC	KS	3010	1916

Locomotives stored off site

CORNWALL

INDUSTRIAL SITES

DRILLSERVE LTD, PLANT YARD, ROSCROGGAN MILL, near CAMBORNE TR14 0BA

Locomotives for resale are occasionally present. www.drillserve.com **SW 648418**

IMERYS MINERALS LTD, IMERYS CLAY, ROCKS WORKS, near BUGLE

Gauge 4ft 8½in www.imerys.com **SX 025586**

10150	P406D	ISAAC	0-6-0DH		S	10150	1963
01571	P405D	ALEX	0-6-0DH		TH	261V	1976
				reb	YEC	L124	1996

SOUTH WESTERN MINING & TUNNELLING LTD, PLANT DEPOT, COTTONWOOD, NANSTALLON, BODMIN PL30 5LQ

Gauge 2ft 0in www.swmtltd.co.uk **SX 021677**

–	0-4-0BE	WR	G7174	1967

SUMMERCOURT SCRAPYARD LTD, SCRAP METAL MERCHANTS & VEHICLE DISMANTLERS, TREFULLOCK, SUMMERCOURT, NEWQUAY TR8 5BY

Gauge 1524mm (Closed) www.summercourtscrapyard.co.uk **SW 897563**

1103	2-8-0	OC	LO	141	1943	Pvd

T. WARE & SONS, HEATHER BANK, UNITED ROAD, CARHARRACK, near REDRUTH TR16 5HT

Gauge 4ft 8½in www.twareandsons.co.uk **SW 741417**

TO 9362	4wDM	RH	349041	1953	OOU

WHEAL JANE GROUP, EARTH SCIENCE PARK, WHEAL JANE, BALDHU, near TRURO TR3 6EH

Gauge 1ft 10in www.wheal-jane.co.uk SW 777422

–	4wBE	CE	Pvd

PRESERVATION SITES

BODMIN & WENFORD RAILWAY plc

Locomotives are kept at :- Bodmin General Station PL31 1AQ SX 074664
Bodmin PL30 4BB SX 110640

Gauge 4ft 8½in www.bodminrailway.co.uk

4247			2-8-0T	OC	Sdn	2637	1916	
4612			0-6-0PT	IC	Sdn		1942	
5552			2-6-2T	OC	Sdn		1928	
6435			0-6-0PT	IC	Sdn		1937	
30120	(120)		4-4-0	IC	9E		1899	
30587	(3298)		2-4-0WT	OC	BP	1412	1874	
	JUDY		0-4-0ST	OC	WB	2572	1937	
75178			0-6-0ST	IC	WB	2766	1944	
No.19			0-4-0ST	OC	WB	2962	1950	
	ALFRED		0-4-0ST	OC	WB	3058	1953	
(D442)	50042	TRIUMPH	Co-CoDE		_(EE (EEV	3812 D1183	1968 1968	
(D1787)	47306		Co-CoDE		BT	549	1964	
D3452			0-6-0DE		Dar		1957	
(D3559)	08444		0-6-0DE		Derby		1958	
(D6527)	33110		Bo-BoDE		BRCW	DEL119	1960	
(D6842)	37142		Co-CoDE		_(EE (EEV	3317 D816	1963 1963	
(W51947)			2-2w-2w-2DMR		DerbyC&W		1961	OOU
121020	W55020		2-2w-2w-2DMR		PSteel		1960	
	PETER		0-4-0DM		JF	22928	1940	a
NDS 3	"BRIAN"		4wDM		RH	443642	1960	
P403D	DENISE		4wDH		S	10029	1960	

a carries works plate JF 4000001/1945

CORNWALL COUNTY COUNCIL, GEEVOR TIN MINES MUSEUM, PENDEEN, near ST.JUST

TR19 7EW

Gauge 1ft 6in www.geevor.com **SW 375346**

3		4wBE	CE	5739	1970	
–		4wBE	CE	B1501	1977	
–		4wBE	CE	B1592A	1978	
–		4wBE	CE	B1592B	1978	
13		4wBE	CE	B1851A	1978	
B3606A		4wBE	CE	B3606A	1989	
–		4wBE	CE	B3606C	1989	
59		4wBE	CE	B3606D	1989	
1		0-4-0BE	Geevor			
2	80	0-4-0BE	Geevor			
4		0-4-0BE	Geevor			Pvd
11		0-4-0BE	Geevor			
13		0-4-0BE	Geevor			
15		0-4-0BE	Geevor			
19		0-4-0BE	Geevor			

6		0-4-0BE		Geevor		
16		0-4-0BE		Geevor		
–		4wBE		WR		
3		4wBE		WR		

The Geevor-built locomotives are based on, and use parts of, WR or CE locomotives.

Two of the un-numbered CE locomotives are numbered 2 and 8, but which works numbers these pair up with is unknown.

DAVIDSTOW AIRFIELD & CORNWALL AT WAR MUSEUM, NOTTLES PARK, DAVIDSTOW, near CAMELFORD — PL32 9YF

Gauge 2ft 0in www.cornwallatwarmuseum.co.uk

–		4wDM		MR	8882	1944

THE DELABOLE SLATE COMPANY LTD, PENGELLY, DELABOLE, CAMELFORD — PL33 9AZ

Gauge 1ft 11in www.delaboleslate.co.uk **SX 074836**

No.2		4wDM		MR	3739	1925

HELSTON RAILWAY PRESERVATION COMPANY LTD, HELSTON RAILWAY, PROSPIDNICK HALT, SITHNEY — TR13 0RY

Gauge 4ft 8½in www.helstonrailway.co.uk **SW 647305, 645309**

6	KILMERSDON	0-4-0ST	OC	P	1788	1929
2	WILLIAM MURDOCH	0-4-0ST	OC	P	2100	1949
97649		0-4-0DM		RH	327974	1954
–		0-4-0DM		RH	395305	1956
W50413		2-2w-2w-2DMR		PRoyal	B38848	1958

KING EDWARD MINE LTD, KING EDWARD MINE MUSEUM, TROON, near CAMBORNE — TR14 9HW

Gauge 1ft 10in www.kingedwardmine.co.uk **SW 664389**

60		4wBE		SouthCrofty		1992

LAPPA VALLEY STEAM RAILWAY & COUNTRY LEISURE PARK, BENNY HALT, ST.NEWLYN EAST, near NEWQUAY — TR8 5LX

Gauge 1ft 3in www.lappavalley.co.uk **SW 838574, 839564**

No.2	MUFFIN	0-6-0	OC	Berwyn		1967
			reb	Tambling		1991
	RUBY	0-4-2T	OC	ESR	302	1997
	ELLIE	0-4-2T	OC	ESR	331	2006
No.1	ZEBEDEE	0-6-2T	OC	SL	7434	1974
	rebuilt as	0-6-4T	OC	Tambling		1990
D5905	CITY OF DERBY	4w-4wDM		BrownJ		1995

4	ARTHUR	4wDM		L	20698	1942	
			reb	MalyanG		2011	
3	GLADIATOR	4w-4DH		Minirail		c1960	OOU

LAUNCESTON STEAM RAILWAY CO, ST. THOMAS ROAD, LAUNCESTON PL15 8DA

Gauge 600mm www.launcestonsr.co.uk **SX 328850**

	LILIAN	0-4-0STT	OC	HE	317	1883	
	VELINHELI	0-4-0ST	OC	HE	409	1886	
	COVERTCOAT	0-4-0STT	OC	HE	679	1898	
	DOROTHEA	0-4-0ST	OC	HE	763	1901	
89		2w-2VBT	VC	ParmenterC		2004	
38		2w-2-2-2wRE		EEDK	761	1930	
42		2w-2-2-2wRE		EEDK	806	1930	Dsm
–		4wDM		FH	1896	1935	Dsm
–		4wDM		MR	5646	1933	
	"ELECTRIC DILLY"	2w-2BER		Bowman		c1986	
	rebuilt as	4wBE		Bowman			
–		4w-4DER		Bowman		2003	
2	"THE DILLY"	4wPER		Bowman		2003	
	rebuilt as	2w-2DER		Bowman		2004	
–		4wPE		ShooterA		2014	

MOSELEY N.G. INDUSTRIAL TRAMWAY & MUSEUM, TUMBLY DOWN FARM, TOLGUS MOUNT, REDRUTH

Visitors by appointment only.

Gauge 2ft 0in www.tumblydownfarm.co.uk **SW 686428**

	THE LADY D	4wDM		MR	8934	1944
AD40	LOD 758366	4wDM		RH	202000	1940
	SMELTER	4wDM	s/o	RH	229647	1943
15		4wBE		CE	B3132A	1984
–		4wBE		GB	2345	1951
	"CATHODE"	4wBE		GB	2960	1959
	ANODE	4wBE		GB	420172	1969
1298	LITTLE GEORGE	0-4-0BE		WR	1298	1938
20	DIODE	0-4-0BE		WR	L1021	1983

Gauge 1ft 11in

–		4wBE		?		
			reb	SouthCrofty		

Gauge 1ft 10in

–		4wBE		(CE	B2944D	1982)?
–		4wBE		(CE	B2930B	1981)?
–		4wBE		CE		Dsm
1	MACCA	4wBE		CE		
			reb	SouthCrofty		
6		4wBE		CE		
			reb	SouthCrofty		
(66)	LEWIS	4wBE		SouthCrofty		1996

-	4wBE		(SouthCrofty?)		
-	0-4-0BE		WR		

Gauge 1ft 6in

–	4wWE		Saxton		1988

M.P.S. METAL PROTECTION SERVICES,
GREAT WESTERN RAILWAY YARD, near ST. AGNES — **TR5 0PD**

Gauge 4ft 8½in — **SW 721492**

YARD No.5200	4wDM		FH	3776	1956

NATIONAL TRUST,
LEVANT MINE, LEVANT ROAD, TREWELLARD, PENDEEN — **TR19 7SX**

Gauge 1ft 6in www.nationaltrust.org.uk/levant-mine — **SW 368346**

–	4wBE		CE	B3606B	1989

PARADISE RAILWAY, PARADISE PARK, 16 TRELISSICK ROAD, HAYLE — **TR27 4HB**

Gauge 1ft 3in www.paradisepark.org.uk — **SW 555365**

No.3 ZEBEDEE	4wDM		L	10180	1938

POLDARK MINE, TRENEAR, WENDRON, HELSTON — **TR13 0ER**

Gauge 4ft 8½in www.poldarkmine.org.uk — **SW 682315**

–	0-4-0ST	OC	P	1530	1919

Gauge 900mm

–	0-4-0WT	OC	OK	5102	1912

ROSEVALE MINING HISTORICAL SOCIETY, ZENNOR

Gauge 2ft 0in www.rosevalemine.com — **SW 458380**

–	0-4-0BE		WR		
		reb	SouthCrofty		

ROYAL CORNWALL AGRICULTURAL ASSOCIATION,
ROYAL CORNWALL SHOWGROUND, WADEBRIDGE — **PL27 7JE**

Gauge 600mm www.royalcornwallshow.org

–	0-4-0BE		(WR?)		

JOHN SPENCELEY, TREVAYLOR FARM

Locomotives stored at private location with no public access.

Gauge 2ft 0in

–	0-4-0VBT	VC	Roanoke		2004
(LOD/758221)	4wDM		MR	8886	1944

ST. AUSTELL CHINA CLAY MUSEUM LTD, WHEAL MARTYN MUSEUM & COUNTRY PARK, WHEAL MARTYN, CARTHEW, near ST. AUSTELL PL26 8XG

Gauge 4ft 6in www.wheal-martyn.com **SX 004555**

	LEE MOOR No.1	0-4-0ST	OC	P	783	1899	

Gauge 2ft 6in

–		4wDM	RH	244558	1946	

BRENDON SWEET, GREAT POLGOOTH TIN MINE, POLGOOTH

Gauge 1ft 11½in

–		4wPM	Sweet

CUMBRIA

INDUSTRIAL SITES

AYLE COLLIERY CO LTD, QUARRY DRIFT, ALSTON

The site of the colliery is actually located in Northumberland, though the original Ayle East Drift and related workings were located in Cumbria.

Gauge 2ft 6in — Surface **NY 728498**

15/PLB6	152183	No.5261	4wBEF		CE	5074	1965	
15PLB15	15/19	No.5097	4wBEF		CE	5097	1966	
	15/27	No.B0909B	4wBEF		CE	B0909B	1976	OOU
–			4wBEF		CE	B3168	1985	
–			4wBEF		CE	B3482B	1988	
	6/43		4wBE		WR	6297	1960	Dsm

Gauge 2ft 0in — Surface & Underground

EL 16	5667		4wBE		CE	5667	1969	Dsm
No.2			4wBE		CE	5667	1969	
–			4wBE		CE	5667	1969	Dsm
	263 022		4wBE		CE	5955A	1972	Dsm
–			4wBE		CE	(5955C?	1972)	
1520			4wBEF		CE	B0445	1975	
	263 075		4wBE		CE	B3642A	1990	
				reb	CE	B3766A	1991	
	263 073		4wBE		CE	B3642B	1990	
				reb	CE	B3766B	1991	
	263 074		4wBE		CE	B3766D	1991	
–			4wBEF		GB	2382	1953	Dsm
–			4wDMF		HE	3496	1947	Dsm
–			4wDMF		HE	4569	1956	OOU
–			0-4-0DMF		HE	4991	1955	
		rebuilt as	4wDMF		Ayle		1977	OOU
–			0-4-0DM		HE	5222	1958	
LE/12/75	P17271		4wBE		WR	E6807	1965	

LM03	4wBE	WR	N7607	1973	
LM 5 1	4wBE	WR		1973	a
7664	0-4-0BE	WR	P7664	1975	

Gauge 1ft 7½in — Surface

–	0-4-0BE	WR	D6754	1964	Dsm

Gauge 1ft 6in — Surface

(14) 77 4	4wBE	CE	5712	1969	
16 82	4wBE	CE	B3132B	1984	

a one of WR N7606, WR N7620 or WR N7621

CUMBRIA COUNTY COUNCIL, THE PORT OF WORKINGTON, PRINCE OF WALES DOCK, WORKINGTON CA14 2JH

Gauge 4ft 8½in www.portofworkington.co.uk **NX 993294**

–	0-6-0DH		HE	8976	1979
		reb	HE	9306	1992
		reb	HE	6706	2000
–	0-6-0DH		HE	8977	1979
		reb	HE	9307	1992
		reb	HE	6707	2000

DIRECT RAIL SERVICES LTD, KINGMOOR DEPOT, ETTERBY ROAD, ETTERBY, CARLISLE CA3 9NZ

Gauge 4ft 8½in www.directrailservices.com **NY 385575**

Locomotives from British Nuclear Fuels plc, Sellafield occasionally present for repair/overhaul.

EGREMONT MINING CO LTD, FLORENCE IRON ORE MINE, EGREMONT CA22 2NR

Gauge 2ft 6in (Closed) www.florenceartscentre.com **NY 018103**

7	4wBE	WR	6218	1961	OOU

HONISTER SLATE MINE LTD, HONISTER SLATE QUARRY, HONISTER PASS, BORROWDALE CA12 5XN

Gauge 2ft 0in www.honister.com **NY 225136**

–	0-4-0DM	s/o	Bg	3236	1947	Pvd
–	4wBE		PWR	AO.296V.02	1992	a
–	4wBE		PWR	AO.296V.03	1992	

a currently stored off site

MINISTRY OF DEFENCE, DEFENCE MUNITIONS, LONGTOWN DEPOT, near CARLISLE

See Section 6 for details.

NUCLEAR DECOMMISSIONING AGENCY, SELLAFIELD SITE, SELLAFIELD, near SEASCALE — CA20 1PG

(operated by Sellafield Ltd, a subsidiary of Nuclear Management Partners Ltd)

Gauge 4ft 8½in — **NY 025034**

B.N.F.L.1	PB 1056.860	0-4-0DH		HE	7426	1982
			reb	HAB	6480	1997
2		0-6-0DH		S	10111	1963
			reb	TH		1987
			reb	HAB	6479	1997
3		0-4-0DH		HE	7406	1977
			reb	HE	9200	1983
			reb	HAB	6478	1997
4		0-4-0DH		HE	7427	1982
			reb	HE	9384	2014
(5)	YARD No.12228	0-4-0DH		HE	6975	1968
6		0-6-0DH		HE	9000	1983
			reb	HE	9288	1987
			reb	YEC	L117	1992
No.6	H0344	4wDH	R/R	(Minilok	158	1991
				(YEC	L105	1992

"Pile No.2", Building B12, Windscale, Sellafield Site, Sellafield

Gauge 2ft 5½in — **NY 030043**

PV 7	2w-2BE	WR

F.G.SHEPHERD, FLOW EDGE SCRAP YARD, MIDDLE FELL, ALSTON

Locomotives stored at unknown location

Gauge 2ft 0in — **NY 734442**

–	4wBEF	CE	B3084	1984
3	0-4-0BE	WR		
9	0-4-0BE	WR		

STOBART RAIL, BLACKDYKE ROAD, KINGSTOWN, CARLISLE — CA3 0PJ

Road-rail vehicles are present in this yard between use on contracts.

Gauge 4ft 8½in www.stobartrail.com — **NY 389594**

W 031	99709 977011-4	4wDM	R/R	(Unimog	160572	1990
				(Zweiweg	1291	1990
W 020	99709 977012-2	4wDM	R/R	Unimog	166228	c1991
W 061	99709 970010-6	4wDM	R/R	(Unimog	195311	2000
				(Zweiweg	1850	2000

STORY RAIL LTD, BURGH ROAD INDUSTRIAL ESTATE, CARLISLE — CA2 7NA

Road-rail vehicles are present in this yard between use on contracts.

Gauge 4ft 8½in www.storycontracting.com/rail — **NY 377562**

SR 924	99709 977013-0	4wDM	R/R	(Unimog	130580	1986
				(Zweiweg	1143	1986

TATA STEEL EUROPE, SHAPFELL LIMESTONE QUARRIES, SHAP, PENRITH

(part of the Tata Group)

Gauge 4ft 8½in www.tatasteeleurope.com **NY 571134**

272	GROSMONT	6wDE		GECT	5470	1978	
278		6wDE		GECT	5468	1977	
(No.403)		0-6-0DH		HE	7543	1978	
(404)		0-6-0DH		HE	8978	1979	

PRESERVATION SITES

BLENNERHASSET WATERMILL, near ASPATRIA

Gauge 2ft 0in Private site **NY 184419**

–		4wDM		FH	3756	1955	
–		4wDM		RH	273525	1949	
–		0-4-0BE		WR	(4149	1949?)	
			reb	ShepherdFG			
NVR No.1		4wBE	s/o	ShepherdFG		c1973	
			reb	‡		1988	
41		4wBER		HallamP			

‡ rebuilt by S.Frogley, Nent Valley Rly.

Gauge 1ft 3in

–		4wPH		CurleA		c1985	Dsm

EDEN VALLEY RAILWAY TRUST, WARCOP STATION WARCOP CA16 6PR

Gauge 4ft 8½in www.evr-cumbria.org.uk **NY 753156**

(D1654	47070 47620 47835) 47799	Co-CoDE		Crewe		1965	
(D6742)	37042	Co-CoDE		_(EE	3034	1962	
				(VF	D696	1962	
ND 3815		0-4-0DM		HE	2389	1941	
21		0-4-0DH		JF	4220045	1967	
	DARLINGTON	0-6-0DH		_(RSHD	8343	1962	
				(WB		1962	
(ARMY 244	88 EQ)	0-4-0DH		TH	130c	1963	
	a rebuild of	0-4-0DM		JF	22971	1942	
(227)		0-4-0DM		_(VF	5262	1945	
				(DC	2181	1945	
205009	60108	4-4wDER		Afd/Elh		1957	
2315	61798	4w-4wRER		Afd/Elh		1961	
2315	61799	4w-4wRER		Afd/Elh		1961	
2311	61804	4w-4wRER		Afd/Elh		1961	
2311	61805	4w-4wRER		Afd/Elh		1961	
9003	(S68003 931093)	4w-4RE/BER		Afd/Elh		1960	
(9005)	S68005 931095	4w-4RE/BER		Afd/Elh		1960	
9010	68010 (931090)	4w-4RE/BER		Afd/Elh		1961	
(68045)		2w-2PMR		Wkm	730	1932	DsmT
DE 320468	(950042 753)	2w-2PMR		Wkm	1724	1934	

HAIG COLLIERY MINING MUSEUM, SOLWAY ROAD, WHITEHAVEN **CA28 9BG**
Gauge 2ft 6in **(Closed)** **NX 966175**

–		0-4-0BE		WR	(5931	1958 ?)	Dsm

LAKESIDE & HAVERTHWAITE RAILWAY CO LTD, HAVERTHWAITE **LA12 8AL**
Gauge 4ft 8½in www.lakesiderailway.co.uk **SD 349843**

42073		2-6-4T	OC	Bton		1950
42085		2-6-4T	OC	Bton		1951
46441		2-6-0	OC	Crewe		1950
–		0-6-0T	OC	AB	1245	1911
1	DAVID	0-4-0ST	OC	AB	2333	1953
	"REPULSE"	0-6-0ST	IC	HE	3698	1950
	PRINCESS	0-6-0ST	OC	WB	2682	1942
	VICTOR	0-6-0ST	OC	WB	2996	1951
(D2072	03072)	0-6-0DM		Don		1959
(D2117)		0-6-0DM		Sdn		1959
(D8314)	20214 AUSTIN MAHER - CHAIRMAN LAKESIDE & HAVERTHWAITE RAILWAY 1970-2006					
		Bo-BoDE		_(EE	3695	1967
				(EEV	D1090	1967
(7120)	AD601	0-6-0DE		Derby		1945
	"RACHEL"	4wPM		MR	2098	1924
(M52071)		2-2w-2w-2DMR		BRCW		1962
(M52077)		2-2w-2w-2DMR		BRCW		1961

NATURAL ENGLAND, UNIT 2, KIRKBRIDE AIRFIELD, KIRKBRIDE, WIGTON
Gauge 2ft 0in **NT 222551**

–		4wDM	MR	9231	1947

OLD HALL FARM ENGINEERING & RESTORATION,
OLD HALL FARM, WEAR BRIDGE, BOUTH, ULVERSTON **LA12 8JA**
www.oldhallfarmbouth.com **SD 325855**
Railway vehicles under restoration occasionally present

H. POTTS, APPLEBY
Gauge 2ft 0in

5		4wBE	WR

RAVENGLASS & ESKDALE RAILWAY CO LTD, RAVENGLASS **CA18 1SW**
Gauge 1ft 3¼in www.ravenglass-railway.co.uk **SD 086967, NY 137000**

	RIVER MITE	2-8-2	OC	Clarkson	4669	1966
No.3	RIVER IRT	0-8-2	OC	DB	3	1894
			reb	Ravenglass		1927
	RIVER ESK	2-8-2	OC	DP	21104	1923
	WHILLAN BECK	4-6-2	OC	KraussM	8457	1929

	THE FLOWER OF THE FOREST	2w-2VBT	VCG	Ravenglass	5	1985	
No.10	NORTHERN ROCK	2-6-2	OC	Ravenglass	10	1976	
No.1	BLACOLVESLEY	4-4-4PM	s/o	BL		1909	
–		4wBE		GB	2782	1957	
I.C.L.9	CYRIL	4wDM		L	4404	1932	
			reb	Ravenglass		1986	
–		4wDM		L	40009	1954	a DsmT
21	LES	4wDM		LB	51721	1960	
	PERKINS	4w-4DM		Ravenglass		1933	
	a rebuild of	4wPM		MH	NG39A	1929	
	LADY WAKEFIELD	4w-4wDH		Ravenglass		1980	
	ANITA	4wDM		RH	277273	1949	b
	SHELAGH OF ESKDALE	4-6-4DH		SL		1969	
I.C.L. No.11	DOUGLAS FERRIERA	4w-4wDH		TMA	28800	2005	
136	(126 RC1)	4w-4wDHR		Crow		1975	c Dsm
137	(RC 2)	4w-4wDHR		Ravenglass		1983	c Dsm

a converted to a flat wagon
b in use as a non-self propelled flail mower
c converted from coaching stock; since reverted to coaching stock

RAVENGLASS RAILWAY MUSEUM TRUST, RAVENGLASS RAILWAY MUSEUM, RAVENGLASS — CA18 1SW

Gauge 1ft 3in www.ravenglassrailwaymuseum.co.uk **SD 086967**

	ELLA	0-6-0T	OC	DB	2	1881	a
	KATIE	0-4-0T	OC	DB	4	1896	
			reb	StationRoad		2016	
	SYNOLDA	4-4-2	OC	BL	30	1912	
	(QUARRYMAN)	4wPM		MH	2	1926	
I.C.L.No.1		4-4wPM		Ravenglass		1925	
	(SCOOTER)	2-2wPMR		Ravenglass		1971	Dsm

a side frames only; currently stored elsewhere

SOUTH TYNEDALE RAILWAY PRESERVATION SOCIETY, ALSTON STATION, ALSTON — CA9 3JB

Gauge 2ft 0in www.south-tynedale-railway.org.uk **NY 717467**

10	NAKLO	0-6-0WTT	OC	Chrz	3459	1957	
	GREEN DRAGON	0-4-2T	OC	HE	1859	1937	
740		0-6-0T	OC	OK	2343	1907	
	BARBER	0-6-2ST	OC	TG	441	1908	
	CARLISLE	4wBE		CE	B4427B	2006	
			reb	AK	96R	2015	
			reb	SBR	271	2016	
	NEWCASTLE	4wBE		CE	B4427C	2006	
			reb	SBR	272	2017	
NG 25		4wBE		BD	3704	1973	
			reb	AB	6526	1987	
	(2103/35)	4wBEF		_(EE	2519	1958	
				(Bg	3500	1958	a Dsm
4	NAWORTH STR 04	0-6-0DMF		HC	DM819	1952	
–		0-6-0DMF		HC	DM1169	1960	Dsm

1247	OLD RUSTY STR 18	0-6-0DMF		HC	DM1247	1961	
9	(2403/50)	0-4-0DMF		HE	4109	1952	
			reb	STRPS		1992	
4110	2403/51	0-4-0DMF		HE	4110	1953	b Pvd
11	CUMBRIA	4wDM		HE	6646	1967	
21		4wDM		Moës			
20		4wDM		MR	5880	1935	c Dsm
51		2w-2PMR		Wkm		193x	d DsmT

a chassis used as a frame for a mobile crane
b plinthed outside the Hub Museum, Station Road, Alston (NY 717467)
c frame only.
d converted to weedkiller wagon

STAINMORE RAILWAY COMPANY, KIRKBY STEPHEN EAST STATION, KIRKBY STEPHEN — CA17 4LA

Gauge 4ft 8½in www.kirkbystepheneast.co.uk **NY 769075**

No.910		2-4-0	IC	Ghd		1875	
68009		0-6-0ST	IC	HE	3825	1954	
	F.C. TINGEY	0-4-0ST	OC	P	2084	1948	
	LYTHAM ST.ANNE'S	0-4-0ST	OC	P	2111	1949	
–		4wDH		FH	3958	1961	
No.305		0-4-0DH		YE	2952	1965	
	STANTON No.50	0-6-0DE		YE	2670	1958	
(PWM 2222	B154W)	2w-2PMR		Wkm	4139	1947	
–		2w-2PMR		Wkm			DsmT

THRELKELD QUARRY & MINING MUSEUM, THRELKELD QUARRY, near KESWICK — CA12 4TT

Locomotives are also kept at Hilltop Quarry www.threlkeldquarryandminingmuseum.co.uk

Gauge 4ft 8½in **NY 321231, 318215, 327245**

	"ASKHAM HALL"	0-4-0ST	OC	AE	1772	1917	

Gauge 3ft 2¼in

–		4wDM		RH	320573	1951	Dsm

Gauge 2ft 6in

–		0-4-0DM		HE	2248	1940	Dsm
–		0-4-0DM		HE	2267	1940	Dsm

Gauge 2ft 0in

	HELEN KATHRYN	0-4-0WT	OC	Hen	28035	1948	
	SIRTOM	0-4-0ST	OC	WB	2135	1925	
–		0-4-0DMF		HC	DM752	1949	Dsm
–		0-4-0DMF		HE	3149	1945	a
	SILVERBAND	4wDM		HE	3595	1948	
–		4wDM		LB	51651	1960	
3	K 11143	4wDM		MR	7191	1937	Dsm
–		4wDM		MR	8627	1941	
–		4wDM		MR	8698	1941	
–		4wDM		RH	217993	1943	

ND 6456	4wDM		RH	221626	1943	
–	4wDM		RH	223744	1944	
ND 6440	4wDM		RH	242918	1947	
2 B173106 AA 1 L.R. 10227 BR(E) No.1067	4wBE		Red(T)		1980	b
7 TAMAR	4wBE		WR			
LLECHWEDD	4wBE		WR	C6766	1963	
–	4wBE		WR	P7624	1975	

a carries worksplate HE 2254/1940
b carries worksplate MR 461/1917

DERBYSHIRE

INDUSTRIAL SITES

ALLSOP (PLANT & HAULAGE) LTD, HOLLY MOUNT, HEANOR ROAD, SMALLEY, ILKESTON — DE7 6DW

Gauge 4ft 8½in www.allsop-plant.co.uk **SK 412451**

44 MITCHELL	0-6-0DH		HE	7396	1974	
		reb	Wilmott		2000	

BOMBARDIER TRANSPORTATION UK LTD, LITCHURCH LANE, DERBY — DE24 8AD

Gauge 9ft 10in www.bombardier.com **SK 364345**

(TRAVERSER No.4)	0-4-0WE		DerbyC&W		1985	

Gauge 4ft 8½in

(D3769 08602) 004	0-6-0DE		Derby		1959	OOU
(D3849 08682) LIONHEART	0-6-0DE		Hor		1959	
(D4014 08846) 003	0-6-0DE		Hor		1961	a
–	4wDH	R/R	Zephir	2677	2016	
–	4wDH	R/R	Zephir	2733	2017	

a currently at Railway Support Services Ltd, Wishaw, Warwickshire

BREEDON CEMENT LTD, HOPE CEMENT WORKS, PINDALE ROAD, HOPE — S33 6RP

Gauge 4ft 8½in www.breedongroup.com **SK 167823**

(D3536 08421) 09201	0-6-0DE		Derby		1958	a
(D3790) 08623	0-6-0DE		Derby		1959	a
(D3881) 08714	0-6-0DE		Crewe		1960	a
(D4033) 08865 GILLEY	0-6-0DE		Dar		1960	a
(D4135) 08905	0-6-0DE		Hor		1962	a
(D8066 20066) 82	Bo-BoDE		(EE	2972	1961	
			(RSHD	8224	1961	a
(D8168 20168) 2 SIR GEORGE EARLE	Bo-BoDE		(EE	3639	1966	
			(EEV	D1038	1966	a

(D8319	20219 20906) 3	Bo-BoDE		_(EE	3700	1968		
				(EEV	D1095	1968	a	
4		0-6-0DH		AB	616	1977		
			reb	HN(R)	DH L102	2008	a	
5		0-6-0DH		AB	613	1977		
			reb	AB		1986		
			reb	HN(R)	DH L101	2008	a	
20/109/89	41	0-6-0DH		AB	647	1979	a	Dsm
01570	BLUE JOHN	B-B DH		HAB	773	1990		OOU
	PEVERIL	0-6-0DH		S	10087	1963		
			reb	AB	6140	1989		OOU
	DERWENT	0-6-0DH		S	10156	1963		
			reb	AB	6004	1988		OOU
	62	6wDH		TH	V316	1987	a	OOU

a property of Harry Needle Railroad Co Ltd, Derbyshire

PETER BRIDDON, LOCOMOTIVE ENGINEER, GEOFFREY BRIDDON BUILDING, STATION ROAD, DARLEY DALE, near MATLOCK DE4 2EQ

www.petebriddon.co.uk SK 273625

Locos for overhaul or repair usually present

BROOKSIDE ENGINE CO LTD, BROOKSIDE WORKS, NEW ROAD, WHALEY BRIDGE SK23 7JG

Gauge 4ft 8½in Private Site - no visitors without prior permission

32		0-4-0ST	OC	AB	1659	1920	

Gauge 2ft 0in

–		0-6-0WT	OC	OK	9239	1921	
7		4wBE		WR		1973	a
	"PANTHER WAGON"	2-2wPMR		Brookside		2017	

Gauge 600mm

5662		0-4-0WTT	OC	OK	5662	1912

a one of WR N7606, WR N7620 or WR N7621

W.H. DAVIS LTD, LANGWITH ROAD, LANGWITH JUNCTION NG20 9RN

Gauge 4ft 8½in www.whdavis.co.uk **SK 529683**

P22		0-4-0DH	AB	499	1965
(3052	Car No.91) 291	4w-4wRER	MetCam		1932
3051	Car No.88 (288)	4w-4wRER	MetCam		1932

DONFABS & CONSILLIA LTD, THE OLD IRON WAREHOUSE, THE WHARF, SHARDLOW, near DERBY DE72 2GH

www.trackgeometry.co.uk **SK 442303**

New Consillia railcars under construction or repair occasionally present

LEANDER ARCHITECTURAL, FLETCHER FOUNDRY, HALLSTEADS CLOSE, DOVE HOLES, near BUXTON — SK17 8BP

Locomotives usually present for restoration or overhaul.
[visitors welcome – but first phone T. McAvoy on 01298 814941]

Gauge 2ft 0in www.leanderarchitectural.co.uk **SK 076784**

No.	Name	Type		Builder	Works No.	Year	Note
	PHOENIX	0-4-0ST	OC	Ferndale	21	2001	a
23		4wDM		L	52031	1960	b
87009		4wDM		MR	4572	1929	
22		4wDM		MR	8756	1942	
6	TURBO TED	4wDM		MR	9543	1950	
–		4wDM		RH	264252	1952	
–		4wDM		RH		1958	c
	FAULD	4wDM		RH	444208	1961	
	WASP	2w-2PM		WilsonAJ		1969	
			reb	WilsonAJ		1979	

a carries worksplate H.K. Porter 18635 1896
b plate reads 25031
c one of RH 418676 or RH 418776

LORAM UK LTD, RTC BUSINESS PARK, LONDON ROAD, DERBY — DE24 8UP

Locomotives for repair usually present.

Gauge 4ft 8½in www.loram.co.uk **SK 364349**

No.	Name	Type	Builder	Works No.	Year	Note
(D3755) 08588	H 047	0-6-0DE	Crewe		1959	a
(D6898) 37198		Co-CoDE	(EE	3376	1964	
			(EES	8419	1964	

a property of British American Railway Services Ltd, Stanhope, Co. Durham

MOORSIDE MINING CO LTD, PLANT YARD, ROTHERSIDE ROAD, ECKINGTON — S21 4HL

Plant depot with locomotives for resale occasionally present. **SK 436800**

HARRY NEEDLE RAILROAD COMPANY LTD, BARROW HILL ROUNDHOUSE, CAMPBELL DRIVE, BARROW HILL, STAVELEY — S43 2PR

Gauge 4ft 8½in **SK 412754**

No.	Name	Type	Builder	Works No.	Year	Note
D2853 (02003)		0-4-0DH	YE	2812	1960	
(D2996) 07012		0-6-0DE	RH	480697	1962	
(D3543) 08428		0-6-0DE	Derby		1958	
(D3927 08759 09106) 6		0-6-0DE	Hor		1961	
(D3933 08765)		0-6-0DE	Hor		1961	
(D3950) 08782		0-6-0DE	Derby		1960	
(D3954) 08786		0-6-0DE	Derby		1960	
(D3992) 08824	IEMD 01	0-6-0DE	Derby		1960	
(D4045 08877)	WIGAN 1 REVENGE	0-6-0DE	Dar		1961	
(D4047) 08879		0-6-0DE	Dar		1961	
D4092		0-6-0DE	Dar		1962	
(D8096) 20096		Bo-BoDE	(EE	3002	1961	
	IAN GODDARD 1938 - 2016		(RSHD	8254	1961	a

(D8101	20101) 20901	Bo-BoDE		_(EE	3007	1961		
				(RSHD	8259	1961		
(D8102	20102) 20311	Bo-BoDE		_(EE	3008	1961		
				(RSHD	8260	1961	a	
(D8107)	20107	Bo-BoDE		_(EE	3013	1961		
				(RSHD	8265	1961	a	
(D8117	20117) 20314	Bo-BoDE		_(EE	3023	1962		
				(RSHD	8275	1962	a	
(D8118)	20118	Bo-BoDE		_(EE	3024	1962		
	SALTBURN-BY-THE-SEA			(RSHD	8276	1962	a	
(D8121	20121)	Bo-BoDE		_(EE	3027	1962		
				(RSHD	8279	1962		
(D8132)	20132	Bo-BoDE		_(EE	3603	1966		
	BARROW HILL DEPOT			(EEV	D1002	1966		
(D8166	20166)	Bo-BoDE		_(EE	3637	1966		
				(EEV	D1036	1966		
(D8325	20225) 20905	Bo-BoDE		_(EE	3706	1967		
				(EEV	D1101	1967	a	
–		0-6-0DH		AB	612	1976		Dsm
L127	BILL	0-6-0DH		EEV	D1199	1967		
			reb	YEC	L127	1996		
L149	BEN	0-6-0DH		EEV	D1200	1967		
			reb	YEC	L149	1996		
10		0-6-0DH		EEV	D1228	1967		
–		0-6-0DH		GECT	5365	1972		
(11)	"VALIANT"	0-6-0DH		RR	10213	1964		
			reb	TH		1988		
01515	(304)	4wDH		TH	V321	1987		
			rep	LH Group	76629	2002		

a currently on hire to GB Railfreight Ltd (Mainline)

See also entry for Barrow Hill Engine Shed Society in preservation section

STORED ENERGY TECHNOLOGY LTD.
ATLAS WORKS, LITCHURCH LANE, DERBY — DE24 8AQ

SET locomotives under construction/repair occasionally present. www.set-gb.com **SK 362348**

TARMAC plc - A CRH Company,
TUNSTEAD QUARRY, GREAT ROCKS, TUNSTEAD, BUXTON — SK17 8TG

Gauge 4ft 8½in www.tarmac.com/tunstead **SK 101743, 097755**

PATRICK D. DUGGAN	0-6-0DH		HE	9386	2015	
		rebuild of	EEV	D1226	1967	
GRAHAM LEE JNR	0-6-0DH		HE	9387	2015	
		rebuild of	EEV	_(D1249	1968	
				(3947	1968	
DOVEDALE	0-6-0DH		RR	10284	1969	
		reb	TH		1974	
HIGH PEAK	6w-6wDH		Vollert	02/013	2003	

WALKER & PARTNERS LTD, INKERSALL ROAD ESTATE, STEPHENSON ROAD, STAVELEY, CHESTERFIELD **S43 3JN**

Dealers yard, with locomotives for resale occasionally present.

Gauge 2ft 0in www.walkerandpartners.co.uk **SK 436743**

1524	2143	4wBEF		CE	B3611	1989

S.E. WARD ENGINEERING LTD, STATION ROAD, KILLAMARSH

Locomotives for restoration occasionally present **SK 448810**

PRESERVATION SITES

BARROW HILL ENGINE SHED SOCIETY, BARROW HILL ROUNDHOUSE RAILWAY CENTRE, CAMPBELL DRIVE, BARROW HILL, STAVELEY **S43 2PR**

Site also encompasses some commercial operations including Rampart Engineering Ltd, and Harry Needle Railroad Company, along with other short term contractors.

Gauge 4ft 8½in www.barrowhill.org **SK 414755**

5164		2-6-2T	OC	Sdn		1930	
(41000)	1000	4-4-0	3C	Derby		1902	
41708		0-6-0T	IC	Derby		1880	
(62660)	No.506 BUTLER-HENDERSON	4-4-0	IC	Gorton		1920	
(65567)	No.8217	0-6-0	IC	Str		1905	
68006		0-6-0ST	IC	HE	3192	1944	
			reb	HE	3888	1964	
	HENRY	0-4-0ST	OC	HL	2491	1901	
	E.B. WILSON	0-4-0ST	OC	MW	1795	1912	
(D67)	45118 THE ROYAL ARTILLERYMAN	1Co-Co1DE		Crewe		1962	
(D86)	45105	1Co-Co1DE		Crewe		1961	
(D100)	45060 SHERWOOD FORESTER	1Co-Co1DE		Crewe		1961	
(D212)	40012 AUREOL	1Co-Co1DE		_(EE	2668	1959	
				(VF	D429	1959	
(D2066)	03066	0-6-0DM		Don		1959	
D2868		0-4-0DH		YE	2851	1961	
(D5300)	26007	Bo-BoDE		BRCW	DEL45	1958	
(D5386	27103) 27066	Bo-BoDE		BRCW	DEL229	1962	
D5910		Bo-BoDE		BabyDeltic		2013	
	a rebuild of	Co-CoDE		_(EE	3337	1963	
(D6859	37159 37372)			(EES	8390	1963	a
(D6521)	33108	Bo-BoDE		BRCW	DEL113	1960	
(E3003)	81002	Bo-BoWE		_(BRCW		1960	
				(BTH	1085	1960	
E3035	(83012)	Bo-BoWE		_(EE	2941	1960	
				(VF	E277	1960	
(E3054)	82008	Bo-BoWE		_(BP	7892	1961	
				(AEI/MV	1029	1961	
(E3061)	85006 (85101)	Bo-BoWE		Don		1961	
89001	(AVOCET)	Co-CoWE		_(Crewe		1986	
				(BT	875	1986	

12589	HARRY	0-4-0DM		_(RSHN	7922	1957	
				(DC	2589	1957	
(3905)	62266	4w-4wRER		York		1969	
(3918	62321)	4w-4wRER		York(BRE)		1970	
1499	62364	4w-4wRER		York(BRE)		1971	
(1393)	62384	4w-4wRER		York(BRE)		1971	
–	(99709 909133-9)	2w-2PMR		Geismar ST/04/01		2004	

a carries worksplate EEV 4004/D1281

Deltic Preservation Society

Gauge 4ft 8½in www.thedps.co.uk **SK 413756**

(D9009)	55009 ALYCIDON	Co-CoDE	_(EE	2914	1961	
			(VF	D566	1961	
D9015	(55015) TULYAR	Co-CoDE	_(EE	2920	1961	
			(VF	D572	1961	
(D9019)	55019 ROYAL HIGHLAND FUSILIER	Co-CoDE	_(EE	2924	1961	
			(VF	D576	1961	

ANDREW BRIDDON, GEOFFREY BRIDDON BUILDING, STATION ROAD, DARLEY DALE, MATLOCK DE4 2EQ

Gauge 4ft 8½in www.andrewbriddonlocos.co.uk SK 273625

(D2128	03128) 03901	0-6-0DH		BriddonA		2011		
	a rebuild of	0-6-0DM		Sdn		1960		
D9500	(9312/92 No.1)	0-6-0DH		Sdn		1964		
(D9524)	14901	0-6-0DH		Sdn		1964	a	
(64)		0-6-0DE		BT	803	1978		
	"PLUTO"	4wDM		FH	3777	1956		
	LUDWIG MOND	6wDE		GECT	5578	1980	b	
(D1)	"ASHDOWN"	0-6-0DM		HC	D1186	1959		
			reb	HE	8526	1977		
	"GRACE"	0-4-0DH		HC	D1345	1970		
D1388	6 "CLAIRE"	0-4-0DH		HC	D1388	1970	c	
(875)		0-4-0DH		HE	9222	1984		Dsm
	"CORONATION"	0-4-0DH		NBQ	27097	1953		
107		0-6-0DH		NBQ	27932	1959		Dsm
	"TOM"	0-6-0DH		S	10180	1964		
"RS 8"		0-4-0DH		SCW		1960		
	a rebuild of	0-4-0ST	OC	AE	1913	1923	d	
	(CHEEDALE)	4wDH		TH	284V	1979	c	
(6)	(CHARLIE)	4wDH		TH	265V	1976		
72229		0-4-0DM		_(VF	5265	1945		
				(DC	2184	1945		
			reb	YEC	L120	1993	e	
	"DONCASTER"	0-4-0DE		YE	2654	1957	f	
"3"	"JACK"	0-4-0DH		YE	2679	1962		
H 051	(LIBBY)	0-6-0DH		YE	2940	1965		
2	JAMES	0-4-0DH		YE	2675	1961		
RRM 15	1382	0-6-0DE		YE	2872	1962	f	
61287	4311	4w-4wRER		Afd/Elh		1959		

DR 98307		4w-4wDHR	Geismar	825	1998	f
–		2w-2DMR	Wkm	(6607	1953?)	

a currently at Colne Valley Railway, Castle Hedingham, Essex
b currently at Rutland Railway Museum, Cottesmore, Rutland
c currently at UK Rail Leasing Ltd, Leicester, Leicestershire
d currently at Tarmac plc, Tunstead Quarry, Derbyshire
e currently at East Anglian Railway Museum, Essex
f privately owned

CRICH TRAMWAY VILLAGE, CRICH, near MATLOCK — DE4 5DP

Gauge 4ft 8½in www.tramway.co.uk **SK 345549**

47		0-4-0VBTram		BP	2464	1885	
–		0-4-0VBTram		BP	2734	1886	a Dsm
–		4wWE		EEDK	717	1927	
–		4wDM		RH	223741	1944	b
	G.M.J.	4wDM		RH	326058	1952	c
058	(53.0692-3)	2w-2DMR		SolHütte	7808	1978	
			reb	SolHütte	692	1999	

a currently stored at Clay Cross store
b rebuilt in 1963 from 600mm gauge
c rebuilt in 1969 from 3ft 3in gauge

Gauge 1000mm

–	4wDM		RH	373363	1954	Dsm

Gauge 1ft 2½in

46	0-4-0VBTram	OC	Pritchard		1941	

Cliff Quarry, Wake Bridge Station

Gauge 2ft 0in www.pdmhs.co.uk **SK 341555**

4	4wBE	WR	3492	1946	a

a property of Peak District Mines Historical Society

TERRY GIBSON, DERBY

Gauge 1ft 3in

	KING GEORGE	4-4-2	OC	BL	22	1915
6100	ROYAL SCOT	4-6-0	OC	Carland	[6100]	1950

GULLIVERS KINGDOM LTD, MATLOCK BATH — DE4 3PG

Gauge 1ft 9½in www.gulliversfun.co.uk **SK 289578**

OLD TIMER	4w-4RE	s/o	Sharon		1988

MIKE HART, PRIVATE STORAGE SITE

Gauge 4ft 8½in

9120	4wDMR	BD	3709	1975

LITTLE MILL INN, ROWARTH, near NEW MILLS — SK22 1EB

Gauge 4ft 8½in www.thelittlemillinn.co.uk **SK 011890**

(3051	Car No.89 289)	4w-4wRER		MetCam	289	1932

MIDLAND RAILWAY – BUTTERLEY (MIDLAND RAILWAY TRUST LTD), BUTTERLEY HILL, RIPLEY — DE5 3QZ

Locomotives are kept at :-

Butterley	SK 403520
Swanwick Junction	SK 412519
Hammersmith	SK 397519
Diesel Locomotive Shed, Swanwick Junction	SK 414519

Gauge 4ft 8½in www.midlandrailway-butterley.co.uk

158A		2-4-0	IC	Derby		1866	
(47327	16410) 23	0-6-0T	IC	NBH	23406	1926	
47357		0-6-0T	IC reb	NBQ Derby	23436	1926 1973	
(47445)		0-6-0T	IC	HE	1529	1927	Dsm
(47564)		0-6-0T	IC	HE	1580	1928	Dsm
73129		4-6-0	OC	Derby		1956	
	"STANTON No.24"	0-4-0CT	OC	AB	1875	1925	
No.2	PN 8292	0-4-0F	OC	AB	2008	1935	
68067	ROBERT	0-6-0ST	IC	HC	1752	1943	
	"GLADYS"	0-4-0ST	OC	Mkm	109	1894	
	"OSWALD"	0-4-0ST	OC	NW	454	1894	
	WHITEHEAD	0-4-0ST	OC	P	1163	1908	a
	VICTORY	0-4-0ST	OC	P	1547	1919	
CASTLE DONINGTON POWER STATION 1		0-4-0ST	OC	RSHN	7817	1954	
D4	(44004) GREAT GABLE	1Co-Co1DE		Derby		1959	
(D40)	45133	1Co-Co1DE		Derby		1961	
D182	(46045) (97404)	1Co-Co1DE		Derby		1962	
D1048	WESTERN LADY	C-C DH		Crewe		1962	
(D1500)	47401 NORTH EASTERN	Co-CoDE		BT	342	1962	
D1516	(47417)	Co-CoDE		BT	358	1963	
(D1619	47564) 47761	Co-CoDE		Crewe		1964	
D2138		0-6-0DM		Sdn		1960	
D2858		0-4-0DH		YE	2817	1960	
(D3401)	08331	0-6-0DE		Derby		1957	
(D3757)	08590 RED LION	0-6-0DE		Crewe		1959	
(D5522)	31418 BOADICEA	A1A-A1A DE		BT	121	1959	
(D5526)	31108	A1A-A1A DE		BT	125	1959	
D5814	(31414 31514)	A1A-A1ADE		BT	315	1961	
(D6890)	37190 (37314)	Co-Co DE		_(EE (EES	3368 8411	1963 1963	
D7671	(25321)	Bo-BoDE		Derby		1967	
(D)8001	(20001) VULCAN PIONEER	Bo-BoDE		_(EE (VF	2348 D376	1957 1957	
(D8007)	20007	Bo-BoDE		_(EE (VF	2354 D382	1957 1957	b
(D8048)	20048	Bo-BoDE		_(EE (VF	2770 D495	1959 1959	

(D8142)	20142	Bo-BoDE		_(EE	3614	1966	
	SIR JOHN BETJEMAN			(EEV	D1013	1966	b
(D8189)	20189	Bo-BoDE		_(EE	3670	1966	
				(EEV	D1065	1966	b
(D8305)	20205	Bo-BoDE		_(EE	3686	1967	
				(EEV	D1081	1967	b
(D8327)	20227	Bo-BoDE		_(EE	3685	1967	
	SHERLOCK HOLMES			(EEV	D1080	1967	b
12077		0-6-0DE		Derby		1950	
(E)27000	(ELECTRA)	Co-CoWE		Gorton	1065	1953	
(No.2)		0-4-0DM		AB	416	1957	
	rebuilt	0-4-0DH		AB		1980	
441		0-4-0DH		AB	441	1959	
	ALBERT	0-6-0DM		HC	D1114	1958	
(No.20)	MANTON	0-6-0DM		HC	D1121	1958	
16038	ANDY	0-4-0DM		JF	16038	1923	
–		2-2wDM		Mercury	5337	1927	
"RS 12"		4wDM		MR	460	1918	
"RS 9"		4wDM		MR	2024	1920	
–		0-4-0DE		RH	384139	1955	
E50019		2-2w-2w-2DMR		DerbyC&W		1957	
51118		2-2w-2w-2DMR		GRC&W		1957	
M51591	(M55966)	2-2w-2w-2DHR		DerbyC&W		1959	
(51937	977806 LO 905)	2-2w-2w-2DMR		DerbyC&W		1960	
E50015	(55929 977775)	2-2w-2w-2DMR		DerbyC&W		1956	
(M 51610)	M 55967	2-2w-2w-2DHR		DerbyC&W		1959	
M51625	(M55976)	2-2w-2w-2DHR		DerbyC&W		1959	
M51907	LO 262	2-2w-2w-2DMR		DerbyC&W		1960	
55513	141113	4wDHR		_(BRE(D)		1984	
				(Leyland		1984	
			reb	AB	761	1989	
55533	141113	4wDHR		_(BRE(D)		1984	
				(Leyland		1984	
			reb	AB	760	1989	
F540 YCK		4wDM	R/R	_(Multicar	16305	1988	
				(Perm		1988	

PWM 3949		2w-2PMR		Wkm	6934	1955	DsmT
DX 68062	(TR34 DB 965566 PT52P)	2w-2PMR		Wkm	8272	1959	

a based here, but visits other locations
b property of Class 20189 Ltd, based here for maintenance

Golden Valley Light Railway
Gauge 2ft 0in www.gvlr.org.uk **SK 412519, 414519**

	JOAN	0-4-2IST	OC	Civil	No.1	1997
2		0-4-0WT	OC	OK	7529	1914
BD 3753	DARCY	4wDH		BD	3753	1980
Dtz	10249	4wDM		Dtz	10248	1931
–		0-6-0DMF		HC	DM1117	1957
AD 34		4wDH		HE	7009	1971
–		4wDM		HE	7178	1971
–		4wDM		L	3742	1931
			reb	FMB		1993

–		4wDM		L	10994	1939
–		4wDM		LB	53726	1963
LOD 758228	G.V.L.R. 15 TUBBY	4wDM		MR	8667	1941
(T2)	PIONEER	4wDM		MR	8739	1942
LOD 758028	LO 3009	4wDM		MR	8855	1943
	HOLWELL CASTLE	4wDM		MR	11177	1961
15		4wDM		MR	11246	1963
	CAMPBELL BRICKWORKS	4wDM		MR	60S364	1968
AD 41	(LOD 758263) LYDDIA	4wDM		RH	191646	1938
	BERRY HILL	4wDM		RH	222068	1943
(L203N)	U84	4wDM		RH	7002/0567/6	1967
	ELLISON	4wDH		SMH	101T020	1979
40SD529		4wDM		SMH	40SD529	1984
NG 24		4wBE		BD	3703	1973
			reb	AB		1986
19	BABY JAYNE	0-4-0BE		WR		

Princess Royal Class Locomotive Trust, Swanwick Junction

Gauge 4ft 8½in www.prclt.co.uk **SK 410520**

44767		4-6-0	OC	Crewe		1947
46203	PRINCESS MARGARET ROSE	4-6-2	4C	Crewe	253	1935
(46233)	6233 DUCHESS OF SUTHERLAND	4-6-2	4C	Crewe		1938
80080		2-6-4T	OC	Bton		1954
80098		2-6-4T	OC	Bton		1954
	GEORGE	0-4-0ST	OC	RSHN	7214	1945

Gauge 1ft 9in

6201	PRINCESS ELISABETH	4-6-2DH	s/o	HC	D611	1938
6203	PRINCESS MARGARET ROSE	4-6-2DH	s/o	HC	D612	1938

PEAK DISTRICT MINES HISTORICAL SOCIETY, PEAK DISTRICT MINING MUSEUM, TEMPLE MINE, TEMPLE ROAD, MATLOCK BATH (DE4 3NR)

Gauge 1ft 5in www.peakdistrictleadminingmuseum.co.uk **SK 293583**

–		4wBE		GB	1445	1936

PEAK RAIL plc, ROWSLEY SOUTH STATION, off HARRISON WAY, near MATLOCK DE4 2LF

Gauge 4ft 8½in www.peakrail.co.uk **SK 262640**

5224		2-8-0T	OC	Sdn		1924
5553		2-6-2T	OC	Sdn		1928
6634		0-6-2T	IC	Sdn		1928
(RRM 28	No.65)	0-6-0ST	IC	HE	3889	1964
WD 150	ROYAL PIONEER	0-6-0ST	IC	RSHN	7136	1944
			reb	HE	3892	1969
No.72	2235/72	0-6-0ST	IC	VF	5309	1945
D8	(44008) PENYGHENT	1Co-Co1DE		Derby		1959
(D172)	46035 (97403) IXION	1Co-Co1DE		Derby		1962

(D429)	50029 RENOWN	Co-CoDE	_(EE (EEV	3799 D1170	1968 1968
(D430)	50030 REPULSE	Co-CoDE	_(EE (EEV	3800 D1171	1968 1968
(D3998)	08830	0-6-0DE	Derby		1960
(D5800)	31270 ATHENA	A1A-A1ADE	BT	301	1961
(D6852)	37152	Co-CoDE	_(EE (EEV	3327 D826	1963 1963
D7659	(25309 25909)	Bo-BoDE	BP	8069	1966
58022		Co-CoDE	Don		1984
E1	CASTLEFIELD	0-6-0DM	HC	D1199	1960
01531	H4323	0-6-0DH reb	HE HAB	7018 6576	1971 1999
	(CYNTHIA)	4wDM	RH	412431	1957
	"TONY"	4wDMR	Wkm	9688	1965

Gauge 1ft 11½in

109		2-6-2+2-6-2T 4C	BP	6919	1939

Heritage Shunters Trust, Rowsley South — DE4 2LF

Gauge 4ft 8½in www.heritageshunters.com **SK 261640**

(D2027	03027)	0-6-0DM	Sdn		1958
(D2099)	03099	0-6-0DM	Don		1960
(D2113)	03113	0-6-0DM	Don		1960
D2139		0-6-0DM	Sdn		1960
(D2180)	03180	0-6-0DM	Sdn		1962
D2199		0-6-0DM	Sdn		1961
D2205		0-6-0DM	_(VF (DC	D212 2486	1953 1953
(D2229)		0-6-0DM	_(VF (DC	D278 2552	1955 1955
D2272	2272 "ALFIE"	0-6-0DM	_(RSHD (DC	7914 2616	1958 1958
D2284		0-6-0DM	_(RSHD (DC	8102 2661	1960 1960
(D2289)	3945	0-6-0DM	_(RSHD (DC	8122 2669	1960 1960
D2337	"DOROTHY"	0-6-0DM	_(RSHD (DC	8196 2718	1961 1961
D2420	(06003 97804)	0-4-0DM	AB	435	1959
D2587		0-6-0DM reb	HE HE	5636 7180	1959 1969
D2854		0-4-0DH	YE	2813	1960
(D2866)		0-4-0DH	YE	2849	1961
D2953		0-4-0DM	AB	395	1955
(D2985)	07001	0-6-0DE	RH	480686	1962
(D3000)	13000	0-6-0DE	Derby		1952
(D3023)	08016	0-6-0DE	Derby		1953
(D3665)	09001	0-6-0DE	Dar		1959
D9525		0-6-0DH	Sdn		1965
PWM 650	97650	0-6-0DE	RH	312990	1952
PWM 654	(97654)	0-6-0DE	RH	431761	1951
319284	FARADAY	0-4-0DM	RH	319284	1952

	BIGGA		0-4-0DH		TH	102C	1960
		a rebuild of	0-4-0DM		JF	4200019	1947
B.S.C.2			0-4-0DE		YE	2480	1950

Ashover Light Railway Society, Rowsley South — DE4 2LF

Gauge 2ft 0in — **SK 261642**

–		4wVBT	VC	Jaywick		1939	
NCB 24	LINBY	4wDHF		HE	8917	1980	
	BOLTON FELL	4wDM		LB	52726	1961	
–		4wDM		RH	260712	1948	
85051		4wDM		RH	404967	1957	
–		4wDM		RH	487963	1963	
No.1	SPONDON	4wBE		Spondon		1926	

PLEASLEY PIT TRUST, PLEASLEY COLLIERY, PIT LANE, PLEASLEY, MANSFIELD — NG19 7PH

Gauge 3ft 0in www.pleasleypittrust.org.uk **SK 498643**

E15	P6	4wBEF		CE	B3045B	1983	

STEEPLE GRANGE LIGHT RAILWAY, STEEPLEHOUSE JUNCTION, WIRKSWORTH

Gauge 2ft 0in www.steeplegrange.co.uk **SK 288554**

No.1881	No.7	4wPM		FH	1881	1934	
–		4wPM		FH	3424	1949	
–		4wDM		L	37366	1951	a

Gauge 1ft 6in

ZM32	No.11 HORWICH	4wDM		RH	416214	1957	
551		4wBE		BEV	551	1924	a
No. 4	LIZZIE	4wDM		ClayCross		1973	b
No.9	4008 ULLR	4wDM		FH	4008	1963	
(No.6)		4wBE		GB	2493	1953	
No.3	"GREENBAT"	4wBE		GB	6061	1961	
L 10		4wBE		CE	5431	1968	
L 13		4wBE		CE	5965B	1973	
L 16	PEGGY	4wBE		CE	B0109B	1973	
No.14	PETER	4wBE		CE	B0922B	1975	
4	"COSTAIN"	4wBE		Plymouth			a
No.2	(HUDSON)	2-2wPMR		SGLR		1988	

a currently under renovation elsewhere
b constructed from parts supplied by Lister

WYVERNRAIL plc, ECCLESBOURNE VALLEY RAILWAY, WIRKSWORTH STATION, STATION ROAD, WIRKSWORTH — DE4 4FB

Gauge 4ft 8½in www.e-v-r.com **SK 289541**

	HENRY ELLISON	0-4-0ST	OC	AB	2217	1947	
No.3	BRIAN HARRISON	0-4-0ST	OC	AB	2360	1954	
–		0-6-0T	OC	HC	1884	1955	

68012		0-6-0ST	IC	WB	2746	1944	
(D3772	08605) G. R. WALKER	0-6-0DE		Derby		1959	
(D3871)	08704	0-6-0DE		Hor		1960	
(D5609	31186) 31601	A1A-A1ADE		BT	209	1960	
(D5630)	31206	A1A-A1A DE		BT	230	1960	
(D6514)	33103 SWORDFISH	Bo-BoDE		BRCW	DEL106	1960	
D9537	ERIC	0-6-0DH		Sdn		1965	
(E6022	73116) 73210 SELHURST	Bo-BoDE/RE		_(EE	3584	1966	
				(EEV	E354	1966	
402803		0-4-0DE		RH	402803	1956	
	L.J. BREEZE	6wDH		RR	10275	1969	
	"MEGAN"	4wDH		TH	103C	1960	
	a rebuild of	4wVBT	VCG	S	9390	1949	
RRM 134	"TOM"	4wDH		TH	188C	1967	
	a rebuild of	4wVBT	VCG	S	9597	1955	
E50170	(53170)	2-2w-2w-2DMR		MetCam		1957	
E50253	(53253)	2-2w-2w-2DMR		MetCam		1957	
(E50599)	E53599 L990	2-2w-2w-2DMR		DerbyC&W		1958	
W51073	L594	2-2w-2w-2DMR		GRC&W		1959	
E51505		2-2w-2w-2DMR		MetCam		1959	
(51567)	977854	2-2w-2w-2DMR		DerbyC&W		1959	
W55006		2-2w-2w-2DMR		GRC&W		1958	
960302	977975 (55027)	2-2w-2w-2DMR		PSteel		1960	
960303	977976 (55031)	2-2w-2w-2DMR		PSteel		1960	
(79018	975007)	2-2w-2w-2DMR		DerbyC&W		1954	
(975 010)	M 79900 IRIS	2-2w-2w-2DMR		DerbyC&W		1956	
68500	(S61269) 9101	4w-4RER		Afd/Elh		1959	
			reb	Elh		1983	a
(68506	S61292 9107)	4w-4RER		Afd/Elh		1957	
			reb	Elh		1984	a
	(L263 MNU)	4wDMR	R/R	_(Multicar		1993	
				(Perm		1993	
	R447 XRA	4wDM	R/R	_(Landrover		1998	
				(Perm		1998	
	R482 JGG	4wDM	R/R	_(Landrover		1998	
				(Perm		1998	
	MURIEL	2w-2DER		EVRA		2000	
	99709 901209-5	2w-2PMR		Lesmac	LMA001		

a unpowered hauled coaching stock

Gauge 2ft 0in — Stone Line

	LESLEY THE LISTER	4wDM		L	26288	1944

DEVON

INDUSTRIAL SITES

AGGREGATE INDUSTRIES UK LTD, MELDON QUARRY, near OKEHAMPTON — EX20 4LT

Gauge 4ft 8½in (Closed) www.aggregate.com **SX 568927**

FLYING FALCON	0-4-0DH		JF	4220016	1962

BABCOCK INTERNATIONAL GROUP plc, DEVONPORT ROYAL DOCKYARD, DEVONPORT — (PL1 4SG)

Gauge 4ft 8½in www.babcockinternational.com **SX 449558**

RTU 1	DENNIS	4wDM		CE	B4314A	2000
RTU 2	HENRY	4wDM		CE	B4314B	2000
			reb	CE	B4622	2016

FIRST GREATER WESTERN LTD, t/a GREAT WESTERN RAILWAY, LAIRA DEPOT, EMBANKMENT ROAD, PLYMOUTH — PL4 9JN

(part of the First Group plc)

Gauge 4ft 8½in www.gwr.com **SX 504556**

See Section 7 and Appendix 1 for details

MINISTRY OF DEFENCE, OKEHAMPTON RANGES TARGET RAILWAY, OKEHAMPTON

Gauge 2ft 6in **SX 586932**

RTT/767138 CAPTAIN	2w-2PM		Wkm	3284	1943

PRESERVATION SITES

ASHBRITTLE LIGHT RAILWAY, HOLE FARM, HOCKWORTHY

Gauge 2ft 0in Private Site

YARD No.1073	EARL OF MORTAIN	4wDH	HE	7450	1976
34	SO 34 LR 10756	4wDM	MR	5879	1935
25		4wDM	RH	375694	1954

BICTON WOODLAND RAILWAY, BICTON PARK BOTANICAL GARDENS, EAST BUDLEIGH — EX9 7BJ

Gauge 1ft 6in www.bictongardens.co.uk **SY 074862**

	SIR WALTER RALEIGH	0-4-0DH	s/o	AK	61	2000
	CLINTON	4wDM		HE	2290	1941
B.W.R. 2	BICTON	4wDM		RH	213839	1942
	reb	4wDH	s/o	AK	75R	2007

BIDEFORD HERITAGE RAILWAY CENTRE C.I.C., BIDEFORD STATION, BIDEFORD HILL, BIDEFORD EX39 4BB

Gauge 4ft 8½in www.bidefordrailway.co.uk **SS 457262**

	KINGSLEY	4wDM		FH	3832	1957

C. BURGES, EXETER & TEIGN VALLEY RAILWAY, G.W.R.(CHRISTOW), SHELDON LANE, DODDISCOMBSLEIGH, EXETER EX6 7YT

Gauge 4ft 8½in www.teignrail.co.uk **SX 839868**

	PERSEUS	0-4-0DM		_(VF	D98	1949
				¯(DC	2269	1949
DX 90011		4wDMR		Matisa	PV6 620	1967
(DX 68004) (PWM)2831 (B)194(W)		2w-2PMR		Wkm	5009	1949

'BYGONES' VICTORIAN EXHIBITION STREET AND RAILWAY MUSEUM, FORE STREET, ST.MARYCHURCH, TORQUAY TQ1 4PR

Gauge 4ft 8½in www.bygones.co.uk **SX 922658**

No.5	PATRICIA	0-4-0ST	OC	HC	1632	1929

COMBE MARTIN WILDLIFE AND DINOSAUR PARK, HIGHER LEIGH MANOR, LEIGH ROAD, COMBE MARTIN, near ILFRACOMBE EX34 0NG

Gauge 1ft 3in www.cmwdp.co.uk **SS 600452**

–		2-8-0PH	s/o	SL	70.5.87	1987

DART VALLEY RAILWAY plc, DARTMOUTH STEAM RAILWAY & RIVERBOAT CO, DARTMOUTH STEAM RAILWAY

Locomotives are kept at :-

Churston TQ5 0LL SX 896564
Kingswear TQ6 0AA SX 884515
Park Siding, Paignton SX 889606

Gauge 4ft 8½in www.dartmouthrailriver.co.uk

4277	HERCULES	2-8-0T	OC	Sdn	2857	1920
4555	WARRIOR	2-6-2T	OC	Sdn		1924
5542		2-6-2T	OC	Sdn		1928
7827	LYDHAM MANOR	4-6-0	OC	Sdn		1950
75014		4-6-0	OC	Sdn		1951
D2192	TITAN	0-6-0DM		Sdn		1961
D2371	(03371)	0-6-0DM		Sdn		1958
D3014	SAMSON	0-6-0DE		Derby		1952
(D)6975	(37275)	Co-CoDE		_(EE	3535	1965
				¯(EEV	D964	1965
(DB 966030)	CE 68200	2w-2DMR		Plasser	419	1975

DARTMOOR RAILWAY CIC (subsidiary of British American Railway Services Ltd)
Locomotives are kept at :- Okehampton EX20 1EJ SX 591944
Meldon Quarry EX20 4LT SX 567927

Gauge 4ft 8½in www.dartmoorrailway.com

–		0-6-0T	OC	HC	1864	1952	Dsm
(D1966	47266 47629) 47828	Co-CoDE		Crewe		1965	
D4167	(08937)	0-6-0DE		Dar		1962	
205028	60146	4-4wDER		Afd/Elh		1962	
1132	S60150 (205032)	4-4wDER		Afd/Elh		1962	
	S61743	4w-4wRER		Afd/Elh		1960	
98303		4wDMR		Geismar	G.780.004	1985	
(DX 68086)	SM 02	2w-2DMR		Wkm	10841	1975	
(DX 68088)	SM 01	2w-2DMR		Wkm	10842	1975	DsmT
–		2w-2PMR		Wkm			

DEVON RAILWAY CENTRE,
BICKLEIGH MILL, CADLEIGH STATION, TIVERTON **EX16 8RG**
Gauge 4ft 8½in www.devonrailwaycentre.co.uk **SS 938076**

	"BORIS"	0-4-0DM		Bg	3357	1952	

Gauge 2ft 0in

No.14	REBECCA	0-4-0WT	OC	OK	5744	1912	
	"PLANET"	4wDM		FH	2201	1939	
–		4wDM		L	34025	1949	a
	IVOR	4wDM	s/o	MR	8877	1944	
	EDEN	4wDM		MR	20058	1949	
	SIR TOM	4wDM		MR	40S273	1966	
	HORATIO	4wDM	s/o	RH	217967	1942	
	"RUSTON"	4wDM		RH	418770	1957	
1300	S14	0-4-0BE		WR		c1950	a

a currently under restoration off site

EXMOOR STEAM RAILWAY, CAPE OF GOOD HOPE FARM, BRATTON FLEMING
Private Site. Locomotives under construction and repair are usually present.
Gauge 2ft 0in **SS 662383**

77	2-6-2+2-6-2T	4C	Hano	10629	1928	
115	2-6-2+2-6-2T	4C	BP	6925	1937	
135	2-8-2	OC	AFB	2685	1952	
-	0-6-0WT+T	OC	OK	7122	1914	
1	4wDH		BD	3783	1984	
2	4wDH		BD	3784	1984	
–	4wDH		HE	9333	1994	
–	4wDH		HE	9335	1994	
LOD 758035	4wDM		MR	8856	1944	

C. GROVE, THE TAMAR BELLE HERITAGE GROUP, BERE FERRERS STATION

PL20 7LT

Gauge 4ft 8½in www.tamarbelle.co.uk **SX 452635**

	HILDA	0-4-0ST	OC	P	1963	1938	
PD&SWJR 3	A.S. HARRIS	0-4-0DM		HE	2642	1941	
PD&SWJR 4	EARL OF MOUNT EDGCUMBE	0-4-0DM		HE	3133	1944	
PD&SWJR 5	LORD ST. LEVAN	0-4-0DM		HE	3395	1946	

Gauge 2ft 0in

–		4wDM		RH	186318	1937	

LYNTON & BARNSTAPLE RAILWAY TRUST, WOODY BAY STATION, MARTINHOE CROSS, near PARRACOMBE

EX31 4RA

Gauge 2ft 0in www.lynton-rail.co.uk **SS 682464**

E 762	LYN	2-4-2T	OC	AK	92	2017	
	AXE	0-6-0T	OC	Gartell		2006	a
8	FAITH	0-4-2T	OC	UphillJ		2016	
	CHARLES WYTOCK	4-4-0T	OC	WB	2819	1946	
	ISAAC	0-4-2T	OC	WB	3023	1953	
D2393	PILTON	0-6-0DM		Bg/DC	2393	1952	
D6652		4wDH		HE	6652	1965	
–		4wDH		HE	6660	1965	

a locomotive built with new frames, incorporating parts from KS 2451

THE MILKY WAY ADVENTURE PARK, THE MILKY WAY RAILWAY, DOWNLAND FARM, near CLOVELLY

EX39 5RY

Gauge 2ft 0in www.themilkyway.co.uk **SS 327229**

–		4w-4wDH		SL	23	1973	

D. MOORE, BRATTON FLEMING

Gauge 2ft 0in

	"CHRIS"	4wDH		HLH	001	1996	

THE MORWELLHAM & TAMAR VALLEY TRUST, MORWELLHAM QUAY, near TAVISTOCK

PL19 8JL

Gauge 2ft 0in www.morwellham-quay.co.uk **SX 448699**

–		4wBE		AK	67	2003	
1	GEORGE S259	4wBE		WR	H7197	1968	
(2)	BERTHA	4wBE		WR	6298	1960	
5494 (3)	CHARLOTTE	4wBE		WR	G7124	1967	
4	LUDO	4wBE		WR	6769	1964	
(5)	WILLIAM	4wBE		WR	C6770	1964	
(6)	MARY	4wBE		WR	5665	1957	

No.7	HAREWOOD	4wBE		WR	D6800	1964	
8		4wBE		WR			Dsm

PLYM VALLEY RAILWAY COMPANY LTD, MARSH MILLS STATION, COYPOOL ROAD, PLYMPTON, PLYMOUTH PL7 4NW

Gauge 4ft 8½in www.plymrail.co.uk **SX 520571**

705		0-4-0ST	OC	AB	2047	1937	
	ALBERT	0-4-0ST	OC	AB	2248	1948	
5374	VANGUARD	0-6-0T	OC	Chrz	5374	1959	
	BYFIELD No.2	0-6-0ST	OC	WB	2655	1942	
–		0-4-0F	OC	WB	3121	1957	
(D2046)		0-6-0DM		Don		1958	
			reb	HE	6644	1967	
(D3002	11 DULCOTE) 13002	0-6-0DE		Derby		1952	
–		4wDM		FH	3281	1948	
10077		4wDH		S	10077	1961	
–		4wDH		TH	125V	1963	
W51365	T304	2-2w-2w-2DMR		PSteel		1960	
W51407	T304	2-2w-2w-2DMR		PSteel		1960	
–		2w-2PMR		Wkm	3366	1943	Dsm
–		2w-2PMR		Wkm	4154	1949	Dsm
–		2w-2PMR		Wkm	4992	1949	
			reb	PVRA		1989	DsmT
–		2w-2PMR		Wkm	5002	1949	
			reb	PVRA		1989	DsmT
–		2w-2PMR		Wkm	7139	1955	

PRIVATE OWNER, WHITCHURCH, TAVISTOCK

Gauge 2ft 0in

–		4wDM	L	11410	1939
	SIMON	4wDM	MR	7126	1936

PRIVATE SITE, Near PLYMOUTH

Gauge 4ft 8½in

–		2w-2PMR	Wkm	7973	1958

W.L.A. PRYOR, LYNTON RAILWAY STATION, STATION HILL, LYNTON

Gauge 2ft 0in **SS 719488**

No.3	BRUNEL 42	4wDM	RH	179880	1936

SOUTH DEVON RAILWAY TRUST, THE PRIMROSE LINE

Locomotives are kept at :-

Buckfastleigh TQ11 0QZ	SX 747663
Staverton Bridge	SX 785638
Totnes Littlehempston	SX 804613

Gauge 7ft 0¼in www.southdevonrailway.org

151	TINY	0-4-0VBWT	VCG	Sara		1868	

Gauge 4ft 8½in

1369		0-6-0PT	OC	Sdn		1934	
1420		0-4-2T	IC	Sdn		1933	
2873		2-8-0	OC	Sdn	2779	1918	
3205		0-6-0	IC	Sdn		1946	
3803		2-8-0	OC	Sdn		1939	
4920	DUMBLETON HALL	4-6-0	OC	Sdn		1929	
5526		2-6-2T	OC	Sdn		1928	
5786	L92	0-6-0PT	IC	Sdn		1930	
6412		0-6-0PT	IC	Sdn		1934	
	GLENDOWER	0-6-0ST	IC	HE	3810	1954	
47	CARNARVON	0-6-0ST	IC	K	5474	1934	
	LADY ANGELA	0-4-0ST	OC	P	1690	1926	
1	ASHLEY	0-4-0ST	OC	P	2031	1942	
(D402)	50002 SUPERB	Co-CoDE		_(EE	3771	1967	
				(EEV	D1142	1967	
D2246	(11216)	0-6-0DM		_(RSHN	7865	1956	
				(DC	2578	1956	
D2271		0-6-0DM		_(RSHD	7913	1958	
				(DC	2615	1958	
D3721	(09010)	0-6-0DE		Dar		1959	
D6501	(33002)	Bo-BoDE		BRCW	DEL93	1960	a
(D)6737	(37037 37321)	Co-CoDE		_(EE	2900	1961	
				(VF	D616	1961	
D7541	(25191)	Bo-BoDE		Derby		1965	
D7535	(25185) HERCULES	Bo-BoDE		Derby		1965	
D7612	(25262, 25901)	Bo-BoDE		Derby		1966	
W55000	(122 100)	2-2w-2w-2DMR		GRC&W		1958	
	MFP No.4	0-4-0DM		JF	4210141	1958	
2745		0-6-0DE		YE	2745	1960	
DR 98210		4wDMR		Plasser	52765A	1985	
–		2w-2PMR		Wkm	4149	1947	Dsm
(PWM 3767)	DS 3321 ADRIAN	2w-2PMR		Wkm	6646	1953	
(PWM 3773)	B 12 W	2w-2PMR		Wkm	6652	1953	
–		2w-2PMR		Wkm	8198	1958	
THE ADDICK	BERNARD WICKHAM	2w-2PMR		Wkm	11717	1976	

Gauge 4ft 6in

	LEE MOOR No.2	0-4-0ST	OC	P	784	1899	

a currently at Epping Ongar Railway, Ongar, Essex

P. SPENCER, BRIXHAM

Stored at a private location.

Gauge 1ft 3in

3205	EARL OF DEVON	4-4-0	IC	Prestige		2001

TARKA VALLEY RAILWAY C.I.O.,
TORRINGTON STATION, off STATION HILL, TORRINGTON **EX38 8JD**

Gauge 4ft 8½in www.tarkavalleyrailway.co.uk **SS 479198**

No.1	PROGRESS	0-4-0DM		JF	4000001	1945
544998	TORRINGTON CAVALIER	0-4-0DE		RH	544998	1969

TIVERTON MUSEUM SOCIETY,
TIVERTON MUSEUM, ST. ANDREW STREET, TIVERTON **EX16 6PJ**

Gauge 4ft 8½in www.tivertonmuseum.org.uk **SS 955124**

1442		0-4-2T	IC	Sdn		1935

MARTIN TURNER,
c/o BRIGHTLYCOTT COTTAGE, SHIRWELL ROAD, BARNSTAPLE

Gauge 4ft 8½in **SS 571354**

–		4wDM		FH	2893	1944

WORLD OF COUNTRY LIFE,
WEST DOWN LANE, SANDY BAY, near EXMOUTH **EX8 5BY**

Gauge 1ft 6in www.worldofcountrylife.co.uk **SY 035808**

–		4-2-2	OC	RegentSt		1898
	SIR FRANCIS DRAKE	4-6-0	OC	Scarrott		1988

DORSET

INDUSTRIAL SITES

ALASKA ENVIRONMENTAL CONTRACTING LTD,
STOKEFORD FARM, EAST STOKE, WAREHAM **BH20 6AL**

Locomotive may be present in yard between contracts.

Gauge 2ft 0in www.alaska.ltd.uk **SY 873872**

		4wDM		MR	8614	1941

MINISTRY OF DEFENCE, LULWORTH RANGES, EAST LULWORTH

See Section 6 for details. **SY 863822**

SOUTH COAST STEAM CO LTD,
UNIT 9, TRADECROFT INDUSTRIAL ESTATE, PORTLAND **DT5 2LN**

Locomotives usually present for maintenance, repair and /or restoration

Gauge 4ft 8½in **SY 685723**

–	4wDH	RR	10252	1966

FIRST MTR SOUTH WESTERN TRAINS LTD,
T/A SOUTH WESTERN RAILWAY, BOURNEMOUTH TRAINCARE DEPOT,
NELSON ROAD, WESTBOURNE, BOURNEMOUTH **BH4 9JA**

(A first Group and MTR Company)

Gauge 4ft 8½in www.southwesternrailway.com **SZ 066917**

See section 7 for details

VITACRESS SALADS LTD, WATERCRESS GROWERS,
DODDINGS FARM, BERE REGIS

Gauge 1ft 6in www.vitacress.com **SY 852934**

VITACRESS	4wPH	Jesty		1948

PRESERVATION SITES

AVON CAUSEWAY HOTEL, HURN, near BOURNEMOUTH **BH23 6AS**

Gauge 4ft 8½in www.avoncauseway.co.uk **SZ 136976**

–	0-4-0DM	JF	22871	1939

G.A. & L.M. FELDWICK, 22A ROPERS LANE, WAREHAM **BH20 4QT**

Gauge 2ft 0in **SY 921875**

Private location with locomotives occasionally present. Visitors by appointment only.

JEFF LEAKE, 154 WAREHAM ROAD, LYTCHETT MATRAVERS

Gauge 1ft 3in **SY 945950**

–	4w-4BER	Leake		1992
–	0-4-0BE	Leake		1992
–	0-4-0DM	Leake		1992

MICKY FINN RAILWAY, near POOLE

Gauge 2ft 0in Private site

8	4wDM	L	28039	1945
–	4wDM	MR	26007	1964

MOORS VALLEY RAILWAY, MOORS VALLEY COUNTRY PARK, HORTON ROAD, ASHLEY HEATH, RINGWOOD — BH24 2ET

Gauge 2ft 0in www.moorsvalleyrailway.co.uk **SZ 104061**

	EMMET		0-4-0T		Moors Valley	20	1995	
		a rebuild of	0-4-0DM		OK	21160	1938	a

a currently at the Old Kiln Light Railway, near Farnham, Surrey

Gauge 1ft 3in

	KATIE		0-4-0T	OC	FMB	002	1997	
		rebuilt as	0-4-2T	OC	HaylockJ			

THE NORTH DORSET RAILWAY TRUST, THE SHILLINGSTONE STATION PROJECT, SHILLINGSTONE STATION, SHILLINGSTONE — DT11 0SA

Gauge 4ft 8½in www.shillingstone-railway-project.co.uk **ST 825116**

(62-669)	30075	0-6-0T	OC	DDak	669	1960	
(62-521)	30076	0-6-0T	OC	DDak	521	1954	
DS 1169	LITTLE EVA	4wDM		RH	305302	1951	
TP57P		2w-2PMR		Wkm	8267	1959	

PRIVATE OWNER, PRIVATE LOCATION

Gauge 4ft 8½in

47160	CUNARDER	0-6-0T	OC	HE	1690	1931	
92207	MORNING STAR	2-10-0	OC	Sdn		1959	
(PWM 4302)		2w-2PMR		Wkm	7505	1956	

SWANAGE RAILWAY SOCIETY

Locomotives are kept at :-

Swanage Station BH19 1HB SZ 026789
Harmans Cross BH19 3EB SY 983800
Herston Railway Works BH19 1AH SZ 018793
Norden BH20 5DW SY 955828, 954831

Gauge 4ft 8½in www.swanagerailway.co.uk

(30053)53	0-4-4T		IC	9E		1905		
(31625)	5 "JAMES"		2-6-0	OC	Afd		1929	a
31806	(806)		2-6-0	OC	Afd		1928	
		rebuild of	2-6-4T	OC	Bton		1926	
31874			2-6-0	OC	Woolwich		1925	a
34010	SIDMOUTH		4-6-2	3C	Bton		1945	
34028	EDDYSTONE		4-6-2	3C	Bton		1946	
34072	257 SQUADRON		4-6-2	3C	Bton		1948	
80104			2-6-4T	OC	Bton		1955	
(D3551)	08436		0-6-0DE		Derby		1958	
D3591	(08476)		0-6-0DE		Crewe		1958	
D6515	(33012)	LT. JENNY LEWIS RN	Bo-BoDE		BRCW	DEL107	1960	
(D6528)	33111		Bo-BoDE		BRCW	DEL120	1960	
W51346			2-2w-2w-2DMR		PSteel		1960	b
W51388			2-2w-2w-2DMR		PSteel		1960	b
M51933			2-2w-2w-2DMR		DerbyC&W		1960	

W55028	(960.012 L128 977860)	2-2w-2w-2DMR		PSteel		1960	b
2054	BERYL	4wPM		FH	2054	1938	
	MAY	0-4-0DM		JF	4210132	1957	
	99709 909140-4	2w-2PMR		Geismar	ST/04/09	2004	

a currently stored off site
b currently at Arlington Fleet Group Ltd, Eastleigh, Hampshire

Purbeck Mineral & Mining Group, Corfe Castle
Goods Shed, Corfe Castle
Gauge 2ft 8in www.purbeckminingmuseum.org **SZ 962821**

	(SECUNDUS)	0-6-0WT	OC	B&S		1874

Norden Park & Ride, Corfe Castle
BH20 5DW
Gauge 2ft 0in **SZ 957829**

–		4wVBT	VCG	DonnellyN		2018
	SNAPPER	4wDM		RH	283871	1950
	NORDEN	4wDM		RH	392117	1956
–		4wDM		MR	8994	1943

P.C. VALLINS, near BLANDFORD
Private Location
Gauge 2ft 0in

–	4wDM	L	9256	1937
–	4wPM	L	18557	1942
–	4wDM	LB	51917	1960

DURHAM

OUR DEFINITION OF "COUNTY DURHAM" INCLUDES THOSE LOCATIONS WHICH WERE SITUATED WITHIN THE PART OF THE FORMER COUNTY OF TYNE & WEAR WHICH IS SOUTH OF THE RIVER TYNE AND NOW COMPRISES A NUMBER OF UNITARY AUTHORITIES. LOCOMOTIVES WHICH ARE LOCATED WITHIN THE COUNTY BOROUGHS OF HARTLEPOOL AND STOCKTON-ON-TEES ARE LISTED UNDER A SEPARATE HEADING OF TEESSIDE.

INDUSTRIAL SITES

BRITISH AMERICAN RAILWAY SERVICES LTD, t/a RMS LOCOTEC, STANHOPE STATION, BONDISLE, STANHOPE, BISHOP AUCKLAND
(A subsidiary of Iowa Pacific Holdings) Office address.

Locomotives are maintained at Weardale Railway Community Interest Company, Wolsingham Depot, Co. Durham and Tata Steel Colours, Shotton Works, Flintshire. The following is a FLEET LIST of locomotives owned by this contractor.

Gauge 4ft 8½in www.rmslocotec.com

(D3378)	08308		0-6-0DE	Derby	1957	a
(D3460)	08375	21	0-6-0DE	Dar	1957	b
(D3538)	08423	H 011 14	0-6-0DE	Derby	1958	c
(D3685)	08523	H 061	0-6-0DE	Don	1958	d
(D3740)	08573		0-6-0DE	Crewe	1959	a

(D3755)	08588 H 047	0-6-0DE		Crewe		1959	e	
(D3780)	08613 H 064	0-6-0DE		Derby		1959	c	
(D3789)	08622 H 028 19	0-6-0DE		Derby		1959	b	
(D3815)	08648	0-6-0DE		Hor		1959	d	
(D3918)	08750	0-6-0DE		Crewe		1960	a	
(D3922)	08754 H 041	0-6-0DE		Hor		1961	a	
(D3924)	08756 H 039	0-6-0DE		Hor		1961	a	
(D3930)	08762	0-6-0DE		Hor		1961	a	
(D3956)	08788	0-6-0DE		Derby		1960	f	
(D3977)	08809	0-6-0DE		Derby		1960	f	
(D4015)	08847	0-6-0DE		Hor		1961	g	
(D4038	08870) H 024	0-6-0DE		Dar		1960	a	
(D4039)	08871 H 074	0-6-0DE		Dar		1960	h	
(D4042)	08874	0-6-0DE		Dar		1960	a	
(D4115)	08885 H 042 18	0-6-0DE		Hor		1962	a	
(D4166)	08936 H 075	0-6-0DE		Dar		1962	a	
(625 690)	H 043 16	0-6-0DE		_(EE	2122	1956		
				(VF	D312	1956	j	
(653)	H 050	0-6-0DE		_(EE	2150	1956		
				(VF	D340	1956	a	
11		0-6-0DH		EEV-AEI	3994	1970		
			rep	YEC	L180	2000	a	
(01573)	H 006 15	0-6-0DH		HE	6294	1965	a	
(H 015)		0-6-0DH		HE	7410	1976	a	
H 032	PETE GANNON LOCOMOTIVE ENGINEER 1948-2001							
		0-6-0DH		HE	7541	1976	a	
–		4wDH		RR	10232	1965	a	
H 014		0-6-0DH		RR	10262	1967	f	+
H 058		4wDH		RR	10280	1968	k	
H 055	VINCENT DE RIVAZ	4wDH		S	10037	1960	l	
H 003	ROSEDALE	4wDH		S	10070	1961	f	
H 057	DL2	4wDH		S	10177	1964	m	
TNS 105		2-2wDMR		Robel				
					56.27-10-AG36	1982	n	
TNS 107		2-2wDMR		Robel				
					56.27-10-AG38	1983	o	
(DX 68086)	SM 02	2w-2DMR		Wkm	10841	1975	p	
(DX 68088)	SM 01	2w-2DMR		Wkm	10842	1975	p	Dsm

a currently at Weardale Railway, Wolsingham, Co. Durham
b currently at Heidelberg Cement Group, t/a Hanson Cement, Ketton Works, Rutland
c currently at PD Ports Teesport, Tees Dock
d currently at Abellio Scotrail Ltd, Inverness T&RSMD, Highland, Scotland
e currently at Loram UK Ltd, Derby, Derbyshire
f currently at Tata Steel Europe, Shotton Works, North Wales
g currently at Mid-Norfolk Railway, Dereham, Norfolk
h currently at Bombadier Transportation, Ilford Depot, Greater London
j currently at Llanelli & Mynydd Mawr Railway Co Ltd, Cynheidre
k currently at EDF Energy plc, Hartlepool Power Station, Co.Durham
l currently at EDF Energy plc, Heysham Power Stations, Lancashire
m currently at Ford Motor Co Ltd, Bridgend, South Wales
n currently at EDF Energy (UK) Ltd, Cottam Power Station, Notts
o currently at Flixborough Wharf Ltd, Scunthorpe, Lincolnshire
p currently at Dartmoor Railway, Meldon Quarry, Devon
\+ carries plate RR 10287

Gauge 3ft 0in

LM 363		4wDH		DunEW	LM363	c1984	a
(LM 370)		4wDH		DunEW			a
LM 373		4wDH		DunEW		1984	
			reb	BoothWKelly		2008	b +
H 048	BERTIE	0-6-0DM		RH	281290	1949	a

a currently at Weardale Railway, Wolsingham, Co. Durham
b location not known
\+ rebuilt as non-self propelled generator and compressor unit

D.C. ENGINEERING, UNIT 8A, HACKWORTH INDUSTRIAL PARK, SHILDON — DL4 1HS

Gauge 4ft 8½in — **NZ 224255**

(DX 68003 DB 965331)	2w-2PMR	Wkm	10179	1968	

vehicles under restoration usually present

GRANGE IRON COMPANY LTD 56 BROOMSIDE LANE, CARRVILLE, DURHAM — DH1 2QT

Gauge 2ft 0in www.grangeironcompany.weebly.com NZ 309439

1	ROSIE	0-4-0BE	WR

new minetubs under construction usually present

HITACHI RAIL EUROPE, TRAIN MANUFACTURING FACILITY, IEP MERCHANT PARK, MILLENNIUM WAY, HEIGHINGTON, NEWTON AYCLIFFE — DL5 6xx

Gauge 4ft 8½in www.hitachirail-eu.com **NZ 266221**

CRAB 1	NAY 1000075	4wBE	R/R	Zephir	2630	2016	
–		4wDH	R/R	Zephir	2631	2016	
	NAY 1000185	4wBE	R/R	Zephir	2632	2016	
CRAB 2	NAY 1000371	4wBE	R/R	Zephir	2678	2016	
(D3599)	08484 CAPTAIN NATHANIEL DARELL	0-6-0DE		Hor		1958	a

a property of Railway Support Services Ltd, Wishaw, Warwickshire

new multiple unit type rolling stock under construction/repair usually present

NBR ENGINEERING SERVICES LTD, UNIT 6, 2 BARTON STREET, DARLINGTON — DL1 2LP

Gauge 2ft 0in www.nbres.co.uk NZ 300155

–	0-4-0ST	OC	NBRES	2017	a

a incomplete loco

locos under construction, refurbishment, overhaul or repair usually present

TYNE & WEAR FIRE & RESCUE SERVICE, BARMSTON MERE TRAINING CENTRE, NISSAN WAY, WASHINGTON — SR5 3QY

Gauge 4ft 8½in www.twfire.gov.uk **NZ 329570**

3721		4w-4wRER		MetCam	1983	OOU

used as a static training aid

U.K. MINING VENTURES LTD, ROGERLEY QUARRY

Gauge 2ft 0in www.ukminingventures.com

1291	2848	4wBE	CE	(5858	1971?)	a

a battery box carries worksplate CE B0476

PRESERVATION SITES

A1 LOCOMOTIVE TRUST, DARLINGTON LOCOMOTIVE WORKS, STOCKTON & DARLINGTON RAILWAY CARRIAGE SHED, HOPETOWN LANE, DARLINGTON — DL3 6RQ

Gauge 4ft 8½in www.a1steam.com **NZ 288156**

60163	TORNADO	4-6-2	3C	Darlington	2195	2008
	Based here for maintenance					

BEAMISH – THE LIVING MUSEUM OF THE NORTH, BEAMISH, near STANLEY — DH9 0RG

Gauge 4ft 8½in www.beamish.org.uk **NZ 215548, 215545, 218543, 221543**

	"PUFFING BILLY"	4wG	VC	AK	71	2006	
E No.1		2-4-0VBCT	OC	BH	897	1887	
	STEAM ELEPHANT	6wG		(Dorothea		2001	
				(AK		2001	
–		0-4-0	VC	GS(H)		c1852	
–		0-4-0VBT	VC	HW		1871	
			reb	Wilton(ICI)		1984	a
17		0-4-0VBT	OC	HW	33	1873	
18		0-4-0T	OC	Lewin	683	1877	
	NEWCASTLE	0-6-0ST	IC	MW	1532	1901	
	MAY	0-4-0ST	OC	P	1370	1915	
–		0-6-0ST	IC	P	2000	1942	
No.15	ROKER	0-4-0CT	OC	RSHN	7006	1940	
No. 5		0-4-0ST	OC	SDSI(S)		1900	
	VULCAN	0-4-0ST	OC	VF	3272	1918	
14		0-4-0DE		AW	D21	1933	
2		4wWE		Siemens	455	1908	

a carries plate T.H. Head, Engineer, 50 Cannon St, London 1871

Gauge 2ft 0in

5		0-4-0T	OC	AB	988	1903
	SAMSON	0-4-0WTG	OC	Beamish	BM2	2014
No.2		0-4-0WT	OC	KS	721	1901
	L.R. 3098	4wPM		MR	1377	1918

Gauge 1ft 11½in

	GLYDER	0-4-0WT	OC	AB	1994	1931
	EDWARD SHOLTO	0-4-0ST	OC	HE	996	1909
	ASHOVER	4wDM		FH	3307	1948

BOWES RAILWAY CO LTD,
SPRINGWELL ROAD, SPRINGWELL, GATESHEAD — NE9 7QJ

Gauge 4ft 8½in www.bowesrailway.uk **NZ 285589**

No.22	No.85	0-4-0ST	OC	AB	2274	1949
	W.S.T.	0-4-0ST	OC	AB	2361	1954
101		4wDM		FH	3922	1959
–		0-4-0DH		HE	6263	1964
	PERKY	0-4-0DE		RH	395294	1956
	PINKY	0-4-0DE		RH	416210	1957
	REDHEUGH	4wDM		RH	476140	1963
			reb	Wilton(ICI)		1982

Gauge 3ft 0in

No.1	9307/110	4wBE		CE	5921	1972
No.2	20/270/34	4wBE		CE	B3060	1983

Gauge 2ft 6in

B03		4w-4wDHF		_(HE	8515	1981
				(AB	651	1981
No.8		4wBE		CE	B1840	1978

Gauge 2ft 0in

20.123.945	SER No.2476	4wBEF		_(EE	2476	1958
	PLANT No.2207/456			(RSHN	7980	1958
–		0-6-0DMF		HC	DM842	1954

CLASS G5 LOCOMOTIVE COMPANY LTD,
UNIT 8 S, HACKWORTH INDUSTRIAL PARK, SHILDON — DL4 1HS

Gauge 4ft 8½in www.g5locomotiveltd.co.uk **NZ 224255**

"1759"	0-4-4T	IC	GNS/G5 Loco Co		a

a new build locomotive under construction

DARLINGTON RAILWAY PRESERVATION SOCIETY,
STATION ROAD, HOPETOWN, DARLINGTON — DL3 6ST

Gauge 4ft 8½in www.drps.synthasite.com **NZ 290157, 285160**

	NORTHERN GAS BOARD No.1	0-4-0ST	OC	P	2142	1953
No.39		0-6-0T	OC	RSHD	6947	1938
	PATONS	0-4-0F	OC	WB	2898	1948

185	DAVID PAYNE	0-4-0DM		JF	4110006	1950
–		0-4-0DM		JF	4200018	1947
–		4wDM		RH	279591	1949
–		0-4-0DE		RH	312988	1952
–		0-4-0DM		(RSHN	7925	1959
				(DC	2592	1959
1		4wWE		GEC		(1928?)
(DB 965096)		2w-2PMR		Wkm	7611	1957
(DX 68003)	68/007 DB965951 MPP 0007	2w-2PMR		Wkm	10647	1972

Gauge 1ft 8in

–		4wDM		RH	476124	1962
(DL 354013)		4wDM		RH	354013	1953

HEAD OF STEAM, DARLINGTON RAILWAY MUSEUM, NORTH ROAD STATION, STATION ROAD, HOPETOWN, DARLINGTON — DL3 6ST

Gauge 4ft 8½in — **NZ 288157**

(63460)	901	0-8-0	3C	Dar		1919
No.1463		2-4-0	IC	Dar		1885
25	DERWENT	0-6-0	OC	Kitching		1845
	LOCOMOTION	0-4-0	VC	RS	1	1825

KYNREN, FLATTS FARM, TORONTO, BISHOP AUCKLAND — DL14 7SF

(operated by Eleven Arches charity) — NZ 211307

Gauge 5ft 0in www.kynren.com

	LOCOMOTION	0-4-0BE	s/o			c2016

LOCOMOTION – THE NATIONAL RAILWAY MUSEUM AT SHILDON, DALE ROAD INDUSTRIAL ESTATE, SHILDON — DL4 2RE

Gauge 4ft 8½in www.locomotion.org.uk — **NZ 239255**

	"SANS PAREIL"	0-4-0	VC	Hackworth		1829
	LOCOMOTION	0-4-0	VC	LocoEnt	No.1	1975
	SANS PAREIL	0-4-0	VC	BRE(S)		1980
No.251		4-4-2	OC	Don	991	1902
790	HARDWICKE	2-4-0	IC	Crewe	3286	1892
1621		4-4-0	IC	Ghd		1893
34051	WINSTON CHURCHILL	4-6-2	3C	Bton		1946
(45000)	5000	4-6-0	OC	Crewe	216	1934
49395		0-8-0	IC	Crewe	5662	1921
(60800)	4771 GREEN ARROW	2-6-2	3C	Don	1837	1936
65033		0-6-0	IC	Ghd		1889
	IMPERIAL No.1	0-4-0F	OC	AB	2373	1956
2253		2-8-0	OC	BLW	69496	1944
	JUNO	0-6-0ST	IC	HE	3850	1958
44	"CONWAY"	0-6-0ST	IC	K	5469	1933
No.15	"EUSTACE FORTH"	0-4-0ST	OC	RSHN	7063	1942
No.77	NORWOOD	0-6-0ST	OC	RSHN	7412	1948
	DELTIC	Co-CoDE		EEDK	2007	1955

D2090	(03090)	0-6-0DM		Don		1960	
(D4141)	08911 MATEY	0-6-0DE		Hor		1962	
E5001	(71001) 89403	Bo-BoRE/WE		Don		1959	
(26500)	No.1	Bo-BoWE/RE		BE		1905	
(DS 75)	75S	4wRE		_(Siemens ¯(HC	6	1898 1898	c
–		4wPM		MR	4217	1931	
H 001		4wDH		S	10003	1959	
2090	S10656	4w-4wRER		_(Lancing ¯(Elh		1937 1937	
3131	11179	4w-4RER		EE/Elh		1938	
4308	61275	4w-4wRER		Afd/Elh		1959	
65217	306 017	4w-4wWER		MetCam		1949	
APT-E	PC1/TC1/TC2/PC2	4w-4w-4-4w-4wArticGTE		Derby		1972	
(960209)		2w-2PMR		Wkm	899	1933	

c carries worksplate Siemens 7/1898

Gauge 3ft 6in

390		4-8-0	OC	SS	4150	1896	

Gauge 3ft 0in

No. 14	No.4 9306/108	0-6-0DMF		HC	DM1274	1961	

Gauge 2ft 6in

No.14	SLR 85	2-6-2T	OC	HE	3815	1954	+

\+ currently on road tour of UK

Gauge 1ft 3in

1	LOCOMOTION 1825	0-4-0	VC	(Dar?)		1875	

Soho Engine Shed, Soho Street, Shildon **DL4 1PQ**
Gauge 4ft 8½in (open during special event days) **NZ 233257**

"NELSON"	(BRADYLL)	0-6-0	OC	(Hackworth?)		c1835	a

a possibly built by Fossick & Hackworth, Stockton-on-Tees

NORTH EASTERN LOCOMOTIVE PRESERVATION GROUP, HOPETOWN CARRIAGE WORKS, DARLINGTON **DL3 6RQ**

Gauge 4ft 8½in www.nelpg.org.uk **NZ 288157**

62005		2-6-0	OC	NBQ	26609	1949
69023	JOEM	0-6-0T	IC	Dar	2151	1951

SUNDERLAND CITY COUNCIL, TYNE & WEAR JOINT MUSEUMS SERVICE, WASHINGTON "F" PIT MUSEUM, WASHINGTON NEW TOWN **NE37 1BN**

Gauge 2ft 0in www.seeitdoitsunderland.co.uk/washington-f-pit **NZ 303574**

–	0-4-0DMF		RH	392157	1956

TANFIELD RAILWAY PRESERVATION SOCIETY, MARLEY HILL, near STANLEY — NE16 5ET

Gauge 4ft 8½in www.tanfield-railway.co.uk **NZ 207573**

–		0-6-0ST	OC	AB	1015	1904	
No.6		0-4-2ST	OC	AB	1193	1910	
	WELLINGTON	0-4-0ST	OC	BH	266	1873	
RENISHAW IRONWORKS No.6		0-6-0ST	OC	HC	1366	1919	
	IRWELL	0-4-0ST	OC	HC	1672	1937	
38		0-6-0T	OC	HC	1823	1949	
–		0-4-0ST	OC	HL	2711	1907	
			reb	DL		1956	
No.2		0-4-0ST	OC	HL	2859	1911	
L.& H.C. 14		0-4-0ST	OC	HL	3056	1914	
			reb	‡		1980	
	STAGSHAW	0-6-0ST	OC	HL	3513	1927	
	COAL PRODUCTS No.3	0-6-0ST	OC	HL	3575	1923	
No.13		0-4-0ST	OC	HL	3732	1928	
–		0-6-0F	OC	HL	3746	1929	
No.3	TWIZELL	0-6-0T	IC	RS	2730	1891	
–		0-4-0CT	OC	RSHN	7007	1940	
	LYSAGHT'S	0-6-0ST	OC	RSHD	7035	1940	
49		0-6-0ST	IC	RSHN	7098	1943	
	SIR CECIL A.COCHRANE	0-4-0ST	OC	RSHN	7409	1948	
No.44	9103/44	0-6-0ST	OC	RSHN	7760	1953	
38		0-6-0ST	OC	RSHN	7763	1954	
21		0-4-0ST	OC	RSHN	7796	1954	
–		0-6-0ST	OC	RSHN	7800	1954	
No.48		0-6-0ST	OC	RSHN	7944	1957	
(No.3)		0-4-0ST	OC	RWH	2009	1884	
(No.4)		4wVBT	VCG	S	9559	1953	
No.20		0-6-0ST	IC	WB	2779	1945	
No.2		0-4-0DE		AW	D22	1933	
–		4wDM		FH	3716	1955	
	(2111-125)	0-6-0DH		HE	6612	1965	
No.6		0-6-0DH		JF	4240010	1960	
T.I.C.No.35		0-4-0DE		RH	418600	1958	
–		0-6-0DM		RSHN	7697	1953	
–		0-4-0DM		RSHN	6980	1940	
–		0-6-0DM		RSHN	7746	1954	
–		0-4-0DM		RSHN	7901	1958	
(9)		Bo-BoWE		AEG	1565	1913	
–		4wWE		GB	2509	1955	Dsm
2	DEREK SHEPHERD	Bo-BoWE		HL	3872	1936	a
	rebuilt as	Bo-BoBE		Riley		1993	
	(KEARSLEY No.3)	Bo-BoWE		RSHN	7078	1944	
E10		4wWE		Siemens	862	1913	
MPP 10157		2w-2BER		Bance	097	2000	
–		2w-2DHR		Bg	3565	1962	

DB 965097	68044	2w-2PMR		Wkm	7612	1957	
MNN 0005		2w-2PMR		Wkm			DsmT

a carries plate RSH 3872
‡ rebuilt by Clark Hawthorn Ltd, Northumberland Engine Works, Wallsend-on-Tyne

Gauge 3ft 6in

(M2)		4-6-2	OC	RSHD	7430	1951

Gauge 2ft 6in

–		4wBEF		CE	B1886B	1980

Gauge 600mm

No.11	ESCUCHA	0-4-0T	OC	BH	748	1883

Gauge 2ft 0in

4	DM1067	0-6-0DMF		HC	DM1067	1959
2305/54	TYNESIDE GEORGE	0-6-0DMF		HC	DM1119	1958
No.5	2201/266	0-6-0DMF		HC	DM1170	1960
–		4wDM		HE	2577	1942
2	AYLE	4wDM		HE	2607	1942
No.1		4wDHF		HE	7332	1974
–		4wDM		LB	53162	1962
–		4wDM		LB	54781	1965
–		4wDM		RH	244487	1946
–		4wDM		RH	323587	1952
No.2		4wBEF		CE	B3141B	1984
–		4wBEF		_(EE	2848	1960
				(RSHN	8201	1960
–		0-4-0BE		WR		1972
–		0-4-0BE		WR		

THORPE LIGHT RAILWAY, WHORLTON, near BARNARD CASTLE

Private site with occasional public open days

Gauge 1ft 3in www.thorpelightrailway.co.uk **NZ 106146**

	"BESSIE"	4wDM		Eddy/Knowell		2002	
–		2-8-0DH	s/o	SL	73.35	1973	
	reb	2-6-0DH	s/o			1994	
	reb	2-8-0DH	s/o	MRWRS		c2009	a

a property of MRW Railways Ltd, Sheffield, South Yorkshire

WEARDALE RAILWAY COMMUNITY INTEREST COMPANY, WOLSINGHAM RAILWAY CENTRE, off DURHAM ROAD, WOLSINGHAM

(Operated by Weardale Railway Heritage Services Ltd, a subsidiary of Weardale Railway Trust)

No public access to depot

Gauge 4ft 8½in www.weardale-railway.org.uk **NZ 081370**

No.40	68692 9312/40	0-6-0T	OC	RSHN	7765	1954	
(D3378)	08308	0-6-0DE		Derby		1957	a
(D3740)	08573	0-6-0DE		Crewe		1959	a

(D3918)	08750	0-6-0DE		Crewe		1960	a	OOU
(D3922)	08754 H 041	0-6-0DE		Hor		1961	a	
(D3924)	08756 H 039	0-6-0DE		Hor		1961	a	
(D3930)	08762	0-6-0DE		Hor		1961	a	
(D4038	08870) H 024	0-6-0DE		Dar		1960	a	
(D4042)	08874	0-6-0DE		Dar		1960	a	
(D4115)	08885 H 042 18	0-6-0DE		Hor		1962	a	OOU
(D4166)	08936 H 075	0-6-0DE		Dar		1962	a	OOU
(D5524)	31106	A1A-A1ADE		BT	123	1959		
(D5637	31213) 31465	A1A-A1ADE		BT	237	1960	b	
(653)	H 050	0-6-0DE		_(EE	2150	1956		
				(VF	D340	1956	a	
11		0-6-0DH		EEV-AEI	3994	1970		
			rep	YEC	L180	2000	a	
(01573)	H 006 15	0-6-0DH		HE	6294	1965	a	
LH 005		0-6-0DH		HE	6295	1965		
(H 015)		0-6-0DH		HE	7410	1976	a	
H 032	PETE GANNON LOCOMOTIVE ENGINEER 1948-2001	0-6-0DH		HE	7541	1976	a	
–		0-6-0DH		RR	10187	1964		
–		4wDH		RR	10232	1965	a	OOU
M50980		2-2w-2w-2DMR		DerbyC&W		1959		
M52054		2-2w-2w-2DMR		DerbyC&W		1960		
E55012		2-2w-2w-2DMR		GRC&W		1959		
–	(DB 965950)	2w-2PMR		Wkm	10646	1972		
(DX 68080	DB 965993) POINTLESS 3	2w-2DMR		Wkm	10706	1974	DsmT	
DX 68090		2w-2DMR		Wkm	10843	1975		
MPP 9832		2w-2BER		Bance	00101	1996	Dsm	

Gauge 3ft 0in – (locomotives for storage)

LM 363		4wDH	DunEW	LM363	c1984	a	
(LM 370)		4wDH	DunEW			a	
H 048	BERTIE	0-6-0DM	RH	281290	1949	a	

a property of British American Railway Services Ltd, Stanhope, Co. Durham
b property of Harry Needle Railroad Co Ltd, Barrow Hill, Derbyshire

ESSEX

INDUSTRIAL SITES

HALTERMANN CARLESS LTD, HARWICH REFINERY, REFINERY ROAD, PARKESTON, HARWICH — CO12 4QG

Gauge 4ft 8½in www.haltermann-carless.com **TM 232323**

Q240 JBV	4wDM	R/R	Unimog	092692	1982	OOU

ROGER HARVEY, WALTON-ON-THE-NAZE

Gauge 4ft 8½in

–	4wWE	KS	1269	1912	Dsm

INDUSTRIAL CHEMICALS LTD, TITAN WORKS, TITAN INDUSTRIAL ESTATE, HOGG LANE, GRAYS — RM17 5DU

Gauge 4ft 8½in www.icgl.co.uk **TQ 613783**

–	4wDH	TH	144V	1964	OOU

QINETIQ LTD, ENVIRONMENTAL TEST CENTRE, FOULNESS ISLAND, SHOEBURYNESS

Gauge 2ft 6in www.QinetiQ.com **TQ 980889**

RAMBO	4wBEF	CE	B0483	1976
TERMINATOR	4wBEF	CE	B0483	1976

TRADITIONAL TRACTION LTD, SIBLE HEDINGHAM

Adminstration address only. The following is a FLEET LIST of locomotives owned by this contractor

Gauge 4ft 8½in

(D8035	20035)	Bo-BoDE	(EE	2757	1959	
2001	AT3 DJ 053		(VF	D482	1959	a
(D8063	20063)	Bo-BoDE	(EE	2969	1961	
2002	AT3 DJ 054		(RSHD	8221	1961	b

a currently located at Gloucestershire-Warwickshire Railway, Gloucestershire

b currently located at Battlefield Line, Shackerstone, Leicestershire

PRESERVATION SITES

BRITISH POSTAL MUSEUM, STORE & ARCHIVE, DEBDEN — IG10 3UF

Gauge 2ft 0in www.postalmuseum.org **TQ 444962**

–	4w Atmospheric Car	1861	a

a Cut into two halves, Inventory no.s 31.85/1 and 31.85/2

CHELMSFORD CITY COUNCIL, CHELMSFORD MUSEUM, OAKLANDS PARK, MOULSHAM STREET, CHELMSFORD — CM2 9AQ

Gauge 2ft 8½in www.chelmsford.gov.uk/museums **TL 705069**

8	4wRER		BE	1898
		reb	BE	c1911

COLNE VALLEY RAILWAY PRESERVATION SOCIETY LTD, CASTLE HEDINGHAM STATION, YELDHAM ROAD, CASTLE HEDINGHAM, HALSTEAD — CO9 3DZ

Gauge 4ft 8½in www.colnevalleyrailway.co.uk **TL 774362**

35010	BLUE STAR	4-6-2	3C	Elh		1942	
45163		4-6-0	OC	AW	1204	1935	
45293		4-6-0	OC	AW	1348	1936	
1875	BARRINGTON	0-4-0ST	OC	AE	1875	1921	
WD 190		0-6-0ST	IC	HE	3790	1952	
No.24		0-6-0ST	IC	HE	3800	1953	
–		0-4-0ST	OC	HL	3715	1928	
No.60	JUPITER	0-6-0ST	IC	RSHN	7671	1950	
D2041		0-6-0DM		Sdn		1959	
D2184		0-6-0DM		Sdn		1962	
(D9524)	14901	0-6-0DH		Sdn		1964	a
W55033	(977826 T003)	2-2w-2w-2DMR		PSteel		1960	
55508	141 108	4wDMR		_(BRE(D)		1984	
				(Leyland R4.016		1984	
			reb	AB	759	1989	
55528	141 108	4wDMR		_(BRE(D)		1984	
				(Leyland R4.033		1984	
			reb	AB	758	1989	
E79978		4wDMR		ACCars		1958	
	ROF PURITON No.2	0-4-0DM		AB	349	1941	
–		4wDM		FH	3147	1947	
	HENRY	4wDM		Lake&Elliot		c1924	
			reb	FordTTC		1997	
YD No.43		4wDM		RH	221639	1943	
–		0-4-0DM		RH	281266	1950	
–		4wDM	R/R	Unilok(G)	2109	1980	
1	DOUG TOTTMAN	Bo-BoBE		RSHN	7284	1945	
			reb	Kearsley		1982	
HCT 005	XOP 2298	4wDH		Perm	005	1987	
–		2w-2PMR		Wkm	1946	1935	

a property of Andrew Briddon, Darley Dale, Derbyshire

CRAVEN HERITAGE TRAINS LTD, EPPING SIGNALBOX, EPPING — CM16 4HW

Gauge 4ft 8½in www.cravensheritagetrains.co.uk **TL 460013**

(L11)	4w-4wRE	MetCam	1931/1932

ROGER CRAVEN, PRIVATE LOCATION

Gauge 3ft 0in

–		4wDMF	RH	418803	1957

Gauge 750mm

5	BETSY	4wDH	HE	8829	1979
16		4wDH	HE	9079	1984
15		4wDH	HE	9080	1984
(11)		4wDM	Moës		

27		4wDM	Moës		
4		4wDM	Moës		
(3)		4wDMF	RH	375693	1954

Gauge 650mm

2		0-4-0DM	RH	305326	1952

Gauge 600mm

–		4wDM	Coferna	3821	
–		0-4-0DM	Dtz	47069	1950
1		4wDM	Diema	1407	1951
–		4wDM	Jung	7649	1937
17		4wDM	Moës		
–		4wDM	Moës		
–		4wDM	Moës		
–		4wDM	Ruhrthaler	1343	1934
	a rebuild of	4wDM	Ruhrthaler	1256	1933

EAST ANGLIAN RAILWAY MUSEUM, CHAPPEL & WAKES COLNE STATION, STATION ROAD, WAKES COLNE CO6 2DS

Gauge 4ft 8½in www.earm.co.uk **TL 898289**

69621	(7999)	0-6-2T	IC	Str		1924	a
No.11	STOREFIELD	0-4-0ST	OC	AB	1047	1905	
1		0-6-0T	IC	EARM		2008	
	rebuild of	0-6-0ST	IC	RSHN	7031	1941	
	JEFFREY	0-4-0ST	OC	P	2039	1943	
	(JUBILEE)	0-4-0ST	OC	WB	2542	1936	
	LAMPORT No.3	0-6-0ST	OC	WB	2670	1942	
(D2279	11249)	0-6-0DM		_(RSHD	8097	1960	
				(DC	2656	1960	
E51213		2-2w-2w-2DMR		MetCam		1959	
E79963		2w-2DMR		WMD	1268	1958	
A.M.W.No.144	JOHN PEEL	0-4-0DM		AB	333	1938	
7		0-4-0DH		JF	4220039	1965	b
–		4wPM		MR	2029	1920	
WD 72229		0-4-0DM		_(VF	5265	1945	
				(DC	2184	1945	
			reb	YEC	L120	1993	c
XOH 2299	(HCT 008)	4wDH		Perm	008	1988	
(TR 37)	(PWM 2797)	2w-2PMR		Wkm	6896	1954	d

a based here, but visits other locations
b fitted with GER tram bodywork
c property of Andrew Briddon
d currently off site

EPPING ONGAR RAILWAY

Locomotives are kept at :- Ongar Goods Yard, Ongar CM5 9AB TL 551034
North Weald Station, North Weald Bassett CM16 6BT TL 496036

Gauge 1524mm www.eorailway.co.uk

(1008)	DRACULA CASTLE	4-6-2	OC	LO	157	1948

Gauge 4ft 8½in

1		0-4-4T	IC	Neasden	3	1898	
	LORD PHIL	0-6-0ST	IC	HE	2868	1943	
			reb	HE	3883	1963	
	ISABEL	0-6-0ST	OC	HL	3437	1919	
–		0-6-0ST	OC	HL	3837	1934	
(D22)	45132	1Co-Co1DE		Derby		1961	
(D1606	47029) 47635 JIMMY MILNE	Co-CoDE		Crewe		1964	
(D2119)	03119	0-6-0DM		Sdn		1959	
(D2170)	03170	0-6-0DM		Sdn		1960	
(D5557	31139, 31538) 31438	A1A-A1A DE		BT	156	1959	
D6501	(33002)	Bo-BoDE		BRCW	DEL93	1960	
D6729	(37029)	Co-CoDE		_(EE	2892	1961	
				(VF	D608	1961	
–		4wDM		RH	398616	1956	
51342		2-2w-2w-2DMR		PSteel		1960	
51384		2-2w-2w-2DMR		PSteel		1960	
60110	205205	4-4wDER		Afd/Elh		1957	
1031	(1085)	2-2w-2w-2RER		MetCam		1959	
PM002	BADGER	4wDHR		Perm	T002	1988	

GLENDALE FORGE, MONK STREET, near THAXTED — CM6 2NR

Gauge 2ft 0in **(Closed)** **TL 612287**

145	C.P.HUNTINGTON	4w-2-4wPH	s/o	Chance	76-50145-24	1976	
	ROCKET	0-2-2+4wPH	s/o	FRgroup4		1970	
–		4wDM		RH			a Dsm

a either RH 217973 / 1942, or RH 213853 / 1942

MANGAPPS RAILWAY MUSEUM, SOUTHMINSTER ROAD, BURNHAM-ON-CROUCH — CM0 8QG

Gauge 4ft 8½in www.mangapps.co.uk **TQ 944980**

80078		2-6-4T	OC	Bton		1954	
	TOTO	0-4-0ST	OC	AB	1619	1919	
No. 8	FAMBRIDGE	0-4-0ST	OC	AB	2157	1943	
	MINNIE	0-6-0ST	OC	FW	358	1878	
2087	"GIBRALTAR"	0-4-0ST	OC	P	2087	1948	
	"BROOKFIELD"	0-6-0PT	OC	WB	2613	1940	
(D2018	03018)	0-6-0DM		Sdn		1958	
(D2020)	03020 F134L	0-6-0DM		Sdn		1958	a
(D2081)	03081 LUCIE	0-6-0DM		Don		1960	
(D2089)	03089	0-6-0DM		Don		1960	
D2325		0-6-0DM		_(RSHD	8184	1961	
				(DC	2706	1961	
(D2399)	03399	0-6-0DM		Don		1961	
(D5523)	31105	A1A-A1ADE		BT	122	1959	
(D5560)	31233	A1A-A1ADE		BT	260	1960	
(D5695	31265 31530) 31430 SISTER DORA	A1A-A1ADE		BT	296	1961	

(D6530) 33018	Bo-BoDE	BRCW	DEL122	1960	a
ELLAND No.1	0-4-0DM	HC	D1153	1959	
(DS 1169)	4wDM	RH	207103	1941	
–	4wDM R/R	S&H	7502	1966	
11104	0-6-0DM	(VF (DC	D78 2252	1948 1948	
(226)	0-4-0DM	(VF (DC	5261 2180	1945 1945	
1030	2-2w-2w-2RER	MetCam		1960	
22624	2w-2-2-2wRER reb	GRC&W GRC&W	O/2228	1938 1950	 a
W51381	2-2w-2w-2DMR	PSteel		1960	b
PWM 2786 A14W (TR36)	2w-2PMR	Wkm	6885	1954	
LLPW 01 PWM 3951 PW2 MAISIE	2w-2PMR	Wkm	6936	1955	
3700-84	2w-2PMR	Woodings	A466	19xx	

a currently under restoration at Sonic Rail Services Ltd, Burnham-on-Crouch
b converted to non-powered hauled stock

Gauge 3ft 6in

8	4wRER	ACCars	1949

SOUTHEND-ON-SEA BOROUGH COUNCIL, SOUTHEND BOROUGH PARKS DEPARTMENT, CENTRAL NURSERY, BARLING ROAD, SOUTHEND-ON-SEA

Gauge 3ft 6in **TQ 915873**

7	4wRER	ACCars	1949

SOUTHEND-ON-SEA BOROUGH COUNCIL, SOUTHEND PIER RAILWAY, SOUTHEND-ON-SEA SS1 1EE

Gauge 3ft 0in www.southend.gov.uk **TQ 884850**

A SIR JOHN BETJEMAN	4w-4wDH	SL	SE4	1986
B SIR WILLIAM HEYGATE	4w-4wDH	SL	SE4	1986
1835	4wBER	?		1996

SOUTHEND PIER RAILWAY MUSEUM, SOUTHEND-ON-SEA SS1 1EE

Gauge 3ft 6in www.southendpiermuseum.co.uk **TQ 884850**

11	4wRER	ACCars	1949
22	4wRER	ACCars	1949
6	2-2wRER	BE	1890

LYNN TAIT GALLERY, THE OLD FOUNDRY, LEIGH-ON-SEA SS9 1RP

Gauge 3ft 6in www.thelynntaitgallery.com **TQ 889891**

21	4wRER	ACCars	1949

WALTHAM ABBEY ROYAL GUNPOWDER MILLS CO LTD, BEAULIEU DRIVE, WALTHAM ABBEY — EN9 1JY

Gauge 3ft 0in www.royalgunpowdermills.com **TL 376013**

BB 307		2w-2BE		GB	6099	1964

Gauge 2ft 6in

DH 888	(ND 10392)	4wDH		BD	3755	1981
–		4wDH		HE	8828	1979
–		4wDH		Ruhrthaler	3920	1969

Gauge 1ft 6in

BWR 3	CARNEGIE	0-4-4-0DM		HE	4524	1954
	BUDLEIGH	4wDM		RH	235624	1945

GLOUCESTERSHIRE

INDUSTRIAL SITES

FIRE SERVICE COLLEGE LTD, LONDON ROAD, MORETON-IN-MARSH — GL56 0RH

Vehicles are used for static training purposes.

Gauge 4ft 8½in www.fireservicecollege.ac.uk **SP 216329**

H 004		0-4-0DE		YE	2732	1959
64681	508212 (508133)	4w-4wRER		York		1980
64724	508212 (508133)	4w-4wRER		York		1980
69633	(390033)	4w-4wWER		Alstom		2002
69733	(390033)	4w-4wWER		Alstom		2002

THE FLOUR MILL LTD, BREAM, Forest of Dean

Locomotives for repair and restoration are usually present.

Gauge 4ft 8½in www.theflourmill.com **SO 604067**

(5521)	LT 150	2-6-2T	OC	Sdn		1927	a
	"WILLY"	0-4-0WT	OC	KS	3063	1918	a
1		4wDH		TH	133C	1963	
	rebuild of	4wVBT	VCG	S			b

Preserved vehicles currently present :

5538		2-6-2T	OC	Sdn		1928	
No.229		0-4-0ST	OC	N	2119	1876	
No.563		4-4-0	OC	9E		1893	
No.3		0-4-0WT	OC	EB	37	1898	
	THE MEG	0-8-0T	OC	Chrz	5485	1961	c
–		0-6-0ST	IC	HE	3183	1944	
–		0-6-0ST	OC	P	2153	1954	

Gauge 2ft 6in

No.1	CHEVALLIER	0-6-2T	OC	MW	1877	1915	d

Gauge 2ft 0in

	JANET	4wDM		RH	504546	1963	e

a based here, but visits other locations
b carries 133V on plate in error
c carries plate 4939 / 1957
d currently at Welshpool & Llanfair Light Railway Preservation Co Ltd, Welshpool
e currently located elsewhere

MINISTRY OF DEFENCE, DEFENCE RAILWAY EXECUTIVE, DEFENCE STORAGE & DISTRIBUTION CENTRE, ASHCHURCH

See Section 6 for full details. **R.T.C.** **SO 932338**

PRESERVATION SITES

AVON VALLEY RAILWAY CO LTD, BITTON STEAM CENTRE, BITTON STATION, BATH ROAD, BITTON

BS30 6HD

Gauge 4ft 8½in www.avonvalleyrailway.org **ST 670705**

44123		0-6-0	IC	Crewe	5658	1925	
TKh 4015	KAREL	0-6-0T	OC	Chrz	4015	1954	a
	EDWIN HULSE	0-6-0ST	OC	AE	1798	1918	
	LITTLETON No 5	0-6-0ST	IC	MW	2018	1922	
(No.9)		0-6-0T	OC	RSHN	7151	1944	
–		4wVBT	VCG	S	7492	1928	
(D2994)	07010	0-6-0DE		RH	480695	1962	
(D3272)	08202	0-6-0DE		Derby		1956	
(D)5518	(31101)	A1A-A1ADE		BT	89	1958	
(D5548)	31130 CALDERHALL POWER STATION	A1A-A1ADE		BT	147	1959	
70043	GRUMPY	0-4-0DM		AB	358	1941	
	KINGSWOOD	0-4-0DM		AB	446	1959	
–		4wDM		RH	235519	1945	
	RH 252823	4wDM		RH	252823	1947	b Dsm
(429)	RIVER ANNAN	0-6-0DH		RH	466618	1961	
D2	ARMY 610	0-8-0DH		S	10143	1963	
SC52006		2-2w-2w-2DMR		DerbyC&W		1961	
SC52025		2-2w-2w-2DMR		DerbyC&W		1961	
(B8W)	PWM 3769	2w-2PMR		Wkm	6648	1953	
–		2w-2PMR		Geismar			

a carries plate Chrz 4939/1957
b converted to unpowered weed killing unit

DEAN FOREST RAILWAY CO LTD, LYDNEY, Forest of Dean

Locomotives are kept at :–
Norchard Steam Centre GL15 4ET SO 629044
Lydney Riverside Station GL15 5EW SO 634025

Gauge 4ft 8½in www.deanforestrailway.co.uk

5541		2-6-2T	OC	Sdn		1928	

9681		0-6-0PT	IC	Sdn		1949	
2		0-4-0ST	OC	AB	2221	1946	
–		0-6-0ST	IC	HE	2411	1941	
	GUNBY	0-6-0ST	IC	HE	2413	1941	
	WILBERT REV. W. AWDRY	0-6-0ST	IC	HE	3806	1953	
63.000.432	FRED WARRIOR	0-6-0ST	IC	HE	3823	1954	
6	USKMOUTH 1	0-4-0ST	OC	P	2147	1952	
WD 152	RENNES	0-6-0ST	IC	RSHN	7139	1944	
			reb	HE	3880	1961	
13308	08238 (D3308) CHARLIE	0-6-0DE		Dar		1956	
(D3588)	08473	0-6-0DE		Crewe		1958	Dsm
D3937	(08769) GLADYS	0-6-0DE		Derby		1960	
D5634	(31210)	A1A-A1ADE		BT	234	1960	
(D)5662	(31235)	A1A-A1A DE		BT	262	1960	
(E6002)	73002	Bo-BoDE/RE		Elh		1962	
(D6608	37274) 37308	Co-CoDE		_(EE	3568	1965	
				(EEV	D997	1965	
D7633	(25283 25904)	Bo-BoDE		BP	8043	1965	
D9521		0-6-0DH		Sdn		1964	
D9555		0-6-0DH		Sdn		1965	
3947		4wDM		FH	3947	1960	
	DON CORBETT	0-4-0DH		HE	5622	1960	
–		0-4-0DH		HE	6688	1968	
–		0-4-0DM		JF	4210127	1957	
E50619	(53619) B 962	2-2w-2w-2DMR		DerbyC&W		1958	
M51566		2-2w-2w-2DMR		DerbyC&W		1959	
M51914		2-2w-2w-2DMR		DerbyC&W		1960	
(DS 3057)		4wPMR		Wkm	4254	1947	
	(99709 909141-2?)	2w-2PMR		Geismar (ST/04/10 2004?)			

BRIAN FAULKNER

Locomotives kept at private location. contact : irs@railtruck.org

Gauge 2ft 0in

–	4wPM	L	3834	1931	
–	4wDM	L	8022	1936	
–	4wDM	L	33650	1949	
–	4wDM	L	41803	1955	
–	4wDM	LB	56371	1970	
–	2w-2DM	StokesMJ		1986	Dsm a
–	4wBE	CE			

a converted to non-powered passenger coach

GLOUCESTERSHIRE WARWICKSHIRE STEAM RAILWAY plc

Locomotives are kept at :-

Toddington Goods Yard GL54 5DT SP 049321
Winchcombe Carriage Works Yard GL54 5LB SP 026297
Winchcombe Station Yard GL54 5LB SP 025297

Gauge 4ft 8½in www.gwsr.com

2807	2-8-0	OC	Sdn	2102	1905
2874	2-8-0	OC	Sdn	2780	1918

3850		2-8-0	OC	Sdn		1942	
4270		2-8-0T	OC	Sdn	2850	1919	
7820	DINMORE MANOR	4-6-0	OC	Sdn		1950	
7903	FOREMARKE HALL	4-6-0	OC	Sdn		1949	
9642		0-6-0PT	IC	Sdn		1946	a
35006	PENINSULAR & ORIENTAL S.N.CO						
		4-6-2	3C	Elh		1941	
76077		2-6-0	OC	Hor		1956	
–		0-4-0ST	OC	P	1976	1939	
(D135)	45149	1Co-Co1DE		Crewe		1961	
(D)1693	(47105)	Co-CoDE		BT	455	1963	
(D1895)	47376 FREIGHTLINER 1995	Co-CoDE		BT	657	1965	
D2182		0-6-0DM		Sdn		1962	
(D2280)		0-6-0DM		_(RSHD	8098	1960	
				(DC	2657	1960	
(D)5081	(24081)	Bo-BoDE		Crewe		1960	
D5343	(26043)	Bo-BoDE		BRCW	DEL88	1959	
(D6915)	37215	Co-CoDE		_(EE	3393	1963	
				(EEV	D859	1963	
(D6948)	37248 LOCH ARKAIG	Co-CoDE		_(EE	3505	1964	
				(EEV	D936	1964	
(D8035	20035) 2001 AT3 DJ 053	Bo-BoDE		_(EE	2757	1959	
				(VF	D482	1959	Dsm b
D8137	(20137)	Bo-BoDE		_(EE	3608	1965	
				(EEV	D1007	1965	
E6036	(73129)	Bo-BoDE/RE		_(EE	3598	1965	
				(EEV	E368	1965	
21	(MAVIS)	0-4-0DM		JF	4210130	1957	
11230		0-6-0DM		_(RSHN	7860	1956	
				(DC	2574	1956	
372		0-6-0DE		YE	2760	1959	
W51360		2-2w-2w-2DMR		PSteel		1960	
W51363		2-2w-2w-2DMR		PSteel		1960	
W51372		2-2w-2w-2DMR		PSteel		1960	
W51405		2-2w-2w-2DMR		PSteel		1960	
Sc52029		2-2w-2w-2DMR		DerbyC&W		1961	
W55003		2-2w-2w-2DMR		GRC&W		1958	
(9127)		4wDMR		BD	3743	1976	
XLR 8123	S134 TBC	4wDM	R/R	Landrover		1998	

a currently under restoration elsewhere
b property of Railway Support Services Ltd, Wishaw, Warwickshire

GREAT WESTERN RAILWAY MUSEUM, THE OLD RAILWAY STATION, COLEFORD, Forest of Dean — GL16 8RH

Gauge 4ft 8½in www.gwrmuseumcoleford.co.uk **SO 576105**

2	182	0-4-0ST	OC	P	1893	1936

HOPEWELL COLLIERY MUSEUM, LACINDA COALWAY, near COLEFORD, Forest of Dean — GL16 7EL

Gauge 2ft 0in www.hopewellcolliery.com **SO 603114**

	EILEEN	4wDM		RH	432648	1959	

LEA BAILEY LIGHT RAILWAY, NEWTOWN, near ROSS-ON-WYE

Gauge 2ft 0in www.leabaileylightrailway.co.uk **SO 645196**

	WHISTLING PIG	4wCA	G	EIMCO	401-216	1968	
21282		4wDM		MR	21282	1957	
LM 4	3	4wBE		WR	N7605	1973	
3		4wBE		WR	7888R	1977	
–		0-4-0BE		WR	L1009	1979	

Gauge 1ft 6in

(JMLM6)		0-4-0BE		WR	7617	1973	

B NICHOLLS, GOTHERINGTON STATION

Gauge 4ft 8½in **SP 974298**

(TR2)	PWM 2779	2w-2PMR		Wkm	6878	1954	

NORTH GLOUCESTERSHIRE RAILWAY CO LTD, TODDINGTON GOODS YARD, TODDINGTON — GL54 5DT

Gauge 2ft 0in www.toddington-narrow-gauge.co.uk **SP 048318**

1966	TOURSKA	0-6-0T	OC	Chrz	3512	1957	a
	CHAKA'S KRAAL No.6	0-4-2T	OC	HE	2075	1940	
1091		0-8-0T	OC	Hen	15968	1918	
7	JUSTINE	0-4-0WT	OC	Jung	939	1906	
	"IVAN"	4wPM		FH	3317	1948	
	YARD No. A497 ND 3824	4wDM		HE	6647	1967	
2	DFK 538	4wDM		L	34523	1949	
	"SPITFIRE"	4wPM		MR	7053	1937	
6		4wDM		RH	166010	1932	
5		4wDM		RH	354028	1953	
	BRYNEGLWYS	4wDH		SMH	101T023	1985	
			reb	YEC	L145	1995	

a carries plate Fabrika Kotlow Toron 1966 1957

J. RAINBOW, PRIVATE SITE, GLOUCESTER

Gauge 2ft 0in

	FOXHANGER	4wDM		House		c1974	

ROYAL FOREST OF DEAN'S MINING MUSEUM, CLEARWELL CAVES, CLEARWELL, near COLEFORD, Forest of Dean — GL16 8JR

Gauge 2ft 6in www.clearwellcaves.com **SO 576082**

–		0-6-0DMF	(HC (HE	DM1435 8583	1977 1977	

Gauge 2ft 0in

–		0-4-0DMF	HC	DM739	1950	Dsm
K9		0-6-0DMF	HC	DM801	1954	
T42		0-6-0DMF	HC	DM841	1954	
68		0-4-0DMF	HC	DM924	1955	Dsm
R3		0-6-0DMF	(HC (HE	DM1442 8842	1980 1980	 OOU
–		4wDHF	HE	7386	1976	a
7		4wDHF	HE	7446	1975	
–		4wDHF	HE	8985	1981	
–		4wDHF	HE	8986	1981	a b
1		4wBE	WR	7964	1978	

a at Hawthorn Tunnel, near Drybrook
b uses parts and carries worksplate from HE 9053

P. SADDINGTON

Gauge 2ft 0in

–		4wPM	Bg	2095	1936	

TREASURE TRAIN LTD, PERRYGROVE RAILWAY, PERRYGROVE FARM, PERRYGROVE ROAD, MILKWALL, near COLEFORD, Forest of Dean — GL16 8QB

Gauge 2ft 0in www.perrygrove.co.uk **SO 579095**

HLH 003D		4wDH		HE	9352	1994	

Gauge 1ft 3in

	LYDIA	2-6-2T	OC	AK	77	2008	a
	SPIRIT OF ADVENTURE	0-6-0T	OC	ESR	295	1993	
3	ANNE	0-6-2T	OC	ESR	323	2004	
27	SOONY	0-4-0	OC	NemethJ		2012	b
–		4-6wDM		Guest		1960	
4	JUBILEE	4wDH		HE	9337	1994	
	PYLON	4wDM		L	40407	1954	
No.2	WORKHORSE	4wDM		MR	26014	1967	

a worksplate shows build date of 2007
b constructed from parts supplied by Hillcrest Locomotives, USA; construction started by J. Page; carries w/n BLW 15912/1901.

WINCHCOMBE RAILWAY MUSEUM TRUST, 3 GLOUCESTER STREET, WINCHCOMBE, near CHELTENHAM — GL54 5LX

Gauge 2ft 0in (Closed) www.wrmt.org.uk **SP 022282**

	AMOS	2w-2BE	FoxA		c1972	Dsm

VALE OF BERKELEY RAILWAY, THE ENGINE SHED,
SHARPNESS DOCKS, DOCK ROAD, SHARPNESS **GL13 9YA**

Gauge 4ft 8½in www.valeofberkeleyrailway.co.uk **SO 668023**

44027		0-6-0	IC	Derby		1924	
44901		4-6-0	OC	Crewe		1945	Dsm
No.15	EARL DAVID	0-6-0ST	IC	AB	2183	1945	
D2069	(03069)	0-6-0DM		Don		1959	
(7069)		0-6-0DE		HL	3841	1935	
	99709 909139-6	2w-2PMR		Geismar	ST/04/08	2004	
156W	(PWM 2188 TR12)	2w-2PMR		Wkm	4165	1948	
(68066	DB 965564) 23 PWM 4306	2w-2PMR		Wkm	7509	1956	

Gloucestershire Science and Technology Park,
Berkeley Green, Berkeley **GL13 9PB**

Gauge 4ft 8½in **ST 658991**

–		0-4-0F	OC	AB	2126	1942
D9553	54	0-6-0DH		Sdn		1965

Locomotives in storage for Vale of Berkeley Railway

HAMPSHIRE

INDUSTRIAL SITES

ARLINGTON FLEET GROUP LTD,
EASTLEIGH WORKS, CAMPBELL ROAD, EASTLEIGH **SO50 5AD**

Vehicles for repair/overhaul/restoration/storage usually present

Gauge 4ft 8½in www.arlington-fleet.com **SU 457185**

(D3734)	08567	0-6-0DE		Crewe		1959	
01508	(428)	0-6-0DH		RH	466617	1961	
323·539·7	CHEVIOT	4wDH		Gmd	4861	1955	a

Preserved vehicles based here :

35005	CANADIAN PACIFIC	4-6-2	3C	Elh		1941
(D421)	50021 RODNEY	Co-CoDE		(EE	3791	1968
				(EEV	D1162	1968
(D426)	50026 (89426) INDOMITABLE	Co-CoDE		(EE	3796	1968
				(EEV	D1167	1968
(D1946	47503) 47771	Co-CoDE		BT	708	1966
12	SARAH SIDDONS	4w-4wRE		VL		1922
W51346		2-2w-2w-2DMR		PSteel		1960
51356	L702	2-2w-2w-2DMR		PSteel		1960
W51388		2-2w-2w-2DMR		PSteel		1960
51392	117701	2-2w-2w-2DMR		PSteel		1960
W55028	(960012 L128 977860)	2-2w-2w-2DMR		PSteel		1960
7105	S61229 (99229)	4w-4RER		Afd/Elh		1958
7105	S61230 (99230)	4w-4RER		Afd/Elh		1958

a property of Northumbria Rail Ltd, Bedlington, Northumberland

Knights Rail Environmental Services Ltd

Gauge 4ft 8½in

(D2991) 07007	0-6-0DE	RH	480692	1962	

BLUE FUNNEL FERRIES LTD, HYTHE PIER RAILWAY, PROSPECT PLACE, HYTHE — SO45 6AU

Gauge 2ft 0in www.hytheferry.co.uk **SU 423081**

–	4wRE	BE	16302	1917	
–	4wRE	BE	16307	1917	

BRYAN HIRST LTD, BULLINGTON CROSS, SUTTON SCOTNEY — SO21 3FN

Locomotives for scrap or resale occasionally present. www.bryanhirst.com **SU 463420**

LONDON & NORTH WESTERN RAILWAY COMPANY LTD, t/a ARRIVA TRAINCARE EASTLEIGH TRACTION & ROLLING STOCK DEPOT, CAMPBELL ROAD, EASTLEIGH (Arriva - a DB Company) — SO50 5AD

Gauge 4ft 8½in www.arrivatc.com **SU 458179**

(D3557 08442) 0042	0-6-0DE	Derby		1958	OOU
(D3978) 08810 RICHARD J. WENHAM EASTLEIGH DEPOT DECEMBER 1969 – JULY 1999	0-6-0DE	Derby		1960	

PRIVATE OWNER, MALLARDS, BUCKLERS HARD, BEAULIEU

Locomotive used in connection with timber felling at this estate.

Gauge 600mm **SZ 414997**

–	4wBE	WR	D6905	1964	OOU

SIEMENS AG, SIEMENS RAIL SYSTEMS, NORTHAM TRAINCARE FACILITY, RADCLIFFE ROAD, NORTHAM, SOUTHAMPTON — SO14 0PS

Gauge 4ft 8½in www.siemens.com **SU 429126**

–	4wBE	Niteq	B193	2002	

SOLENT GATEWAY LTD, SEA MOUNTING CENTRE, MARCHWOOD MILITARY PORT, CRACKNORE HARD, MARCHWOOD — SO40 4ZG

(A GBA (Holdings) Ltd & David MacBrayne Group Ltd Joint Venture)

Gauge 4ft 8½in www.solentgateway.com **SU 395103**

01523 (259)	4wDH	TH	299V	1981	
01541 (260)	4wDH	TH	300V	1982	
01542 (262)	4wDH	TH	302V	1982	

PRESERVATION SITES

17 PORT & MARITIME REGIMENT, c/o SOLENT GATEWAY LTD, MARCHWOOD MILITARY PORT, MARCHWOOD SO40 4ZG

Gauge 4ft 8½in **SU 395103**

PERCY	0-4-0DM		JF	22503	1938	Pvd

BEAULIEU – NATIONAL MOTOR MUSEUM, BEAULIEU, BROCKENHURST SO42 7ZN

Gauge 4ft 8½in www.beaulieu.co.uk **SU 384024**

XJ12 596	2w-2PMR		Jaguar		1990	a

a road vehicle, converted to rail operation

Gauge monorail **SU 386026**

–	4wDM		Watson&Haig		c1974	

COLIN BILLINGHURST, PRIVATE LOCATION, FAREHAM

Gauge 2ft 0in

S128 263 001	4wBE		CE	5882A	1971	a

a frame used as a replica; private location with no public vantage points

EAST HAYLING LIGHT RAILWAY SOCIETY, HAYLING SEASIDE RAILWAY, EASTOKE, HAYLING ISLAND PO11 9HL

Gauge 2ft 0in www.haylingrailway.com **SZ 729985**

No.3 JACK	0-4-0DH	s/o	AK	23	1988	
No.1 ALAN B	4wDM		MR	7199	1937	
4 ALISTAIR	4wDM		RH	201970	1940	
No.5 EDWIN	4wDM		RH	7002-0967-5	1967	

R. GAMBRILL, PRIVATE LOCATION

Locomotives not available for viewing. Locomotives may occasionally visit galas and other public events.

Gauge 2ft 9in

–	4wDM		RH	435398	1959

Gauge 2ft 0in

–	0-8-0T	OC	OK	8356	1917
PIXIE	0-4-0ST	OC	WB	2090	1919
–	4wPM		FH	1747	1931
–	4wDM		RH	444193	1960

HAMPSHIRE BUILDINGS PRESERVATION TRUST LTD, CENTRE FOR THE CONSERVATION OF THE BUILT ENVIRONMENT, BURSLEDON BRICKWORKS, SWANWICK LANE, LOWER SWANWICK — SO31 7HB

Gauge 2ft 0in www.bursledonbrickworks.org.uk **SU 499098**

	WENDY	0-4-0ST	OC	WB	2091	1919
–		4wDM		FH	3787	1956
	AGWI PET 2	4wPM		MR	4724	1939
	BRAMBRIDGE HALL	4wPM		MR	5226	1930
	SIMPLEX ASHBY	4wDM		MR	8694	1943

HORSEBRIDGE STATION, HORSEBRIDGE, KINGS SOMBOURNE, STOCKBRIDGE — SO20 6PU

Gauge 4ft 8½in www.horsebridgestation.co.uk **SU 343303**

(9119)		4wDMR		BD	3708	1975

MARWELL'S WONDERFUL RAILWAY, MARWELL WILDLIFE, COLDEN COMMON, near WINCHESTER — SO21 1JH

Gauge 1ft 3in www.marwell.org.uk **SU 508216**

	PRINCESS ANNE	2-6-0DH	s/o	SL	75.3.87	1987

PHIL MASON, PRIVATE SITE

Gauge 2ft 0in

T203	0-4-0DM	Dtz	11898	1934	
–	0-4-0DM	Dtz	16392	1938	Dsm
–	4wDM	RH	339209	1952	

Gauge 1ft 11½in

–	4wDM	RH	200748	1940	

MID HANTS RAILWAY plc, "THE WATERCRESS LINE"

Locomotives are kept at :-

New Alresford Station SO24 9JG SU 588325
Ropley Station SO24 0BL SU 629324
Medstead & Four Marks Station GU34 5EN SU 668353

Gauge 4ft 8½in www.watercressline.co.uk

30499		4-6-0	OC	Elh		1920
(30506)	S.R. 506	4-6-0	OC	Elh		1920
30828	(E828) HARRY E.FRITH	4-6-0	OC	Elh		1928
(30850)	E850 LORD NELSON	4-6-0	4C	Elh		1926
(30925)	925 CHELTENHAM	4-4-0	3C	Elh		1934
34007	WADEBRIDGE	4-6-2	3C	Bton		1945
34058	SIR FREDERICK PILE	4-6-2	3C	Bton		1947
34105	SWANAGE	4-6-2	3C	Bton		1950
41312		2-6-2T	OC	Crewe		1952
45379	(5379)	4-6-0	OC	AW	1434	1937
73096		4-6-0	OC	Derby		1955

75079			4-6-0	OC	Sdn		1956	
76017			2-6-0	OC	Hor		1953	
80150			2-6-4T	OC	Bton		1956	
92212			2-10-0	OC	Sdn		1959	
1			0-6-0ST	IC	HE	3781	1952	
		rebuilt as	0-6-0T	IC	MHR Ropley		1994	
(D427)	50027 LION		Co-CoDE		_(EE	3797	1968	
					(EEV	D1168	1968	
(D1778	47183 47793) 47579 JAMES NIGHTALL G.C.							
			Co-CoDE		BT	540	1964	
(D3044)	08032		0-6-0DE		Derby		1954	
D3358	(08288)		0-6-0DE		Derby		1957	
D3462	(08377)		0-6-0DE		Dar		1957	
(D6587)	33202 DENNIS G. ROBINSON		Bo-BoDE		BRCW	DEL158	1962	
D6593	(33208)		Bo-BoDE		BRCW	DEL164	1962	a
12049	(12082 01553)		0-6-0DE		Derby		1950	
205025	S60124		4-4wDER		Afd/Elh		1959	
68081			2w-2PMR		Wkm	7031	1954	DsmT
(TR22	PWM 4312)		2w-2PMR		Wkm	7515	1956	DsmT
(68075)	DB965991		2w-2DMR		Wkm	10707	1974	
(68082)	DB966031		2w-2DMR		Wkm	10839	1975	

a currently at The Battlefield Line, Shackerstone, Leicestershire

MILESTONES – HAMPSHIRE'S LIVING HISTORY MUSEUM, LEISURE PARK, CHURCHILL WAY WEST, BASINGSTOKE — RG22 6PG

Gauge 4ft 8½in www3.hants.gov.uk/milestones **SU 612524**

WOOLMER	0-6-0ST	OC	AE	1572	1910	

PAULTONS PARK LTD, PAULTONS RAILWAY, OWER, near ROMSEY — SO51 6AL

Gauge 1ft 3in www.paultonspark.co.uk **SU 316167**

–	2-8-0DH	s/o	SL	RG.11.86	1987

PRIVATE OWNER, Near ALTON

Gauge 2ft 0in

LITTLE JIM	4wDM		Schöma	1676	1955

Mr SAMSON, BROOKLANDS GORLEY LIGHT RAILWAY, BROOKLANDS FARM, MOCK BEGGAR, near IBSLEY, FORDINGBRIDGE

Gauge 2ft 0in **SU 162107**

–	4wDM		LB	55070	1966
		reb	Gartell	1001	1987

TWYFORD WATERWORKS TRUST,
TWYFORD WATERWORKS, HAZELEY ROAD, TWYFORD **SO21 1QA**

Locomotives present at advertised public open days.

Gauge 2ft 0in www.twyfordwaterworks.co.uk **SU 493248**

–		4wDM		FH	1731	1931	
	a rebuild of	4wPM?		MR			Dsm
(DOE 3983)		4wDM		FH	3983	1962	
–		4wDM		L	3916	1931	
–		4wDM		L	42494	1956	
–		4wDM		LB	52886	1962	
–		4wPM		MR	5355	1932	
No.29 AYALA		4wDM		MR	7374	1939	
–		4wBE		Red(F)		1979	
–		4wDMF		RH	209429	1943	
JMLM2		0-4-0BE		WR	M7550	1972	
–		4wBE		WR		1973	a

a one of WR N7606, WR N7620 or WR N7621

HEREFORDSHIRE

INDUSTRIAL SITES

ALAN KEEF LTD, LIGHT RAILWAY ENGINEERS & LOCOMOTIVE BUILDERS,
LEA LINE, ROSS-ON-WYE **HR9 7LQ**

Locomotives under construction and repair are usually present. The following is understood to be a complete FLEET LIST of locomotives currently owned by (or in the care of) Messrs Keef. Some may be hired out from time to time, as shown by footnotes.

Gauge 4ft 8½in www.alankeef.co.uk **SO 665214**

(2)		4wDMR		Bg/DC	1647	1927	Dsm

Gauge 1000mm

4 RUR		0-4-0Tram	IC	Hen	5276	1899	
	rebuilt as	0-4-0F Tram	IC			1940	

Gauge 3ft 0in

NANCY		0-6-0T	OC	AE	1547	1908	
No.11 MAITLAND		2-4-0T	OC	BP	4663	1905	

Gauge 2ft 6in

DALMUNZIE		4wPM		MR	2014	1920	

Gauge 2ft 0in

TAFFY		0-4-0VBT	VC	AK	30	1994	a
WOTO		0-4-0ST	OC	WB	2133	1924	
SKIPPY		4wDM		AK	2	1976	
PLANT No.34		4wDM		MR	40SD502	1975	
–		4wBE		BEV	323	1921	

SP 8 "BATTY"	4wBE		WR	1393	1939	
		reb	AK		2006	

a worksplate is dated 1990

PAINTER BROTHERS LTD,
HEREFORD STEELWORKS, MORTIMER ROAD, HEREFORD **HR4 9SW**
(part of the Balfour Beatty Group Ltd)
Gauge 2ft 0in www.painterbrothers.com **SO 508413**

–	4wBE	CE	B0142B	1973
–	4wBE	CE	5806	1970

WYE VALLEY GROUP – EASTSIDE 2000 LTD,
FORDSHILL ROAD, ROTHERWAS INDUSTRIAL ESTATE, HEREFORD
Gauge 4ft 8½in www.wyevalleygroup.co.uk **SO 538377**

220	0-4-0DM	AB	359	1941	Pvd

PRESERVATION SITES

D2578 LOCOMOTIVE GROUP, PRIVATE SITE, MORETON ON LUGG
The locomotives are under restoration inside a building on a secure industrial estate and visits are normally not possible. A contact for enquiries is at : hfdned@hotmail.com
Gauge 4ft 8½in www.beerandrail.co.uk/d2578lg_home.htm

(D2145) 03145	0-6-0DM		Sdn		1961
D2302	0-6-0DM		_(RSHD	8161	1960
			¯(DC	2683	1960
D2578	0-6-0DM		HE	5460	1958
		reb	HE	6999	1968

M. DAVIES, STOKE EDITH STATION, TARRINGTON
Gauge 4ft 8½in **SO 614414**

–	4wDM	RH	463150	1961
A162 PWM 2194	2w-2PMR	Wkm	4171	1948

M. DEEM, LAMARO, ECCLES GREEN, NORTON CANON, HEREFORD
Gauge 1ft 3in **SO 374488**

101	2-4wPM	TaylorJ		c1964

R. HUNT, TITLEY JUNCTION STATION, near KINGTON
Private collection; cannot be viewed from any public place. No visitors without prior appointment.
Gauge 4ft 8½in **SO 328581**

–	0-4-0ST	OC	P	1738	1928
D2158 (03158) MARGARET ANN	0-6-0DM		Sdn		1960

–		4wDM		FH	3906	1959	
10	BRESSINGHAM	4wDH		TH	163V	1966	a
A13W	PWM 2801 (TR 3)	2w-2PMR		Wkm	6884	1954	

a carries worksplate 173V 1966 in error

K. MATTHEWS, FENCOTE OLD STATION, HATFIELD, near LEOMINSTER

Gauge 4ft 8½in **SO 601589**

W40	(TR40 PWM 4314)	2w-2PMR		Wkm	7517	1956

OWEN BROS MOTORS, c/o K. JONES, 13 NORBURY PLACE, TUPSLEY, HEREFORD

Gauge 1ft 3in **SO 542373**

No.303		0-6-0PM	s/o	TaylorJ		1967

BOB PALMER, BROMYARD & LINTON LIGHT RAILWAY ASSOCIATION LTD, BROMYARD & LINTON LIGHT RAILWAY, BROADBRIDGE HOUSE, BROMYARD

This railway does not operate public trains.

Locomotives are kept at :- Bromyard SO 657548, 661546
Linton SO 668542

Gauge 2ft 6in

–		4wDM		Bg	3406	1953	Dsm

Gauge 2ft 0in

	MESOZOIC	0-6-0ST	OC	P	1327	1913	Dsm
–		4wPM		MR	6031	1936	
–		4wDM		MR	9382	1948	
1		4wDM		MR	9676	1952	
2		4wDM		MR	9677	1952	
No.7		4wDM		MR	20082	1953	
–		4wDM		MR	102G038	1972	
–		4wDM		RH	187101	1937	
–		4wDM		RH	195849	1939	a
L 10		4wDM		RH	198241	1939	
No.3	"NELL GWYNNE"	4wDM		RH	229648	1944	
No.6	"PRINCESS"	4wDM		RH	229655	1944	
–		4wDH		RH	437367	1959	
–		4wDM		RH	444200	1960	
–		2w-2PM		Wkm	3034	1941	Dsm

a converted for use as a generating unit

THE WOOLHOPE LIGHT RAILWAY, P.J. FORTEY, THE HORNETS NEST, CHECKLEY, MORDIFORD, near HEREFORD

Gauge 1ft 3in **SO 608378**

202	TREVOR	0-6-0PM	s/o	TaylorJ		c1974

HERTFORDSHIRE

INDUSTRIAL SITES

KIER GROUP plc, KIER CONSTRUCTIO, PLANT DEPOT, LISMIRRANE INDUSTRIAL PARK, ELSTREE ROAD, ELSTREE — WD6 3EA

Locomotives are present in this yard between use on contracts.

Gauge 1ft 6in www.kier.co.uk **TQ 166952**

–	2w-2BE	Iso	T42	1973
–	2w-2BE	Iso	T51	1974
–	2w-2BE	Iso	T54	1974
56	2w-2BE	Iso	T56	1974
–	2w-2BE	Iso		
–	2w-2BE	Iso		
–	2w-2BE	Iso		

RUSH GREEN MOTORS, LONDON ROAD, LANGLEY, HITCHIN — SG4 7PQ

Gauge 4ft 8½in www.rushgreenmotors.com **TL 210236**

(C955 YOR?)	4wDM	R/R	Bruff	(514	1986?)	OOU
C966 YOR	4wDM	R/R	Bruff	524	1986	OOU
(C969 YOR?)	4wDM	R/R	Bruff	(527	1986?)	OOU

PRESERVATION SITES

B.LAWSON, BRY RAILWAY, TRING

Locomotives stored at various private locations.

Gauge 4ft 8½in

No.25	GR5091	4wDM		RH	294269	1951	
–		0-4-0DH		RH	418793	1957	
No.28	L5 (50) HERBERT TURNER	0-4-0DH		RH	513139	1967	
No.27		0-4-0DH		RH	518190	1965	
No.26		4wDH		RH	544996	1968	

Gauge 2ft 6in

No.2	6	4wBE		BV	694	1974	
No.3	9	4wBE		BV	696	1974	
No.17	3135B SP 03 426	4wBEF		CE	B3135B	1984	
			reb	CE	B4066RF	1994	
No.16	SP 03.425	4wBEF		CE	B3204A	1985	
			reb	CE	B4139	1995	
No.24	8	4wBE		GB	2920	1958	
No.14	ND 3307 YARD No. B49	4wBE		GB	3546	1948	
	ND 3308 YARD No.B50	2w-2BE		GB	3547	1948	Dsm
No.15		4wBE		GB	3825	c1949	
No.03		4wDHF		HE	7384	1976	

–		4wDM		LB	55870	1968	Dsm
No.16		4wDM		RH	170200	1934	Dsm
No.24		4wDM		RH	247182	1947	
–		4wDMF		RH	353491	1954	
No.14		4wDM		RH	441945	1959	
No.12		0-4-0DH		RH	476133	1964	
No.17		4wDMF		RH	480680	1963	
No.13		4wDM		RH	506415	1964	
No.19		4wDM		RH	7002/0767/6	1967	
No.15		4wDM		RH	7002/0867/3	1967	

Gauge 2ft 3in

No.31	"ROBERT (BOB) MACBETH"	4wDH		RH	476108	1964	

Gauge 2ft 0in

No.29		4wDM		RH	213848	1942	
No.30		4wDM		RH	246793	1947	
LM26		4wDM		RH	248458	1946	
–		4wDM		RH	280866	1949	
No.21	LM 264	4wDM		RH	371535	1954	
No.22	LM 265	4wDM		RH	375696	1954	
No.23	LM 112	4wDM		RH	375699	1954	
No.20		4wDMF		RH	381704	1955	
No.20	1511	4wBEF		CE	5382	1966	
No.25		4wBE		CE	5688/2	1969	
No.18		4wBE		CE	B0107A	1973	
No.21	1517	4wBEF		CE	B0122	1973	
	JMLM25	4wBE		CE	B0145A	1973	
			reb	CE	B3786B	1991	
"No.22"	S232 263025	4wBE		CE	B0459A	1975	
"No.23"	JMLM17 (JM93)	4wBEF		CE	B1547B	1977	
			reb	CE	B3672	1990	
No.27	LM18 (JM95)	4wBE		CE	B1534B	1977	
			reb	CE	B3672	1990	
No.28	LM16 (JM94)	4wBE		CE	B1534A	1977	
			reb	CE	B3672	1990	
SP 204		4wBE		CE	B1808	1978	
			reb	CE	B3214A	1985	
			reb	CE	B3825	1992	
No.26	LM10	4wBE		CE	B3329B	1986	
			reb	CE	B3804	1991	
			reb	CE	B4181	1996	
No.6	F	0-4-0BE		WR	M7544	1972	
No.1	MANDI MIS 47	0-4-0BE		WR	N7639	1973	

Gauge 1ft 6in

No.8	L 12	4wBE		CE	5965A	1973	
No.7	L 15	4wBE		CE	B0109A	1973	
–		2w-2BE		Iso	T6	1972	
–		2w-2BE		Iso	T9	1972	Dsm
–		2w-2BE		Iso	T15	1972	
–		2w-2BE		Iso	T40	1973	
–		2w-2BE		Iso	T53	1974	Dsm
–		2w-2BE		Iso	T71	1975	

–		2w-2BE		Iso	T79	1975	
–		2w-2BE		Iso	T81	1975	
–		2w-2BE		Iso			a
No.19		4wBH		Tunn		1980	
			reb	Tunn		1996	b
No.9	SP 100 35	2w-2BE		WR	L800	1983	
No.10	SP 101 35T005	2w-2BE		WR	L801	1983	
No.11	SP 102	2w-2BE		WR	544901	1984	
No.12	JM 103	2w-2BE		WR	546001	1987	
No.13	SP 104	2w-2BE		WR	546601	1987	
No.4	N7652	0-4-0BE		WR	F7117	1966	
			reb	WR	10142	1985	
No.5	1580	0-4-0BE		WR			

a one of Iso T21 to T38
b one of Tunn TQ121 to TQ126

Gauge monorail

-		2wPH		RM	8111	1959	

C.& D. LAWSON, DORCLIFF RAILWAY, TRING

(Custodians : B. & G. Lawson) Locomotives are currently stored elsewhere.

Gauge 2ft 6in

No.04		4wDM		Diema	3543	1974	
No.05		4wDM		Fisons		1976	
		rebuilt as 4wDH		Lawson		1999	
No.02		4wDM		HE	7366	1974	
No.01		4wDM		HE	7367	1974	
–		4wPM		L	34652	1949	
–		2w-2PH		Lawson	2	1998	
–		2w-2PH		Lawson	3	2000	
–		4wDH		Lawson	4	2001	
–		2w-2DMR		Lawson	5	2004	
–		2w-2DH		Lawson	6	2005	
–		4wDM		LB	53976	1964	
–		4wDM		LB	53977	1964	a
No.1		4wDM		RH	166045	1933	
No.7	ELLEN	4wDM		RH	200069	1939	
No.6		4wDM		RH	224315	1944	
No.9		4wDM		RH	229657	1945	
No.8		4wDM		RH	244559	1946	
No.2	CUCKOO BUSH	4wDM		RH	247178	1947	
No.3		4wDM		RH	297066	1950	
No.4		4wDM		RH	402439	1957	
No.5		4wDM		RH	432654	1959	
No.18	SIMBA	4wDH		RH	432661	1959	
			reb	Swanhaven		c1985	
No.10	TANIA	4wDM		RH	432665	1959	
No.11	SHEEBA	4wDM		RH	466594	1962	
–		2w-2PM		Wkm	3175	1942	
–		2w-2PM		Wkm	3431	1943	
		rebuilt as 2w-2DH		Lawson		1997	

–		2w-2PM		Wkm	3578	1944	
	rebuilt as	2w-2DH		Lawson		2002	

a currently on loan to County Borough of Doncaster Museum & Art Gallery, South Yorkshire

Gauge 2ft 0in

–		4wDM		RH	441944	1960	Dsm

PARADISE WILDLIFE PARK, WHITE STUBBS LANE, BROXBOURNE — EN10 7QA

Gauge 4ft 8½in www.pwpark.com **TL 338067**

No.1	671	0-4-0DH	s/o	RH	512463	1965

THE RAIL TROLLEY TRUST, PRIVATE SITE

Gauge 4ft 8½in

–		2w-2PMR		Syl	14384	

N.V. SILL, PRIVATE SITE, near HEMEL HEMPSTEAD

Gauge 1524mm

1151	1819	2-8-0	OC	Frichs	397	1949
792	HEN	0-6-0T	OC	TK	373	1927

WARNER BROTHERS STUDIO TOUR LONDON, STUDIO TOUR DRIVE, LEAVESDEN, WATFORD — WD25 7LR

Gauge 4ft 8½in www.wbstudiotour.co.uk **TL 092005**

5972	HOGWARTS CASTLE	4-6-0	OC	Sdn		1937

ISLE OF WIGHT

PRESERVATION SITES

ISLE OF WIGHT RAILWAY CO LTD, ISLE OF WIGHT STEAM RAILWAY, HAVENSTREET STATION — PO33 4DS

Gauge 4ft 8½in www.iwsteamrailway.co.uk **SZ 556898**

(32110	110) No.2 YARMOUTH	0-6-0T	IC	Bton		1877
(32640)	W11 (NEWPORT)	0-6-0T	IC	Bton		1878
(32646)	No.8 FRESHWATER	0-6-0T	IC	Bton		1876
41298		2-6-2T	OC	Crewe		1951
41313		2-6-2T	OC	Crewe		1952
No.24	CALBOURNE	0-4-4T	IC	9E		1891
W37	INVINCIBLE	0-4-0ST	OC	HL	3135	1915
(38)	AJAX	0-6-0T	OC	AB	1605	1918
ARMY 92	WAGGONER	0-6-0ST	IC	HE	3792	1953
ARMY 198	ROYAL ENGINEER	0-6-0ST	IC	HE	3798	1953

D2059	(03059)	0-6-0DM		Don		1959	
D2554	(05001 97803) NUCLEAR FRED	0-6-0DM		HE	4870	1956	
ARMY 235	(233)	0-4-0DM		AB	369	1945	
68800		4wDHR		Perm	001	1985	
(DX)68809		4wDMR		Perm	010	1986	
HCT 002		4wDH		Perm	002	1987	
(DS 3320	PWM 3766) "66 532"	2w-2PMR		Wkm	6645	1953	
6944	PWM 3959	2w-2PMR		Wkm	6944	1955	Dsm

KENT

INDUSTRIAL SITES

AVONDALE ENVIRONMENTAL SERVICES LTD, FORT HORSTED, PRIMROSE CLOSE, CHATHAM — ME4 6HZ

Gauge 4ft 8½in www.avondaleuk.com **TQ 751651**

	C959 YOR	4wDMR	R/R	Bruff	517	1986	

Other road/rail vehicles usually present

BRETT AGGREGATES LTD, CLIFFE WHARF, NORTH SEA TERMINAL, SALT LANE, CLIFFE — ME3 7SX

Gauge 4ft 8½in www.brett.co.uk **TQ 720755**

331		4wDE		Werk	868	1950	OOU

CONTRACKED LANDS LTD, GARDEN CLOSE, MAIDSTONE

Administration address only, vehicles stored elsewhere when not working.

Gauge 4ft 8½in www.contrackedlands.co.uk

	DK02 ORX	4wDMR	R/R	(Landrover	609746	2002
				(Contracked		2013
016	BD03 LVH 99709 976031-3	4wDM	R/R	(Landrover	656671	2003
				(Harsco		2003

DEFENCE INFRASTRUCTURE ORGANISATION, DEFENCE TRAINING ESTATES - SOUTH, CINQUE PORTS TRAINING AREA, LYDD, ROMNEY MARSH

Gauge 600mm **TR 033198**

1	5210	2w-2PM	Wkm	11684	1990
3	5208	2w-2PM	Wkm	11685	1990
4	5211	2w-2PM	Wkm	11679	1990
5	1378	2w-2PM	Wkm	11681	1990
6	5209	2w-2PM	Wkm	11678	1990
7	5207	2w-2PM	Wkm	11677	1990
8	1379	2w-2PM	Wkm	11680	1990

EUROTUNNEL GROUP,
CHERITON TERMINAL, ASHFORD ROAD, FOLKESTONE CT18 8XX

These locomotives, used for shunting and maintenance, are also employed in the Channel Tunnel and at Coquelles Terminal, France, as required.

Gauge 4ft 8½in www.eurotunnel.com **TR 185375**

0001	(21901)	Bo-BoDE	MaK	1000.867	1993
0002	(21902)	Bo-BoDE	MaK	1000.868	1993
0003	(21903)	Bo-BoDE	MaK	1000.869	1993
0004	(21904)	Bo-BoDE	MaK	1000.870	1993
0005	(21905)	Bo-BoDE	MaK	1000.871	1993
0006	(6456 21906)	Bo-BoDE	MaK	1200.056	1991
0007	(6457 21907)	Bo-BoDE	MaK	1200.057	1991
0008	(6450 21908)	Bo-BoDE	MaK	1200.050	1991
0009	(6451 21909)	Bo-BoDE	MaK	1200.051	1991
0010	(6447 21910)	Bo-BoDE	MaK	1200.047	1991
0031	FRANCES	4wDH	Schöma	5366	1993
		Incorporates parts of	HE		
0032	ELISABETH	4wDH	Schöma	5367	1993
		Incorporates parts of	HE		
0033	SILKE	4wDH	Schöma	5263	1994
		Incorporates parts of	HE		
0034	AMANDA	4wDH	Schöma	5262	1994
		Incorporates parts of	HE		
0035	MARY	4wDH	Schöma	5269	1994
0036	LAURENCE	4wDH	Schöma	5268	1994
0037	LYDIE	4wDH	Schöma	5264	1994
0038	JENNY	4wDH	Schöma	5266	1994
0039	PACITA	4wDH	Schöma	5401	1994
0040	JILL	4wDH	Schöma	5402	1994
0041	KIM	4wDH	Schöma	5464	1995
0042	NICOLE	4wDH	Schöma	5465	1995
520		4wDH	Cockerill		1996
521		4wDH	Cockerill		1996
522		4wDH	Cockerill		1996

HITACHI RAIL EUROPE LTD, ASHFORD TRAIN MAINTENANCE CENTRE,
ASHFORD DEPOT, ASHFORD TN23 1EZ

Gauge 4ft 8½in www.hitachirail-eu.com **TR 016420**

	FUJI	0-4-0DH	S	10089	1962
–		4wBE	Scul		2007

LONDON & SOUTH EASTERN RAILWAY LTD t/a SOUTH EASTERN
(part of the Govia Group) **and HITACHI RAIL EUROPE LTD,**
RAMSGATE DEPOT, NEWINGTON ROAD, RAMSGATE CT12 6EA

Gauge 4ft 8½in www.southeasternrailway.co.uk **TR 370658**

See Section 7 for details

METROPOLITAN POLICE SPECIALIST TRAINING CENTRE, off MARK LANE, DENTON, GRAVESEND **DA12 2HN**

London underground railcar used for instructional purposes.

Gauge 4ft 8½in **TQ 672742**

1306		2-2w-2w-2RER		MetCam		1961	

RIDHAM SEA TERMINALS LTD, RIDHAM DOCK, SITTINGBOURNE **ME9 8SR**

Gauge 4ft 8½in R.T.C. www.ridhamseaterminals.co.uk **TQ 918684**

–		0-6-0DH		EEV	D1227	1967	OOU

SEACON TERMINALS LTD, TOWER WHARF, NORTHFLEET **DA11 9BD**

Gauge 4ft 8½in R.T.C. www.seacongroup.co.uk **TQ 612752**

–		2w-2BER		PWR	BO.598W.01	1993	OOU

PRESERVATION SITES

A.J.R. BIRCH & SON LTD, HOPE FARM, SELLINDGE, near ASHFORD

Gauge 1524mm **TR 119388**

799		0-6-0T	OC	TK	355	1925
1134		2-8-0	OC	TK	531	1946
(1157)		2-8-0	OC	Frichs	403	1949

Gauge 4ft 8½in

35011	GENERAL STEAM NAVIGATION	4-6-2	3C	Elh		1944
35025	BROCKLEBANK LINE	4-6-2	3C	Elh		1948
68078		0-6-0ST	IC	AB	2212	1946
No.3180	ANTWERP	0-6-0ST	IC	HE	3180	1944
3142	11201	4w-4wRER		Elh		1937
4902	(4002) S13003S	4w-4wRER		Elh		1949
(14573)	6307	4w-4wRER		Lancing/Elh		1959

DAVID BREAKER, STALISFIELD

Currently stored elsewhere

Gauge 4ft 8½in

–		0-6-0ST	IC	MW	1317	1895

BREDGAR & WORMSHILL LIGHT RAILWAY, "THE WARREN", SWANTON STREET, BREDGAR, near SITTINGBOURNE **ME9 8AT**

Gauge 750mm www.bwlr.co.uk **TQ 873585**

No.105	SIAM	0-6-0WT	OC	Hen	29582	1956

Gauge 2ft 0in

7	VICTORY	0-4-2T	OC	Decauville	246	1897

3	LADY JOAN	0-4-0ST	OC	HE	1429	1922
No.10	ZAMBEZI	0-4-2T	OC	JF	13573	1912
9	LIMPOPO	0-6-0WT	OC	JF	18800	1930
2	KATIE	0-6-0WT	OC	Jung	3872	1931
6	EIGIAU	0-4-0WT	OC	OK	5668	1912
8	HELGA	0-4-0WT	OC	OK	12722	1936
1	BRONHILDE	0-4-0WT	OC	Schw	9124	1927
No.4	ARMISTICE	0-4-0ST	OC	WB	2088	1919
NG 50		4wDH		AB	719	1987
			reb	HAB		1996
NG 51		4wDH		AB	720	1987
			reb	HAB		1996
NG 52		4wDH		AB	721	1987
			reb	HAB		1996
NG 53		4wDH		AB	764	1988
			reb	HAB		1996
14	(NG 54) MILSTEAD	4wDH		AB	765	1988
			reb	HAB		1996
5	BREDGAR	4wDH		BD	3775	1983
11	WORMSHILL	0-6-0DMF		HC	DM1366	1965
			reb	STRPS		1997
			reb	Bredgar		2009
13	LYNE	4wDM		MR	7037	1936
	ESK	4wDM		MR	7498	1940
12	BICKNOR	4wDM		MR	9869	1953
	JENNY	4wDH		Schöma	5239	1991

Gauge 1ft 3in

	JACK	0-6-0	OC	LemonB		c1956

BRETT GRAVEL, MILTON MANOR FARM, ASHFORD ROAD, CHARTHAM, near CANTERBURY — CT4 7PP

Gauge 2ft 0in www.brett.co.uk **TR 120558**

–		4wDM		MR	8730	1941

CANTERBURY CITY COUNCIL, CANTERBURY HERITAGE MUSEUM, STOUR STREET, CANTERBURY — CT1 2JR

Gauge 4ft 8½in **TR 146577**

	(INVICTA)	0-4-0	OC	RS	24	1830

CHATHAM HISTORIC DOCKYARD TRUST, THE HISTORIC DOCKYARD, CHATHAM — ME4 4TE

Gauge 4ft 8½in www.thedockyard.co.uk **TQ 758689**

No.8	INVICTA	0-4-0ST	OC	AB	2220	1946
–		0-4-0ST	OC	P	1903	1936
YARD No.361	AJAX	0-4-0ST	OC	RSHN	7042	1941
WD 42	OVERLORD	0-4-0DM		AB	357	1941
YARD No.562	ROCHESTER CASTLE	4wDM		FH	3738	1955

	THALIA	0-4-0DM		_(RSHN (DC	7816 2503	1954 1954	

Gauge 600mm

	LOD 758148	4wDM		RH	226276	1944	

COLONEL STEPHENS RAILWAY MUSEUM, TENTERDEN TOWN STATION, STATION ROAD, TENTERDEN — TN30 6HE

Gauge 4ft 8½in www.hfstephens-museum.org.uk **TQ 882336**

	GAZELLE	0-4-2WT	IC	Dodman		1893	
	EW&BR No.1	2-2wPMR		ShuttC		c2005	

DOVER TRANSPORT MUSEUM, WILLINGDON ROAD, OLD PARK, DOVER CT16 2HQ

Gauge 4ft 8½in www.dovertransportmuseum.org.uk **TR 301444**

	"ST THOMAS"	0-6-0ST	OC	AE	1971	1927	

Gauge 2ft 0in

–		4wPM		FH	3116	1946	
–		4wDM		MR	8606	1941	
–		4wDM		RH	349061	1953	

EAST KENT RAILWAY TRUST, STATION ROAD, SHEPHERDSWELL — CT15 7PD

Gauge 4ft 8½in www.eastkentrailway.co.uk **TR 258483**

	ST DUNSTAN	0-6-0ST	OC	AE	2004	1927	
(D3843)	08676 DAVE 2	0-6-0DE		Hor		1959	a
(D3852)	08685	0-6-0DE		Hor		1959	a
(D3910)	08742	0-6-0DE		Crewe		1960	a
(D3967)	08799 FRED	0-6-0DE		Derby		1960	a
(D3972)	08804	0-6-0DE		Derby		1960	a
(102)	RICHBOROUGH CASTLE	0-6-0DH		EEV	D1197	1967	
	SNOWDOWN	0-4-0DM		JF	4160002	1952	
THE BUFFS, ROYAL EAST KENT REGIMENT, 1572-1961							
	ARMY 427 C4 SA	0-6-0DH		RH	466616	1961	
01530	(269)	4wDH		TH	311V	1984	a
01546	(255)	4wDH		TH	273V	1977	a
42	(KEARSLEY No.1)	Bo-BoWE		HL	3682	1927	
S11161S	3142	4w-4wRER		Elh		1937	
S11187S	3142	2-2w-2w-2RER		EE/Elh		1937	
(60154	1101 205001)	4-4wDER		Afd/Elh		1957	
62385	(1399)	4w-4wRER		York		1971	
67300	7001 (316999)	4w-4wWER a rebuild of		York(BRE) DerbyC&W		1989 1981	 b
	99709 901014-9	2w-2PMR		Lesmac	LMS009	2006	

a property of Harry Needle Railroad Co Ltd, Barrow Hill, Derbyshire
b rebuild of non-motorised Driving trailer

ELHAM VALLEY LINE TRUST, COUNTRYSIDE CENTRE & RAILWAY MUSEUM, PEENE, FOLKESTONE CT18 8AZ

Gauge 4ft 8½in www.elhamvalleylinetrust.org **TR 185377, 185378**

–	0-6-0T	OC	EVM			a

a non-working replica locomotive

Gauge 900mm

RR10	4wDH	RACK	HE	9283	1988	
TU 20 56404	4wDH		Moës			

RAY FAGG, UNKNOWN LOCATION

Gauge 4ft 8½in

–	4wDM		RH	294268	1951	

KEN JACKSON, EYNSFORD LIGHT RAILWAY, EYNSFORD

Private location.

Gauge 2ft 0in

–	4wDM		MR	9711	1952	
AD 22	4wDM		RH	211609	1941	
–	2w-2PM		Westwood			a

a converted ride-on lawnmower with fixed rail wheels for guidance

MICHAEL LIST-BRAIN, PRESTON SERVICES, THE STEAM MUSEUM, COURT LANE, PRESTON, near CANTERBURY CT3 1DH

Gauge 4ft 8½in www.prestonservices.co.uk **TR 244604**

No.1	0-4-0T	OC	N	4444	1892

Gauge 2ft 0in

–	0-10-0	OC	OK	11309	1927

Gauge 600mm

SMT T907	0-10-0T	OC	OK	10956	1925
SMT T912	0-10-0T	OC	OK	10957	1925
SMT T908	0-10-0T	OC	OK	12470	1934

Gauge 1ft 6in

–	4-2-2	OC	WB	1425	1893

LOCOMOTIVE SERVICES LTD, c/o HORNBY HOBBIES LTD, THE HORNBY VISITOR CENTRE, WESTWOOD INDUSTRIAL ESTATE, MARGATE CT9 4JX

Gauge 4ft 8½in www.hornby.com / www.locomotivestorage.co.uk **TR 362686**

(60019) 4464 BITTERN	4-6-2	3C	Don	1866	1937
12795 4732	4w-4wRER		Lancing/Elh		1951
(12796) 4732	4w-4wRER		Lancing/Elh		1951
28690	4w-4wRER		DerbyC&W		c1939

J. MARTIN, THE RICHMOND LIGHT RAILWAY, near HEADCORN

Gauge 2ft 0in **Private site – visits by invitation only**

	CHUQUITANTA	0-4-0T	OC	(Couillet	810	1885	
				(Decauville	36	1885	
	ELIN	0-4-0ST	OC	HE	705	1899	
	"LEARY"	0-4-0VBT	VC	Foulds/Collins		2010	
CFSE 2.1	ANNE-MARIE	0-4-0WT	OC	Jung	2569	1918	
No.3	JENNY	0-4-0WT	OC	Jung	3175	1921	
–		0-6-0T	OC	KS	2442	1915	
No.2	SUSAN	0-4-0WT	OC	OK	3136	1908	
No.1	(NG 49)	4wDH		BD	3701	1973	
–		4wPM		Campagne	903	1925	
"49"	"SAMPSON"	4wDM		FH	1887	1934	
(4)		4wDM		MR	7403	1939	
N.230		4wDM		MR	8828	1943	
–		4wDM		MR	20073	1950	
5		4wDM		RH	235654	1946	
			reb	ALR	5	2002	
–		4wDM		RH	235711	1945	

Gauge 1ft 3in

1904	"CAGNEY"	4-4-0	OC	McGarigle		1904	

ROMNEY HYTHE & DYMCHURCH RAILWAY, NEW ROMNEY STATION, NEW ROMNEY

TN28 8PL

Gauge 1ft 3in www.rhdr.org.uk **TR 074249**

1	GREEN GODDESS	4-6-2	OC	DP	21499	1925	
2	NORTHERN CHIEF	4-6-2	OC	DP	21500	1925	
3	SOUTHERN MAID	4-6-2	OC	DP	22070	1926	
4	THE BUG	0-4-0TT	OC	KraussS	8378	1926	
5	HERCULES	4-8-2	OC	DP	22071	1926	
6	SAMSON	4-8-2	OC	DP	22072	1926	
7	TYPHOON	4-6-2	OC	DP	22073	1926	
8	HURRICANE	4-6-2	OC	DP	22074	1926	
9	WINSTON CHURCHILL	4-6-2	OC	YE	2294	1931	
No.10	DOCTOR SYN	4-6-2	OC	YE	2295	1931	
11	BLACK PRINCE	4-6-2	OC	Krupp	1664	1937	
	"PRINCESS CORONATION"	4-6-2	OC	MaxEng		2008	a
No.12	J. B. SNELL	4w-4wDH		TMA	6143	1983	
14	CAPTAIN HOWEY	4w-4wDH		TMA	2336	1989	
PW3	REDGAUNTLET	4wPM		AK	[1977]	1977	
7		4wDM	s/o	L	37658	1952	
–		4wDM		MR	7059	1938	
	rebuilt as	4wDH		TMA		1988	
(PW2)	SCOOTER	2w-2PM		RHDR		1964	

a incomplete loco

SITTINGBOURNE & KEMSLEY LIGHT RAILWAY LTD, SITTINGBOURNE and KEMSLEY

Locomotives are kept at Kemsley, access to which is by train from Sittingbourne.

Gauge 4ft 8½in www.sklr.net **TQ 904642, 920661**

–		0-4-0F	OC	AB	1876	1925	
	BEAR	0-4-0ST	OC	P	614	1896	
			reb	AB	5997	1941	

Gauge 2ft 6in

	PREMIER	0-4-2ST	OC	KS	886	1905	
	LEADER	0-4-2ST	OC	KS	926	1905	
	MELIOR	0-4-2ST	OC	KS	4219	1924	
	UNIQUE	2-4-0F	OC	WB	2216	1923	OOU
	ALPHA	0-6-2T	OC	WB	2472	1932	OOU
	TRIUMPH	0-6-2T	OC	WB	2511	1934	OOU
	SUPERB	0-6-2T	OC	WB	2624	1940	
	VICTOR	4wDM		HE	4182	1953	
P6495	BARTON HALL	4wDH		HE	6651	1965	
	EDWARD LLOYD	0-4-0DM		RH	435403	1961	

TENTERDEN RAILWAY CO LTD, (KENT & EAST SUSSEX RAILWAY)

Locomotives are kept at :-

Bodiam Station TN32 5UD TQ 783249
Rolvenden Station TN17 4JP TQ 865328
Tenterden Station TN30 6HE TQ 882336
Wittersham Road Station TN17 4QA TQ 866288

Gauge 4ft 8½in www.kesr.org.uk

1638		0-6-0PT	IC	Sdn		1951
4253		2-8-0T	OC	Sdn	2640	1917
5668		0-6-2T	IC	Sdn		1926
6619		0-6-2T	IC	Sdn		1928
30065		0-6-0T	OC	VIW	4441	1943
(30070	DS 238) WD 300	0-6-0T	OC	VIW	4433	1943
(31556)	753	0-6-0T	IC	Afd		1909
32670	(BODIAM)	0-6-0T	IC	Bton		1872
32678		0-6-0T	IC	Bton		1880
No.23	HOLMAN F.STEPHENS	0-6-0ST	IC	HE	3791	1952
No.25	NORTHIAM	0-6-0ST	IC	HE	3797	1953
14	CHARWELTON	0-6-0ST	IC	MW	1955	1917
376	NORWEGIAN	2-6-0	OC	Nohab	1163	1919
No.12	MARCIA	0-4-0T	OC	P	1631	1923
(W20W)		4w-4wDMR		Sdn		1940
D2023		0-6-0DM		Sdn		1958
D2024		0-6-0DM		Sdn		1958
(D)3174	(08108) DOVER CASTLE	0-6-0DE		Derby		1955
D4118	(08888)	0-6-0DE		Hor		1962
D6570	(33052) ASHFORD	Bo-BoDE		BRCW	DEL174	1961
D7594	(25244)	Bo-BoDE		Dar		1964
D9504	(01566)	0-6-0DH		Sdn		1964
M50971		2-2w-2w-2DMR		DerbyC&W		1959
M51571		2-2w-2w-2DMR		DerbyC&W		1960

–		Bo-BoDE		MV		1932	
–		0-4-0DE		RH	423661	1958	
DR 98211A		4wDMR		Plasser	52766A	1985	
L111 EED		4wDM	R/R	Isuzu		2005	
(900393)		2w-2PMR		Wkm	673	1932	DsmT
–		2w-2PMR		Wkm	6603	1953	
(9043)		2w-2PMR		Wkm	6965	1955	
			reb	Wkm		1961	DsmT
7438		2w-2PMR		Wkm	7438	1956	

TUNBRIDGE WELLS & ERIDGE RAILWAY PRESERVATION SOCIETY LTD, SPA VALLEY RAILWAY, WEST STATION, NEVILL TERRACE, TUNBRIDGE WELLS — TN2 5QY

Gauge 4ft 8½in www.spavalleyrailway.co.uk **TQ 579385**

32650	SUTTON	0-6-0T	IC	Bton		1876	
47493		0-6-0T	IC	VF	4195	1927	
68077		0-6-0ST	IC	AB	2215	1947	
2315	LADY INGRID	0-4-0ST	OC	AB	2315	1951	
	"NEWSTEAD"	0-6-0ST	IC	HE	1589	1929	
No.57		0-6-0ST	IC	RSHN	7668	1950	
62	UGLY	0-6-0ST	IC	RSHN	7673	1950	
No.10	TOPHAM	0-6-0ST	OC	WB	2193	1922	
D3489	COLONEL TOMLINE	0-6-0DE		Dar		1958	
(D4114)	09026 CEDRIC WARES	0-6-0DE		Hor		1962	
(D)6583	(33063) R.J. MITCHELL	Bo-BoDE		BRCW	DEL187	1962	
(D6585)	33065 SEALION	Bo-BoDE		BRCW	DEL189	1962	
(D6586)	33201	Bo-Bo DE		BRCW	DEL157	1962	
15224		0-6-0DE		Afd		1949	
(E6047)	73140	Bo-BoRE/DE		(EE	3719	1966	
				(EEV	E379	1966	
2591	SOUTHERHAM	0-4-0DM		(RSHN	7924	1959	
				(DC	2591	1959	
51669		2-2w-2w-2DMR		DerbyC&W		1960	
51849		2-2w-2w-2DMR		DerbyC&W		1960	
S60142	(207017) 1317	4-4wDER		Afd/Elh		1962	
62402	(1497)	4w-4wRER		York(BRE)		1971	
	99709 901125-3	2w-2PMR	R/R	Geismar	M44/075	2010	
	99709 901126-1	2w-2PMR	R/R	Geismar	M44/076	2010	
	99709 901012-3	2w-2PMR		Lesmac	LMS007/2	2006	
	99709 901020-6	2w-2PMR		Lesmac	LMS003/3	2005	Dsm

WOODLANDS RAILWAY

Gauge 1ft 3in Closed

	PAM	4wPM		Mace		1980	OOU
	SIMON	6wPM	s/o	Mace		1985	OOU

LANCASHIRE

INDUSTRIAL SITES

AMEY GROUP plc, t/a BYZAK LTD, PLANT YARD, PLANTATION ROAD, BURSCOUGH INDUSTRIAL ESTATE, BURSCOUGH — L40 8JT

Locomotives are present in the yard between use on contracts.

Gauge 610mm — **SD 426119**

PTL 02	4wBE		CE	B4246A	1998	
		reb	CE	B4381B	2002	
PTL 01	4wBE		CE	B4246B	1998	
		reb	CE	B4381A	2002	

BLACKPOOL TRANSPORT SERVICES LTD, BLACKPOOL

Blundell Street Depot & Works, Blackpool — **FY1 5DD**

Gauge 4ft 8½in www.blackpooltransport.com — **SD 307350**

938	Q204 HFR	4wDM	R/R	Unimog	029065	1981
939	J271 TEC	4wDM	R/R	Unimog	172544	1992
	MX14 LRJ	4wDM	R/R	Hako		2015

Starr Gate Depot, Blackpool — **FY4 1SN**

Gauge 4ft 8½in — **SD 304317**

–	4wBE		Zephir	2372	2011

EDF ENERGY plc, HEYSHAM POWER STATION, PRINCESS ALEXANDRA WAY, HEYSHAM — LA3 2XH

Gauge 4ft 8½in www.edfenergy.com — **SD 401599**

H 055	VINCENT DE RIVAZ	4wDH	S	10037	1960	a

a on hire from British American Railway Services Ltd, Stanhope, Co. Durham

HEIDELBERG CEMENT GROUP, t/a HANSON CEMENT, RIBBLESDALE WORKS, WEST BRADFORD ROAD, CLITHEROE — BB7 4QF

Gauge 4ft 8½in www.heidelbergcement.com — **SD 749434**

10	WINSTON	0-6-0DH	GECT	5396	1975		
9	CHUG CHUG	0-6-0DH	GECT	5401	1975		
(DB 965051 DE 320477 9036T)		2w-2PMR	Wkm	(7574)	1956	DsmT	Pvd

HELICAL TECHNOLOGY LTD, PRECISION SPRING MANUFACTURERS, (LYTHAM MOTIVE POWER MUSEUM), DOCK ROAD, LYTHAM ST ANNES — FY8 5AQ

Preserved locomotives in store, not on public display.

Gauge 4ft 8½in www.helical-technology.com — **SD 381276**

	RIBBLESDALE No.3	0-4-0ST	OC	HC	1661	1936
SNIPEY	HODBARROW No.6	0-4-0CT	IC	N	4004	1890
	GARTSHERRIE No.20	0-4-0ST	OC	NBH	18386	1908

Gauge 2ft 0in

"No.37"	JONATHAN	0-4-0ST	OC	HE	678	1898	
	MIRANDA	4wDM		HE	2198	1940	

RIBBLE RAIL LTD, off CHAIN CAUL ROAD, RIVERSWAY, PRESTON PR2 2PD

Gauge 4ft 8½in **SD 505294**

Operates freight trains between Total UK Ltd, Chain Caul Way, Preston and Strand Road exchange sidings, Preston - see Ribble Steam Railway entry for locomotives

WEST COAST RAILWAY COMPANY LTD, JESSON WAY, CRAG BANK, CARNFORTH LA5 9UR

Other locomotives may be occasionally present for repairs. Closed to the public.

Gauge 4ft 8½in www.westcoastrailways.co.uk **SD 496708**

34016	BODMIN	4-6-2	3C	Bton		1945	
34073	(249 SQUADRON)	4-6-2	3C	Bton		1948	Dsm
35018	BRITISH INDIA LINE	4-6-2	3C	Elh		1945	
45690	(5690) LEANDER	4-6-0	3C	Crewe	288	1936	
			reb	Derby		1973	
45699	GALATEA	4-6-0	3C	Crewe	297	1936	
46115	(6115) SCOTS GUARDSMAN	4-6-0	3C	NBQ	23610	1927	
(46201)	6201 PRINCESS ELIZABETH	4-6-2	4C	Crewe	107	1933	
48151		2-8-0	OC	Crewe		1942	
No.1		0-6-0F	OC	AB	1572	1917	OOU
	W.T.T.	0-4-0ST	OC	AB	2134	1942	
1		0-4-0ST	OC	AB	2230	1947	
	GLAXO	0-4-0F	OC	AB	2268	1949	OOU
–		0-4-0ST	OC	P	2027	1942	Dsm
–		0-4-0	IC	SS	1585	1865	
	rebuilt as	0-4-0ST	IC	BHSC		1873	
	LINDSAY	0-6-0ST	IC	WCI		1887	
D2084	(03084)	0-6-0DM		Don		1959	
(D2196)	03196 40 JOYCE / GLYNIS	0-6-0DM		Sdn		1961	
D2381	03381	0-6-0DM		Sdn		1961	OOU
(D3533)	08418	0-6-0DE		Derby		1958	
(D3600)	08485	0-6-0DE		Hor		1958	
(D3845	08678) 555	0-6-0DE		Hor		1959	
(D6865)	37165	Co-CoDE		_(EE	3343	1963	
				(EES	8396	1963	Dsm
	TRENCHARD	0-4-0DM		AB	401	1956	OOU
	(ESKDALE)	0-6-0DE		YE	2718	1958	OOU

PRESERVATION SITES

BLACKPOOL MINIATURE RAILWAY CO LTD, RIO GRANDE EXPRESS, c/o BLACKPOOL ZOO, ZOOLOGICAL GARDENS, EAST PARK DRIVE, BLACKPOOL — FY3 8PP

Gauge 1ft 3in www.blackpoolzoo.org.uk **SD 335362**

–		2-8-0DH	s/o	SL	7219	1972

FOLDHOUSE PARK LTD, HOLIDAY HOME PARK, FOLD HOUSE, HEAD DYKE LANE, PILLING, near KNOTT-END-ON-SEA — PR3 6SJ

Gauge 4ft 8½in www.foldhouse.co.uk **SD 409477**

11302	THE PILLING PIG	0-6-0ST	OC	HC	1885	1955

C.J. GIBBONS, UNKNOWN LOCATION

Gauge 1ft 3in

42869		2-6-0	OC	GibbonsCL		1993

HERITAGE PAINTING, UNIT 5, WATERY LANE INDUSTRIAL ESTATE, DARWEN — BB3 2EB

www.heritage-painting.com Vehicles for restoration occasionally present **SD 699208**

MERSEYSIDE TRANSPORT TRUST, OSPREY PLACE, off TOLLGATE ROAD, BURSCOUGH INDUSTRIAL ESTATE, BURSCOUGH — L40 8TG

Gauge 4ft 8½in www.class502.org.uk **SD 429108**

(M)28361(M)		4w-4wRER		DerbyC&W		c1939

PLEASURE BEACH RAILWAY, SOUTH SHORE, BLACKPOOL — FY4 1EZ

Gauge 1ft 9in www.blackpoolpleasurebeach.com **SD 305332**

	BARBIE	4wDM		AK	7	1982
4472	MARY LOUISE	4-6-2DH	s/o	HC	D578	1933
4473	CAROL JEAN	4-6-2DH	s/o	HC	D579	1933
	reb	4-6-4DH	s/o	Ravenglass		1988
6200	GEOFFREY THOMPSON OBE DL	4-6-2DH	s/o	HC	D586	1935
			reb	PBR		2004

POULTON & WYRE RAILWAY PRESERVATION SOCIETY, THORNTON STATION

Gauge 4ft 8½in www.pwrs.org

–		0-4-0DM		JF	4210108	1955	a
	99709 909196-6	2w-2PMR		Geismar	ST/04/11	2004	a

a currently stored off site

PRIVATE LOCATION, EARBY, near COLNE

Gauge 2ft 0in

–		4wPM		L	9993	1938	

THE RIBBLE STEAM RAILWAY, off CHAIN CAUL ROAD, RIVERSWAY, PRESTON **PR2 2PD**

Gauge 4ft 8½in www.ribblesteam.org.uk **SD 504295, 500293**

(30072)	72	0-6-0T	OC	VIW	4446	1943	
(52322)	12322 (1300)	0-6-0	IC	Hor	420	1896	a
19	(11243)	0-4-0ST	OC	Hor	1097	1910	
1439		0-4-0ST	IC	Crewe	842	1862	
No.1	GLENFIELD 15	0-4-0CT	OC	AB	880	1902	
	JOHN HOWE	0-4-0ST	OC	AB	1147	1908	
	EFFICIENT	0-4-0ST	OC	AB	1598	1918	
(No.20)	"NIDDRIE"	0-6-0ST	OC	AB	1833	1924	
	ALEXANDER	0-4-0ST	OC	AB	1865	1926	
	"HEYSHAM"	0-4-0F	OC	AB	1950	1928	
	J.N.DERBYSHIRE	0-4-0ST	OC	AB	1969	1929	
No.6		0-4-0ST	OC	AB	2261	1949	b
No.4	BRITISH GYPSUM	0-4-0ST	OC	AB	2343	1953	
	LUCY	0-6-0ST	OC	AE	1568	1909	
	"MDHB No 26"	0-6-0ST	OC	AE	1810	1918	
	"JOAN"	0-6-0ST	OC	AE	1883	1922	
	THE KING	0-4-0WT	OC	EB	(48?)	1906	
	WINDLE	0-4-0WT	OC	EB	53	1909	
–		0-4-0ST	OC	GR	272	1894	
	"KINSLEY"	0-6-0ST	IC	HE	1954	1939	
WD 75105	WALKDEN	0-6-0ST	IC	HE	3155	1944	
1R	"RESPITE"	0-6-0ST	IC	HE	3696	1950	
	"SHROPSHIRE"	0-6-0ST	IC	HE	3793	1953	
No.4	GLASSHOUGHTON No.4	0-6-0ST	IC	HE	3855	1954	
21	"LINDA"	0-6-0ST	OC	HL	3931	1938	
	"DAPHNE"	0-4-0ST	OC	P	737	1899	
	FONMON	0-6-0ST	OC	P	1636	1924	
	CALIBAN	0-4-0ST	OC	P	1925	1937	
	HORNET	0-4-0ST	OC	P	1935	1937	
"NORTH	WESTERN GAS BOARD"	0-4-0ST	OC	P	1999	1941	
	JOHN BLENKINSOP	0-4-0ST	OC	P	2003	1941	
	AGECROFT No.2	0-4-0ST	OC	RSHN	7485	1948	
	"GASBAG"	4wVBT	VCG	S	8024	1929	
	"ST MONANS"	4wVBT	VCG	S	9373	1947	
	COURAGEOUS	0-6-0ST	OC	WB	2680	1942	
D2148		0-6-0DM		Sdn		1960	
(D2189	03189)	0-6-0DM		Sdn		1961	
D2595		0-6-0DM		HE	7179	1969	
		a rebuild of		HE	5644	1959	
D9539		0-6-0DH		Sdn		1965	
–		4wBE		EEDK	788	1930	
(663)		0-6-0DE		(EE	2160	1956	
				(VF	D350	1956	c

–		4wBE		GB	2000	1945	
	"HOTTO"	4wPM		H	965	1930	
	MIGHTY ATOM	0-4-0DM		HC	D628	1943	
D629		0-4-0DM		HC	D629	1945	
	"MARGARET"	0-4-0DM		HC	D1031	1956	
	PERSIL	0-4-0DM		JF	4160001	1952	
	BICC	0-4-0DH		NBQ	27653	1956	
	ENERGY	4wDH		RR	10226	1965	c
	ENTERPRISE	4wDH		RR	10282	1968	c
	PROGRESS	4wDH		RR	10283	1968	c
	STANLOW No.4	0-4-0DH		TH	160V	1966	
D2870		0-4-0DH		YE	2677	1960	
E79960		2w-2DMR		WMD	1265	1958	
(98404)		4wDHR		Perm	MTU 001	1991	c

a based here, but visits other locations
b currently at Cambrian Heritage Railways, Oswestry, Shropshire
d locomotives owned and operated by Ribble Rail Ltd

Furness Railway Trust

Gauge 4ft 8½in www.furnessrailwaytrust.org.uk

4979	(WOOTTON HALL)	4-6-0	OC	Sdn		1930	
5643		0-6-2T	IC	Sdn		1925	a
No.20		0-4-0	IC	SS	1448	1863	
	rebuilt as	0-4-0ST	IC	BHSC		1870	
	rebuilt as	0-4-0	IC	FRT		1998	
	CUMBRIA	0-6-0ST	IC	HE	3794	1953	a
No.2	FLUFF	0-4-0DM		JF	21999	1937	

a based here, but visits other locations

WEST LANCASHIRE LIGHT RAILWAY TRUST, STATION ROAD, HESKETH BANK, near PRESTON

PR4 6SP

Gauge 2ft 0in www.westlancsrailway.org **SD 448229**

–		4-6-0T	OC	BLW	45190	1917	a
(45)		0-6-0T	OC	Chrz	3506	1957	
"No.3"	IRISH MAIL	0-4-0ST	OC	HE	823	1903	
47		0-8-0T	OC	Hen	14676	1917	
48		0-4-2T	OC	JF	15513	1920	a
S1985.0019	(CHEETAL)	0-6-0WT	OC	JF	15991	1923	
	JOFFRE	0-6-0WTT	OC	KS	2405	1915	
"No.35"	No.21 UTRILLAS	0-4-0WT	OC	OK	2378	1907	
"No.34"	No.22 MONTALBAN	0-4-0WT	OC	OK	6641	1913	
	(SYBIL)	0-4-0ST	OC	WB	1760	1906	
"1"	"CLWYD"	4wDM		RH	264251	1951	
"No.2"	TAWD	4wDM		RH	222074	1943	
"No.4"	"BRADFIELD"	4wPM		FH	1777	1931	
"No.5"		4wDM		RH	200478	1940	
"No.7"		4wDM		MR	8992	1946	
10		4wDM		FH	2555	1942	
"11"		4wDM		MR	5906	1934	
12		4wDM		MR	11258	1964	

"No.12"	No.2	4wDM		MR	7955	1945	
			reb	WLLR	No.2	1987	b Dsm
"No.16"		4wDM		RH	202036	1941	Dsm
19		4wBE		BV	613	1972	a c
"No.19"		4wPM		L	10805	1939	
"No.20"		4wPM		Bg	3002	1937	
No.21		4wDM		HE	1963	1939	
25		4wBE		BV	692	1974	a c
"No.25"		4wDM		RH	297054	1950	
"No.26"	8	4wDM		MR	11223	1963	
No.31	"No.27" MILL REEF	4wDM		MR	7371	1939	
36		4wDM		RH	339105	1953	
"No.38"		0-4-0DMF		HC	DM750	1949	
"No.39"	P37829 "BLACK PIG"	4wDM		FH	3916	1959	
40	DAME VERA DUCKWORTH	4wDM		RH	381705	1956	
"No.8"	"PATHFINDER"	4wDM		HE	4478	1953	d
–		0-6-0DMF		HC	DM1393	1967	
640	"No.44" WELSH PONY	4wWE		BẸV	640	1926	
–		4wBE		GB	1840	1942	
–		4wDM		L	29890	1946	
–		4wDM		MR	8995	1946	Dsm
–		4wBE		WR	C6765	1963	
	"BREDBURY"	2w-2PM		Bredbury		c1954	

a currently stored elsewhere
b converted into a brake van
c to be regauged from 2ft 6in
d plate reads 4480/1953

WINDMILL ANIMAL FARM MINIATURE RAILWAY, WINDMILL ANIMAL FARM LTD, FISH LANE, BURSCOUGH L40 1UQ

Gauge 1ft 3in www.windmillanimalfarm.co.uk **SD 427156**

–		4-4-2	OC	Barnes	104	c1927	
	SIÂN	2-4-2	OC	Guest	18	1963	
4	BLUE PACIFIC	4-6-2VB	OC	GuinnessNL		c1935	Pvd
	MOUNTAINEER	0-4-0TT	OC	WVanHeiden		1972	
–		4-4-2	OC	(USA?)		1948	a
–		2-6-2	OC				a
No.4468	DUKE OF EDINBURGH	4-6-2DE	s/o	Barlow		1948	
	PRINCESS ANNE	4-6w-2DE	s/o	Barlow		1948	b
No.2510	PRINCE CHARLES	4-6-2DE	s/o	Barlow		1954	OOU
15		0-6-0DM		G&S		1961	
	rebuilt as	0-6-0PE/BE		MossDW		2013	
	rebuilt as	0-6-0PE		MossAJ		2017	c
–		4w-4wDH		Guest		1957	
	WHIPPIT QUICK	4wPM		L	6502	1935	
	rebuilt as	4w-4PM		Fairbourne		1955	
	rebuilt as	4w-4PMR		Fairbourne		1962	
	rebuilt as	4w-4DMR		Moss AJ			
	GWRIL	4wPM		L	20886	1943	
–		2-2wPMR		MossAJ		1989	
5	BATTISON	2-6-4DE	s/o	MossDW		2012	
	rebuild of	2-6-4DH	s/o	Battison		1958	

	SAFARI EXPRESS	2-8-0DH	s/o	SL	15/2/79	1979
7 278		2-8-0DH	s/o	SL	17/6/79	1979
	"BLACK SMOKE"	2-4-2PM	s/o	SmithEL		c1956
14		2w-2PM		WalkerG		1985

a incomplete loco
b carries incorrect plate dated 1962 (an overhaul date ?)
c carries plates G&S 15/1959

WINFIELDS MEGASTORE,
HAZEL MILL, BLACKBURN ROAD, ACRE, HASLINGDEN, ROSSENDALE BB4 5DD
Gauge 4ft 8½in www.winfieldsoutdoors.co.uk **SD 787250**

–		0-6-0DE		HC	D1075	1959

LEICESTERSHIRE

INDUSTRIAL SITES

AGGREGATE INDUSTRIES UK LTD
Bardon Hill Quarry, Bardon Hill, Coalville LE67 1TD
Gauge 4ft 8½in www.aggregate.com **SK 446129**

No.59	DUKE OF EDINBURGH	6wDH		RR	10273	1968
No.159	BARDON DUCHESS	6wDH		TH	297V	1981

Croft Quarry, Croft LE9 3GS
Gauge 4ft 8½in SP 517960

01572	MS 5482 "KATHRYN"	0-6-0DH		RR	10256	1966
01562	MS 6475 CHUG	0-6-0DH		TH	257V	1975
			reb	RMS	LWO2918	2006

BRUSH TRACTION – a WABTEC COMPANY,
FALCON WORKS, NOTTINGHAM ROAD, LOUGHBOROUGH LE11 1NF
Locomotives under construction, overhaul or repair are usually present.
Gauge 4ft 8½in www.wabtec.com **SK 543207**

PLANT No.11079	SPRITE	0-4-0DH		HC	D1341	1966
	GEORGE TOMS	0-4-0DH		_(WB	3209	1962
				(RSHD	8364	1962
–		4wDM	R/R	Unimog	072555	1981

KING RAIL, KING VEHICLE ENGINEERING LTD,
RIVERSIDE, MARKET HARBOROUGH LE16 7PX
UK agent for Zagro. www.kingrail.co.uk **SP 743876**
Road/Rail vehicles under conversion/repair occasionally present

LOCOMOTIVE MAINTENANCE SERVICES LTD, 14 BAKEWELL ROAD, LOUGHBOROUGH — LE11 5QY

Gauge 4ft 8½in — **SK 526212**

1340	TROJAN	0-4-0ST	OC	AE	1386	1897
4141		2-6-2T	OC	Sdn		1946
4953	PITCHFORD HALL	4-6-0	OC	Sdn		1929

NETWORK RAIL, RAIL INNOVATION & DEVELOPMENT CENTRE

(Operated by SERCO UK and Europe)

Locomotives are kept at :-

Asfordby Test Centre, Asfordby Business Park, Melton Mowbray LE14 3JL — SK 727207

Old Dalby Test Centre, Station Road, Old Dalby LE14 3NQ — SK 680239

Gauge 4ft 8½in

See Section 7 for details

TARMAC plc - A CRH Company, BARROW RAILHEAD, SILEBY ROAD, BARROW-UPON-SOAR — LE12 8LX

Gauge 4ft 8½in www.tarmac.com — **SK 587168**

40905	"JO"	4wDH		DeDietrich	89134	1988	OOU
	RON-DON-CLIVE	0-6-0DH		TH	290V	1980	
			reb	HE	9382	2012	
	DANNY BOY	4wDH	R/R	Zephir	1928	2005	
	TED	4wDH	R/R	Zephir	2136	2008	
–		4wDH	R/R	Zephir	2779	2018	

UK RAIL LEASING LTD, BEAL STREET, LEICESTER — LE2 0AA

Gauge 4ft 8½in — **SK 597044**

(D6571)	33053	Bo-BoDE	BRCW	DEL175	1961	
(D5410	27059)	Bo-BoDE	BRCW	DEL253	1962	
(D6703)	37003	Co-CoDE	_(EE	2866	1960	
			(VF	D582	1960	
58016	050	Co-CoDE	Don		1984	
D1388	6	0-4-0DH	HC	D1388	1970	a
	"CHEEDALE"	4wDH	TH	284V	1979	a

a property of Andrew Briddon, Darley Dale, Derbyshire

Other mainline and preserved locomotives usually present

PRESERVATION SITES

ARMOURGEDDON MILITARY MUSEUM, SOUTHFIELDS FARM, HUSBANDS BOSWORTH, near LUTTERWORTH — LE17 6NW

Gauge 600mm www.militarymuseum.uk — **SP 638861**

(RTT/767149) PRESIDENT	2w-2PM	Wkm	3151	1943	
RTT/767163	2w-2PM	Wkm	3236	1943	Dsm

THE BATTLEFIELD LINE, (THE SHACKERSTONE RAILWAY SOCIETY LTD), SHACKERSTONE STATION, MARKET BOSWORTH

Locomotives are kept at :-	Shackerstone	CV13 0BS	SK 378067, 379064
	Market Bosworth	CV13 0PF	SK 392030

Gauge 4ft 8½in www.battlefieldline.co.uk

	SIR GOMER	0-6-0ST	OC	P	1859	1932	
	RICHARD III	0-6-0T	OC	RSHN	7537	1949	
(D14)	45015	1Co-Co1DE		Derby		1960	OOU
(D1921	47244) 47640 UNIVERSITY OF STRATHCLYDE	Co-CoDE		BT	683	1965	
(D2310)	04110	0-6-0DM		_(RSHD	8169	1960	
				(DC	2691	1960	
(D2867)		0-4-0DH		YE	2850	1961	a
D6508	(33008) EASTLEIGH	Bo-BoDE		BRCW	DEL100	1960	
(D6574)	33019 GRIFFON	Bo-BoDE		BRCW	DEL126	1960	
D6593	(33208)	Bo-BoDE		BRCW	DEL164	1962	b
D7523	(25173)	Bo-BoDE		Derby		1965	
(D8063	20063) 2002 AT3 DJ 054	Bo-BoDE		_(EE	2969	1961	
				(RSHD	8221	1961	c
(56009)	56201	Co-CoDE		Electro		1976	
58012		Co-CoDE		Don		1984	
58023		Co-CoDE		Don		1984	
58048		Co-CoDE		Don		1986	
(E6020)	73114	Bo-BoDE/RE		_(EE	3582	1965	
				(EEV	E352	1965	
12083	(201276 M413)	0-6-0DE		Derby		1950	
	HOTWHEELS	0-6-0DM		AB	422	1958	a
(19)		0-6-0DH		AB	594	1974	
–		4wBE/WE		EEDK	905	1935	
	"MAZDA"	0-4-0DE		RH	268881	1949	
–		4wDM		RH	263001	1949	
–		0-4-0DM		RH	281271	1950	a
M51131		2-2w-2w-2DMR		DerbyC&W		1958	
W51321	(977753) P464	2-2w-2w-2DMR		BRCW		1960	
M55005		2-2w-2w-2DMR		GRC&W		1958	
(S65321	5791 977505 932053)	4w-4RER		Elh		1954	
	C958 YOR	4wDMR	R/R	Bruff	516	1986	
	C951 YOR	4wDMR	R/R	Bruff	519	1986	d
–		2w-2PMR		Geismar	ST/00/16	2000	

a property of Harry Needle Railroad Co Ltd, Derbyshire
b based at Mid-Hants Railway, Hampshire – here for repairs
c property of Railway Support Services Ltd, Wishaw, Warwickshire
d incorporates chassis from Bruff 519 and parts from 518 & 520

"CONKERS", RAWDON ROAD, MOIRA, near ASHBY-DE-LA-ZOUCH **DE12 6GA**

(operated by Planning Solutions Ltd)

Gauge 2ft 0in www.visitconkers.com **SK 312161**

CONKACHOO	4-4-0DH	s/o	SL	2121	2001	
"CONKACHOO TWO"	4w-4wDH	s/o	SL	8204	2011	

EASTWELL HISTORY GROUP - "EASTWELL HERITAGE PROJECTS", EASTWELL

Gauge 3ft 0in

	LORD GRANBY	0-4-0ST	OC	HC	633	1902

A.R. ETHERINGTON, STATION HOUSE, SHACKERSTONE

Gauge 4ft 8½in **SK 378064**

(5)		4wDM		MR	9921	1959

GREAT CENTRAL RAILWAY plc, GREAT CENTRAL STATION, LOUGHBOROUGH

Locomotives are kept at :-

Loughborough Loco Shed	LE11 1RW	SK 543194
Rothley Carriage & Wagon Works	LE7 7LD	SK 569122
Swithland Sidings	LE7 7SL	SK 563132
Quorn and Woodhouse	LE12 8AW	SK 549161

Gauge 4ft 8½in www.gcrailway.co.uk

6990	WITHERSLACK HALL	4-6-0	OC	Sdn		1948
(30777)	777 SIR LAMIEL	4-6-0	OC	NBH	23223	1925
34039	BOSCASTLE	4-6-2	3C	Bton		1946
45305	ALDERMAN A.E. DRAPER	4-6-0	OC	AW	1360	1937
45491		4-6-0	OC	Derby		1943
46521		2-6-0	OC	Sdn		1953
47406		0-6-0T	IC	VF	3977	1926
48305		2-8-0	OC	Crewe		1943
48624	(8624)	2-8-0	OC	Afd		1943
63601	(102)	2-8-0	OC	Gorton		1912
(69523	4744) No.1744	0-6-2T	IC	NBH	22600	1921
70013	OLIVER CROMWELL	4-6-2	OC	Crewe		1951
73156		4-6-0	OC	Don		1956
78018		2-6-0	OC	Dar		1954
78019		2-6-0	OC	Dar		1954
92214	LEICESTER CITY	2-10-0	OC	Sdn		1959
–		0-6-0ST	IC	HE	3809	1954
4	"MEAFORD"	0-6-0T	OC	RSHN	7684	1951
–		4wVBT	VCG	S	9370	1947
(D53)	45041 ROYAL TANK REGIMENT	1Co-Co1DE		Crewe		1962
D123	(45125 89423) LEICESTERSHIRE AND DERBYSHIRE YEOMANRY	1Co-Co1 DE		Crewe		1961
D1705	(47117) SPARROWHAWK	Co-CoDE		BT	467	1965
(D2989	07005)	0-6-0DE		RH	480690	1960
			reb	Resco	L106	1978
(D3101)	13101	0-6-0DE		Derby		1955
D3690	(08528)	0-6-0DE		Dar		1959
(D3861)	08694	0-6-0DE		Hor		1959
(D4067)	10119 MARGARET ETHEL - THOMAS ALFRED NAYLOR	0-6-0DE		Dar		1961
(D4137)	08907	0-6-0DE		Hor		1962
D5185	(25735 25035) CASTELL DINAS BRAN	Bo-BoDE		Dar		1963

(D)5401	(27056)	Bo-BoDE		BRCW	DEL244	1962	
D5830	(31563)	A1A-A1A DE		BT	366	1962	
(D6535	33116)	Bo-BoDE		BRCW	DEL127	1960	
(D6724	37024) 37714	Co-CoDE		_(EE	2887	1961	
	CARDIFF CANTON			(VF	D603	1961	
D8098	(20098)	Bo-BoDE		_(EE	3003	1961	
				(RSHD	8255	1961	
	BARDON	0-4-0DM		AB	400	1956	
	ARTHUR WRIGHT	0-4-0DM		JF	4210079	1952	
(M51616)	ALF BENNEY	2-2w-2w-2DHR		DerbyC&W		1959	
M51622		2-2w-2w-2DHR		DerbyC&W		1959	
(50193)	960 992 977898	2-2w-2w-2DMR		MetCam		1957	
(50203)	960 992 977897	2-2w-2w-2DMR		MetCam		1957	
E50266	(53266)	2-2w-2w-2DMR		MetCam		1957	
E50321	(960 993 977900)	2-2w-2w-2DMR		MetCam		1958	
E51427	(960 993 977899)	2-2w-2w-2DMR		MetCam		1959	
2		2w-2PMR		Bance	002	1995	
3		2w-2PMR		Bance	037	1995	
–		2w-2BER		Bance	104	2002	Dsm

Mountsorrel Railway
Line runs from Swithland Sidings to Mountsorrel, with trains operated by the Great Central Railway.
(See **Mountsorrel & Rothley Community Heritage Centre** entry)

LEICESTER CITY COUNCIL, LEISURE & CULTURE, ABBEY PUMPING STATION - LEICESTER CITY'S MUSEUM OF SCIENCE & TECHNOLOGY, CORPORATION ROAD, LEICESTER — LE4 5PX

Gauge 2ft 0in www.abbeypumpingstation.org **SK 589067**

	LEONARD	0-4-0ST	OC	WB	2087	1919	
	NEW STAR	4wPM		L	4088	1931	
–		4wPM		MR	5260	1931	
–		4wDM		RH	223700	1943	
	PETER	4wDM		SMH	40SD515	1979	

LEICESTERSHIRE COUNTY COUNCIL, former SNIBSTON MINE, ASHBY ROAD, COALVILLE — LE67 3LN

Gauge 4ft 8½in **(Closed)** **SK 420144**

No.2		0-4-0F	OC	AB	1815	1924	a
–		0-4-0ST	OC	BE	314	1906	a
	CADLEY HILL No.1	0-6-0ST	IC	HE	3851	1962	
	MARS II	0-4-0ST	OC	RSHN	7493	1948	
–		0-6-0DH		HE	6289	1966	

a stored off site

Gauge 2ft 6in

(T42.1994)	T15	4wBEF		_(EE	2416	1957	
				(RSHN	7935	1957	
No.5		4wBEF		_(EE	2300	1956	
				(Bg	3436	1956	

–	4wBEF		(EE	2086	1955	
			(Bg	3434	1955	
–	4wBEF		GECT	5424	1976	
–	0-6-0DMF		HC	DM1238	1960	
–	4wDH		HE	8973	1979	

Gauge monorail

A4	2wPH		RM	8253	1959	

MOUNTSORREL & ROTHLEY COMMUNITY HERITAGE CENTRE, MOUNTSORREL RAILWAY, NUNCKLEY HILL QUARRY, SWITHLAND LANE, ROTHLEY — LE7 7SJ

Gauge 4ft 8½in www.heritage-centre.co.uk **SK 569142**

–	4wDM		RH	393304	1956	
(DB 965080)	2w-2PMR		Wkm	7595	1957	
reb to	2w-2DMR				1999	

Nunckley Narrow Gauge Railway
Gauge 2ft 0in

–	4wDM		MR	22070	1960	
85049	4wDM		RH	393325	1952	

PETER THOMAS, BLABY "WAKES" SHOWGROUND, LEICESTER ROAD, BLABY

Gauge 4ft 8½in **SK 567983**

–	0-6-0F	OC	WB	2370	1929	OOU

TWINLAKES PARK, MELTON SPINNEY ROAD, MELTON MOWBRAY — LE14 4SB

Gauge 1ft 3in www.twinlakespark.co.uk **SK 774213**

–	2-6-0DH	s/o	SL	RG11-86	1986	
–	2-6-0DH	s/o	SL	76.3.88	1988	

WELLAND VALLEY VINTAGE TRACTION CLUB, GLEBE ROAD, MARKET HARBOROUGH

Gauge 3ft 0in **SP 742868**

KETTERING FURNACES No.8	0-6-0ST	OC	MW	1675	1906	a

a currently under renovation at a private location

DAVID WHITE, GREENLEA LIGHT RAILWAY, WORKHOUSE LANE, BURBAGE

Gauge 2ft 0in **SP 447915**

SIR GEORGE	4wDH	s/o	AK	12	1984	
C.P.HUNTINGTON	4w-4wDH	s/o	Chance 78-50157-24		1978	
GOLIATH	4wDM		MR	5881	1935	

LINCOLNSHIRE

INDUSTRIAL SITES

WILLIAM BLYTH, OLD TILE WORKS, FAR INGS TILERIES, FAR INGS ROAD, BARTON-UPON-HUMBER — DN18 5RF

Gauge 2ft 0in www.theoldtileworks.com **TA 023233**

No.	Type	Builder	Works No.	Year	Note	Status
–	4wDM	MR	8678	1941		OOU

BRITISH STEEL LTD

(a Greybull Capital LLP company) www.britishsteel.co.uk

Appleby Coke Ovens

Gauge 4ft 8½in **SE 917108**

No.	Type	Builder	Works No.	Year	Note
5	4wRE	Schalke	10-310-0054	1973	
6	4wRE	Schalke	10-310-0055	1973	
7	4wRE	Schalke	10-310-8070	1979	a

a built under licence by Starco Engineering, Winterton Road, Scunthorpe

Appleby-Frodingham Works, Scunthorpe — DN16 1BP

Gauge 4ft 8½in **SE 910110, 915110, 916105**

No.	Name	Type	Builder	Works No.	Year	Note	Status
4	0448-73-04	0-4-0DE	_(BD	3737	1977		
			(GECT	5437	1977		OOU
29		0-6-0DE	YE	2938	1964		Dsm
30	FUSION	Bo-BoDE	CNES		2013	a	OOU
44		0-6-0DE	YE	2768	1960		OOU
51	5	0-6-0DE	YE	2709	1959		
61		6wDH	RR	10277	1968	b	OOU
63		6wDH	TH	V317	1987	b	OOU
70	BIG KEITH	Bo-BoDE	HE	7281	1972		
71		Bo-BoDE	HE	7282	1972		
72		Bo-BoDE	HE	7283	1972		
73		Bo-BoDE	HE	7284	1972		Dsm
74		Bo-BoDE	HE	7285	1972		
75	GENERAL	Bo-BoDE	HE	7286	1972		
76		Bo-BoDE	HE	7287	1973		
77		Bo-BoDE	HE	7288	1973		Dsm
79		Bo-BoDE	HE	7290	1973		Dsm
80		Bo-BoDE	HE	7474	1977		
(D8056	20056) 81	Bo-BoDE	_(EE	2962	1961		
			(RSHD	8214	1961	b	OOU
90		0-6-0DE	YE	2943	1965		OOU
91		0-6-0DE	YE	2944	1965		OOU
92		0-6-0DE	YE	2788	1960		OOU
93	3	0-6-0DE	YE	2902	1963		
94		0-6-0DE	RR	10238	1967		OOU
95		0-6-0DE	YE	2690	1959		OOU
8.701		Bo-BoDE	MaK	1600.001	1996	c	
8.702	92 76 0308702-8	Bo-BoDE	MaK	1600.002	1996	c	

8.703		Bo-BoDE	MaK	1600.003	1996	c
8.704	PAT	Bo-BoDE	MaK	1600.004	1996	c
8.708	92 76 0308708-5	Bo-BoDE	MaK	1600.008	1996	c
8.712		Bo-BoDE	MaK	1600.012	1996	c
8.716		Bo-BoDE	MaK	1600.016	1996	c
(8.717)	817	Bo-BoDE	MaK	1600.017	1996	c
8.718		Bo-BoDE	MaK	1600.018	1996	c
8.719		Bo-BoDE	MaK	1600.019	1996	c
(8.720)	820 POPPY	Bo-BoDE	MaK	1600.020	1996	c

a constructed using parts from HE 7290
b property of Harry Needle Railroad Co, Derbyshire
c operated by GBRF Ltd and maintained by Electro-Motive Diesel Inc

Blast Furnace Highline, Appleby-Frodingham Works, Scunthorpe

Gauge 4ft 8½in **SE 917104**

HL 1	0449-73-01	0-4-0DE	_(BD (GECT	3734 5434	1977 1977
HL 2	0448-73-02	0-4-0DE	_(BD (GECT	3735 5435	1977 1977
No.3	0448/73/03	0-4-0DE	_(BD (GECT	3736 5436	1977 1977
No.5	0448-73-05	0-4-0DE	_(BD (GECT	3738 5438	1977 1977
6	0448-73-06	0-4-0DE	_(BD (GECT	3739 5439	1977 1977
HL 7		0-4-0DE	_(BD (GECT	3740 5440	1977 1977

Dawes Lane Coke Ovens, Scunthorpe

Gauge 4ft 8½in **(Closed)** **SE 921118**

–	4wWE	GB	420383/1	1977	OOU
–	4wWE	GB	420383/2	1977	OOU

Rail Mill, Appleby-Frodingham Works, Scunthorpe

Gauge 4ft 8½in **SE 910113**

DAISY	4wCE	Vollert	2008

Central Engineering Workshops

Gauge 3ft 6in R.T.C.

–	4wBER		GB	2899	1958	
		reb	Lumb			OOU
–	4wBER		GB	2900	1958	
		reb	Lumb			OOU

DEPOT RAIL LTD, MERCURY HOUSE, WILLOUGHTON DRIVE, FOXBY LANE BUSINESS PARK, GAINSBOROUGH DN21 1DY

Gauge 4ft 8½in www.depotrail.co.uk Administration address only

-	4wDH	R/R	Zephir	1470	1997	a

a hire loco, stored elsewhere between contracts

UK agents for Zephir

H.M. DETENTION CENTRE, NORTH SEA CAMP, FREISTON, near BOSTON PE22 0QX

Gauge 2ft 0in www.justice.gov.uk **TF 385405**

–		4wDM	LB	55413	1967	Pvd

INTER TERMINALS IMMINGHAM LTD, ABP PORT OF IMMINGHAM, IMMINGHAM WEST TERMINAL, IMMINGHAM DOCK, IMMINGHAM DN40 2QU

(subsidiary of Inter Pipeline Ltd)

Gauge 4ft 8½in www.interterminals.com **TA 195167**

45	TMC	4wDH	NNM	83506	1984	a

a permanently coupled to a 4w wagon for brake assistance

PHILLIPS 66 LTD, HUMBER REFINERY, SOUTH KILLINGHOLME DN40 3DW

Gauge 4ft 8½in www.phillips66.co.uk **TA 163168**

	EARL OF YARBOROUGH	6wDH		RFSK	V336	1991
			refurbished	HE	9383	2013
(DH60)		6wDH		HE	9372	2010

VICTORIA GROUP HOLDINGS LTD, t/a VICTORIA GROUP, PORT OF BOSTON, BOSTON PE21 6BN

Gauge 4ft 8½in www.victoriagroup.co.uk **TF 329431**

(D4110)	09022	0-6-0DE	Hor		1961

RMS TRENT PORTS, FLIXBOROUGH WHARF, FLIXBOROUGH STATHER, FLIXBOROUGH (part of the RMS Group) DN15 8RS

Gauge 4ft 8½in www.rms-humber.co.uk **SE 859147**

29	0-6-0DH	HE	7017	1971	a
TNS 107	2w-2DMR	Robel			
			56.27-10-AG38	1982	b

a property of Harry Needle Railroad Co, Derbyshire
b on hire from British American Railway Services Ltd, Stanhope, Co. Durham

TOTAL UK LTD, TOTAL LINDSEY OIL REFINERY, EASTFIELD ROAD, NORTH KILLINGHOLME, IMMINGHAM DN40 3LW

Gauge 4ft 8½in www.total.co.uk **TA 160176**

BEAVER	0-6-0DH		AB	630	1978
		reb	YEC	L168	1999
BADGER	0-6-0DH		AB	658	1980
TIGGA	0-6-0DH		TH	285V	1979
		reb	HAB		1992
DUNDERS	0-6-0DH		HE	6971	1968

PRESERVATION SITES

JASON ALLEN, GRIMOLDBY

Gauge 2ft 0in **TF 388877**

–		2w-2PM		AllenJ		c2005	
		a rebuild of		Wkm	4092	1946	

Gauge 1ft 3in

E1	NUCLEAR ELECTRIC	4wBE		HardyK	E1	1992	

APPLEBY-FRODINGHAM RAILWAY PRESERVATION SOCIETY, c/o TATA, APPLEBY-FRODINGHAM WORKS, SCUNTHORPE (DN16 1XA)

Gauge 4ft 8½in www.afrps.co.uk **SE 913109**

(No.8)		0-4-0ST	OC	AB	2369	1955	
	CRANFORD	0-6-0ST	OC	AE	1919	1924	
3138	HUTNIK A11	0-6-0T	OC	Chrz	3138	1954	a
22		0-6-0ST	OC	HE	3846	1956	d
–		0-4-0ST	OC	P	1438	1916	
58		0-6-0DH		HE	7409	1976	
	ARNOLD MACHIN	0-6-0DE		YE	2661	1958	
1		0-6-0DE		YE	2877	1963	

a carries plate Chrz 3140 in error
d rebuilt using parts from HE 3844 and WB 2758; carries worksplate HE 3844

P. CLARK, FULSTOW STEAM CENTRE, CARPENTERS ROW, MAIN STREET, FULSTOW, near LOUTH

Gauge 4ft 8½in Private site with occasional public open days **TF 331972**

No.1		0-4-0ST	OC	RSHN	7680	1950	

CLEETHORPES COAST LIGHT RAILWAY LTD, THE MERIDIAN LINE, LAKESIDE STATION, KINGS ROAD, CLEETHORPES DN35 0AG

Gauge 1ft 3in www.cleethorpescoastlightrailway.com **TA 319070, 321072**

	(PRINCE OF WALES)	4-4-2	OC	BL	11	1908		
(No.1	SUTTON BELLE)	4-4-2	OC	Cannon		1933		
			reb	HuntTG		1953		
No.2	SUTTON FLYER	4-4-2	OC	_(Cannon				
				(HuntTG		1950		
	"SEA BREEZE"	4-6-2	OC	CCLR		(1998)	Dsm	a
111	(YVETTE)	4-4-0	OC	CravenEA		1946		
24		2-6-2	OC	Fairbourne	No.4	1990		
	EFFIE	0-4-0T+T	OC	GNS	11	1999		
	BONNIE DUNDEE	0-4-0WT	OC	KS	720	1901		
	rebuilt as	0-4-2T	OC	Ravenglass		1981		
	rebuilt as	0-4-2	OC	Ravenglass		1996		
	"GEORGE EDWARD"	0-6-4ST		MasseyD			Dsm	a
6284		2-8-0	OC	_(TurnerT/BVR				
				(Crome/Loxley		2009		

DA1	6	4wDM		Bush Mill Rly		1986	
			reb	Bush Mill Rly		c1995	
No.4		4-4wPMR		G&S		1946	

a not yet completed

COLIN COPCUTT, SILVERLEAF POPLAR RAILWAY, SIBSEY ROAD, OLD LEAKE PE22 9QS

Gauge 2ft 0in / 600mm

–		4wPM		MH	110	1925	
R14	JANE	4wDM		MR	8565	1940	
"37"		4wDM		RH	172901	1935	
SP 202		4wBE		CE	B0182C	1974	
–		4wBE		WR	918	1936	

CROWLE PEATLAND RAILWAY SOCIETY, CROWLE PEATLAND RAILWAY, PRIVATE SITE, DOLE LANE, CROWLE DN17 4BL

Gauge 3ft 0in www.peatland.co.uk

H 11023		4wDM		MR	40S302	1967	a
H23		4wDH		Schöma	5129	1990	
			reb	AK	1998		
–		4wDH		Schöma	5131	1990	
			reb	AK	1998	b	
–		4wDH		Schöma	5132	1990	
			reb	AK		1998	b
S21		4wDH		Schöma	5220	1990	
–		4wDH		Schöma	5221	1990	b

a currently at North Lindsey College, Scunthorpe, for restoration
b slave unit for use with a master unit (5129 or 5220)

private site with no public access

LINCOLNSHIRE COAST LIGHT RAILWAY, SKEGNESS WATER LEISURE PARK, WALLS LANE, INGOLDMELLS, SKEGNESS PE25 1JF

Gauge 2ft 0in www.lclr.co.uk **TF 561671**

No.2	JURASSIC	0-6-0ST	OC	P	1008	1903	
(7)	NOCTON	4wDM		MR	1935	1920	
(1)	PAUL	4wDM		MR	3995	1934	
4	WILTON	4wDM		MR	7481	1940	
No.6	GRICER	4wDM		MR	8622	1941	
9	SARK	4wDM		MR	8825	1943	
No.5	MAJOR J.E. ROBINS R.E.	4wDM		MR	8874	1944	
11140		4wDM		MR	8905	1944	Dsm
–		4wDM		MR	9264	1947	a

a currently off site for restoration

LINCOLNSHIRE COUNTY COUNCIL,
MUSEUM OF LINCOLNSHIRE LIFE, BURTON ROAD, LINCOLN — LN1 3LY

Gauge 4ft 8½in www.lincolnshire.gov.uk **SK 972723**

–		4wDM		RH	463154	1961	

Gauge 2ft 6in

–		4wPM		RP	52124	1918	

Gauge 2ft 3in

–		4wDM		RH	192888	1938	

Gauge 2ft 0in

–		4wDM		RH	421432	1959	

LINCOLNSHIRE WOLDS RAILWAY plc,
LUDBOROUGH STATION, STATION ROAD, LUDBOROUGH — DN36 5SQ

Gauge 4ft 8½in www.lincolnshirewoldsrailway.co.uk **TF 309960**

1313		4-6-0	OC	Motala	586	1917	
	SPITFIRE	0-4-0ST	OC	AB	1964	1929	
	LION	0-4-0ST	OC	P	1351	1914	
	FULSTOW	0-4-0ST	OC	P	1749	1928	
	ZEBEDEE	0-6-0T	OC	RSHN	7597	1949	
D3167	(08102)	0-6-0DE		Derby		1955	
–		4wDM		HE	5308	1960	
			reb	Resco	L107	1981	
–		0-4-0DM		JF	4210131	1957	
–		0-4-0DM		JF	4210145	1958	
4		0-4-0DM		RH	375713	1954	
6		0-4-0DM		RH	414303	1957	
7		4wDM		RH	421418	1958	
–	"LITTLE DEBBIE"	0-4-0DE		RH	423657	1958	
	DEBBIE	0-6-0DM		WB	3151	1962	

S. LYON & SON (HAULAGE) LTD,
LINCOLN ROAD, SKELLINGTHORPE, LINCOLN — LN6 5SA

Gauge 4ft 8½in www.slyon.co.uk **SK 930713**

–		4wDM		RH	417889	1958	

M. MAITLAND, THE OLD STATION, FEN ROAD, RIPPINGALE

Gauge 4ft 8½in **TF 115283**

	ELIZABETH	0-4-0ST	OC	AE	1865	1922	
RRM 3	DORA	0-4-0ST	OC	AE	1973	1927	a

a property of J. Scholes

A. NEALE, PRIVATE SITE, Near GAINSBOROUGH

Gauge 1ft 3in

	ANNIE	0-4-0T	OC	FMB	001	1992	a
–		4wPM		L	35811	1950	
–		4wPM		StanhopeT			Dsm

a carries plate DB/1896

NORTH INGS FARM MUSEUM, FEN ROAD, DORRINGTON, near RUSKINGTON — LN4 3QB

Gauge 2ft 0in www.northingsfarmmuseum.co.uk **TF 098527**

No.9	SWIFT	4wVBT	G	HallT	1859401	1994	
	rebuild of [ODDSON]	4wVBT	G	MarshallJ		1970	
–		4wDM		ClayCross		1961	a
1		4wDM		HE	6013	1961	
	BULLFINCH	4wDM		HE	7120	1969	
–		4wDM		MR	7493	1940	
LOD/758022	PENELOPE	4wDM		MR	8826	1943	
–		4wDM		OK		c1932	
–		4wDM		RH	183773	1937	
–		4wDM		RH	200744	1940	
No.1		4wDM		RH	371937	1954	
–		4wDM		RH	375701	1954	Dsm
–		4wDM		RH	421433	1959	

a constructed from parts supplied by Lister

M.C. PALMER, THE GARDEN RAILWAY, METHERINGHAM

Gauge 1ft 3in **Private Site**

No.3	URSULA	0-6-0T	OC	Waterfield		1999	a
No.4	SLUDGE	4wDM		L	41545	1955	

a carries worksplate DB 1916, incorporating parts from original loco

PLEASURE ISLAND FAMILY THEME PARK, KINGS ROAD, CLEETHORPES — DN35 0PL

Gauge 600mm **(Closed)** **TA 323065**

1	ANNABEL	4-4-0DH	s/o	SL	495.10.92	1992

KELVIN SCULLY, LOUTH

Gauge 1ft 3in

2870	CITY OF LONDON	4-6-0DM	s/o	JMR		1978

TONY SINCLAIR, PRIVATE LOCATION, near BROTHERTOFT, BOSTON

Gauge 4ft 8½in

7514	(PWM 4311)	2w-2PMR	Wkm	7514	1956

STATION ROAD STEAM LTD,
UNIT 16, MOORLANDS INDUSTRIAL ESTATE, METHERINGHAM **LN14 3HX**

Miniature locomotives under construction, repair or for resale, usually present.

www.stationroadsteam.com **TF 078614**

TYSDALE FARM, TYDD ST. MARY

Gauge 2ft 0in

–	4wDM		OK	6931	1937
–	4wDM		OK	7734	1938

JAMES WATERFIELD, BOSTON

Gauge 1ft 3in

No.1 EFFIE	0-4-0T	OC	Waterfield		2010

GREATER LONDON
INDUSTRIAL SITES

ALSTOM TRANSPORT (part of Alstom Holdings S.A.)

Locomotives are kept at :-

SPC	Stonebridge Park Carriage Maintenance Depot, Argenta Way, Wembley TQ 191845
SPW	Stonebridge Park Wheel Lathe Depot, Argenta Way, Wembley TQ 191845
W	Wembley Traincare Centre, Argenta Way, Wembley NW10 0RW TQ 192843

Gauge 4ft 8½in www.alstom.com

(D3778) 08611	0-6-0DE		Derby		1959	SPC
(D3863) 08696	0-6-0DE		Hor		1959	W
–	0-4-0DH		HE	9225	1984	
		rep	YEC	L173	2001	W
–	4wDH	R/R	Unilok	4010	2002	SPW

ARRIVA RAIL LONDON LTD

(operating a concession agreement for Transport for London)

New Cross Gate Depot, Juno Way, New Cross **SE14 5RW**

Gauge 4ft 8½in www.arrivaraillondon.co.uk **TQ 359775, 359778**

See Section 7 for details

Willesden Traction & Rolling Stock Maintenance Depot, Station Approach,
Willesden Junction, Willesden, Brent **NW10 4UY**

Gauge 4ft 8½in www.arrivaraillondon.co.uk **TQ 221828**

See Section 7 for details

BAM NUTTALL LTD, CIVIL ENGINEERS, PLANT DEPOT, RAYLAMB WAY, off MANOR ROAD, SLADE GREEN, ERITH — DA8 2LD

Locomotives may be present at this depot between use on contracts.

Gauge 2ft 0in www.bamnuttall.co.uk **TQ 537777**

TO.11	4wBE	CE	B4071.4	1995

BAZALGETTE TUNNEL LTD, t/a TIDEWAY, THAMES TIDEWAY TUNNEL

(25km sewerage storage and stormwater tunnel for Thames Water /2016 to /2024)

Tideway East, Cvjv - Costain Ltd, Vinci Construction Grands Projets and Bachy Soletanche Joint Venture

(Bermondsley to Stratford contract)

Gauge 750mm www.tideway.london

Locomotive identities not yet confirmed

Tideway Central, Ferrovial Agroman UK Ltd and Laing O'Rourke Construction Joint Venture

(Fulham To Blackfriars Contract)

Kirtling Street, Nine Elms, Wandsworth SW8 5BP TQ 293775

Gauge 750mm www.tideway.london

2	4wDH	Schöma	6912	2017
3	4wDH	Schöma	6913	2017
4	4wDH	Schöma	6914	2017
5	4wDH	Schöma	6915	2017
6	4wDH	Schöma	6916	2017
7	4wDH	Schöma	6917	2017
–	4wDH	Schöma	6987	2018
–	4wDH	Schöma	6988	2018
–	4wDH	Schöma	6989	2018
–	4wDH	Schöma	6990	2018

Tideway West, Bam Nuttall Ltd, Morgan Sindall Plc and Balfour Beatty Ltd Joint Venture

(Acton to Fulham Contract)

Locomotive identities not yet confirmed

BOMBARDIER TRANSPORTATION UK LTD, ILFORD DEPOT, LEY STREET, ILFORD — IG1 4BP

Gauge 4ft 8½in www.bombardier.com **TQ 445869**

(D3867) 08700		0-6-0DE		Hor		1960	b
(D4039) 08871	H 074	0-6-0DE		Dar		1960	a
–		4wBE		Niteq	B238	2005	
–		4wBE	R/R	Zephir	2566	2015	
1		4wBE	R/R	Zwiehoff		2017	
2		4wBE	R/R	Zwiehoff		2018	
3		4wBE	R/R	Zwiehoff		2018	

a property of British American Railway Services Ltd, Stanhope, Co. Durham
b property of Harry Needle Railroad Co Ltd, Derbyshire

BOMBARDIER TRANSPORTATION UK LTD,
OLD OAK COMMON DEPOT, OLD OAK COMMON LANE, LONDON NW10 6DW

Gauge 4ft 8½in www.bombardier.com **TQ 217822**

–	4wBE		Zephir	2672	2017	

CROSSRAIL LTD, ATC (Alstom TSO Costain) JV,
PLUMSTEAD SIDINGS, WHITE HART AVENUE, PLUMSTEAD (SE18 1DE)

Gauge 4ft 8½in www.crossrail.co.uk **TQ 456789**

N1	0-6-0DH		Jung	12842	1958	
		reb	Newag	269	2015	
N2	0-6-0DH		Jung	13286	1962	
		reb	Newag		2015	
N3	0-6-0DH		Jung	13289	1962	
		reb	Newag		2015	
4	0-6-0DH		Jung	12347	1956	
		reb	Newag		2016	
(5) PZB-90 98800 170012-5	6wDH		OK	26880	1979	
		reb	Newag	275	2016	
–	4wDH		Schöma	6844	2015	
		a rebuild of	Schöma	6321	2008	
UNIMOG 1 EBA 01D15A 009	4wDM	R/R	_(Unimog	215665	2008	
			(King/Zagro		2015	
UNIMOG 2	4wDM	R/R	_(Unimog	240772		
			(King/Zagro		2015	
3	4wDM	R/R	_(Unimog	240933		
			(King/Zagro	4346	2016	
4	4wDM	R/R	_(Unimog	240938		
			(King/Zagro	4345	2016	
5 16303	4wDM	R/R	_(Unimog	214126	2007	
			(Zagro	3781	2007	a
6	4wDM	R/R	_(Unimog	224139	2010	
			(Zagro			a
7 16305	4wDM	R/R	_(Unimog	226761	c2011	
			(Zagro			a
8	4wDM	R/R	_(Unimog	219528		
			(Zagro			a
CLAYTON 1	4wDH		CE	B4618.1	2016	
CLAYTON 2	4wDH		CE	B4618.2	2016	
CLAYTON 3	4wDH		CE	B4618.3	2016	
CLAYTON 4	4wDH		CE	B4618.4	2016	
CLAYTON 5	4wDH		CE	B4618.5	2016	
CLAYTON 6	4wDH		CE	B4618.6	2016	
CLAYTON 7	4wDH		CE	B4618.7	2016	
E DRA 02	4wDHR		Cometi	1710/1	2005	
		reb	SVI		c2015	
APV 350	4wDHR		SVI	808910	2009	
E DRA 10 GEMINI-01 APV 325	4wDHR		SVI	1602359	2016	
-	2w-2PMR		Bance	278	2016	

a on hire from SAS Lemmonier, Isigny Le Buat, France

EUROSTAR GROUP LTD, EUROSTAR ENGINEERING CENTRE, TEMPLE MILLS DEPOT, 2 ORIENT WAY, LONDON E10 5YA

Gauge 4ft 8½in www.eurostar.com **TQ 372864, 374862**

See Section 7 for details

GOVIA THAMESLINK RAILWAY LTD, t/a THAMESLINK, HORNSEY TRAIN SERVICING CENTRE, HAMPDEN ROAD, HORNSEY N8 0HF

(subsidiary of Govia Ltd, a Go-Ahead Group and Keolis Joint Venture)

Gauge 4ft 8½in www.thameslinkrailway.com **TQ 310892**

See Section 7 for details

GOVIA THAMESLINK RAILWAY LTD, t/a SOUTHERN, SELHURST TRAINCARE DEPOT, SELHURST ROAD, CROYDON SE25 6LJ

(subsidiary of Govia Ltd, a Go-Ahead Group and Keolis Joint Venture)

Gauge 4ft 8½in www.southernrailway.com **TQ 332675**

See Section 7 for details

FIRST MTR SOUTH WESTERN TRAINS LTD, T/A SOUTH WESTERN RAILWAY, WIMBLEDON TRAINCARE DEPOT, DURNSFORD ROAD, WIMBLEDON SW19 8EG

(a First Group and MTR Joint Venture)

Gauge 4ft 8½in www.southwesternrailway.com **TQ 256723**

See Section 7 for details

FORD MOTOR CO LTD, DAGENHAM ENGINE PLANT, THAMES AVENUE, DAGENHAM RM9 6SA

Gauge 4ft 8½in www.corporate.ford.com **TQ 496825, 499827**

1		0-4-0DH		EEV	D1124	1966
2	"HUDSWELL"	0-6-0DH		HC	D1396	1967
			reb	HAB	6385	1996
3	MALCOLM	0-4-0DH		S	10127	1963
			reb	Wilmott		2003
–		4wDH		CE	B4637	2018
GT/PL/1P 260 C		4wDM		Robel	21.11.RK1	1966

HITACHI RAIL EUROPE LTD, NORTH POLE TRAIN MAINTENANCE CENTRE, MITRE WAY, LONDON W3 7DX & W10 6AU

Gauge 4ft 8½in www.hitachirail-eu.com **TQ 219820, 228822**

–	4wBE	Zephir	2508	2014

LOCOMOTIVE SERVICES LTD and WEST COAST RAILWAY COMPANY LTD, SOUTHALL M.P.D., SOUTHALL — UB2 4SE

Locomotives for maintenance, overhaul and repair usually present

Gauge 4ft 8½in www.iconsofsteam.com / www.westcoastrailways.co.uk **TQ 133798**

34067	TANGMERE	4-6-2	3C	Bton		1947
44932		4-6-0	OC	Hor		1945
60009	UNION OF SOUTH AFRICA	4-6-2	3C	Don	1853	1937
(D3948)	08780 FRED	0-6-0DE		Derby		1960
D2447	LORD LEVERHULME	0-4-0DM		AB	388	1953
960 014	977873 (55022) L122	2-2w-2w-2DMR		PSteel		1960
(S65373)	5759	4w-4RER		Elh		1956
S68001	931091	4w-4RE/BER		Afd/Elh		1959
68002	(9002) 931092	4w-4RE/BER		Afd/Elh		1959
(S68008)	(9008) 931098 093	4w-4RE/BER		Afd/Elh		1961
S68009	9009	4w-4RE/BER		Afd/Elh		1961

LONDON NORTH EASTERN RAILWAY LTD, t/a LNER, BOUNDS GREEN MAINTENANCE DEPOT, BRIDGE ROAD, ALEXANDRA PALACE — N22 7SN

Gauge 4ft 8½in www.lner.co.uk **TQ 301907**

See Section 7 for details.

LONDON & SOUTH EASTERN RAILWAY LTD, t/a SOUTHEASTERN SLADE GREEN MAINTENANCE DEPOT, MOAT LANE, ERITH — DA8 2JN

Gauge 4ft 8½in www.southeasternrailway.co.uk **TQ 527759**

See Section 7 for details

J. MURPHY & SONS LTD, PLANT DEPOT, HIGHVIEW HOUSE, HIGHGATE ROAD, KENTISH TOWN — NW5 1TN

Locomotives are present at this depot between use on contracts.

Gauge 2ft 0in www.murphygroup.co.uk **TQ 287855**

PLM11		4wBE		CE	B3329A	1986
			reb	CE	B3791	1991
			reb	CE	B4181	1996
			reb	CE	B4512/3	2010
LM12	JM	4wBE		CE	B3070C	1983
			reb	CE	B3804	1991
			reb	CE	B4512/2	2010
PLM13		4wBE		CE	B3070D	1983
			reb	CE	B3782	1991
			reb	CE	B4181	1996
LM14		4wBE		CE	B3070B	1983
			reb	CE	B3804	1991
			reb	CE	B4181	1996
			reb	CE	B4512/1	2010
PLM15		4wBE		CE	B3070A	1983
			reb	CE	B3782	1991

(LM24)	PLM13	4wBE		CE	B0167	1974	
			reb	CE	B3786A	1991	
			reb	CE	B4512/4	2010	
LM26JM		4wBE		CE	B0145C	1973	
			reb	CE	B3799	1991	
PLM 27		4wDH		Schöma	5694	2001	
PLM 28		4wDH		Schöma	5695	2001	a
PLM 29		4wDH		Schöma	5696	2001	b
PLM 30		4wDH		Schöma	5697	2001	
PLM 31		4wDH		Schöma	5698	2001	
PLM 32		4wDH		Schöma	5699	2001	
PLM 33		4wDH		Schöma	5700	2001	
PLM 34		4wDH		Schöma	5701	2001	
PLM 35		4wDH		Schöma	5702	2001	

a carries worksplate Schöma 5696
b carries worksplate Schöma 5695

PLASSER UK LTD, MANOR ROAD, WEST EALING W13 0PP

New track maintenance vehicles under construction or repair usually present

Gauge 4ft 8½in www.plasser.co.uk **TQ 161809**

HERBERT	0-4-0DM	Bg/DC	2724	1963

PROCAT (PROSPECTS COLLEGE of ADVANCED TECHNOLOGY), TUNNELLING AND UNDERGROUND CONSTRUCTION ACADEMY, ALDERSBROOK SIDINGS, LUGG APPROACH, ILFORD E12 5LN

Gauge 2ft 0in www.tuca.ac.uk **TQ 430862**

2	S238	4wBE	CE	B0471B	1975
–		4wDH	Schöma	5572	1998
–		4wDH	Schöma	6156	2007

ROYAL MAIL LETTERS LTD, THE POST OFFICE UNDERGROUND RAILWAY

(part of London Region of Post Office Letters Ltd)

System closed, on care and maintenance.

Locomotives are kept at various disused sidings and stations throughout the system, including:-

King Edward Building, St Pauls TQ 321816
Mount Pleasant Parcels Office, Clerkenwell TQ 311823
New Western District Office, Rathbone Place TQ 296814

Gauge 2ft 0in

1			4wBE	EEDK	702	1926		
2			4wBE	EEDK	703	1926		
66	[101	104]	2w-2-2-2wRE	_(EE	3335	1962		a
				(EES	8314	1962	OOU	
01	[169	170]	2w-2-2-2wRE	HE	9134	1982	OOU	b
02	[103	104]	2w-2-2-2wRE	GB	420461/2	1980	OOU	
03	[105	106	2w-2-2-2wRE	GB	420461/3	1980	OOU	
04	[107	108]	2w-2-2-2wRE	_(GB	420461/4	1980		
				(HE	9103	1980	OOU	
05	[109	110]	2w-2-2-2wRE	_(GB	420461/5	1980		
				(HE	9104	1980	OOU	

06	[111 112]	2w-2-2-2wRE	(GB	420461/6	1980		
			(HE	9105	1980	OOU	
07	[113 114]	2w-2-2-2wRE	(GB	420461/7	1980		
			(HE	9106	1980	OOU	
08	[115 116]	2w-2-2-2wRE	(GB	420461/8	1980		
			(HE	9107	1980	OOU	
09	[117 118]	2w-2-2-2wRE	(GB	420461/9	1981		
			(HE	9108	1981	OOU	
10	[119 120]	2w-2-2-2wRE	(GB	420461/10	1981		
			(HE	9109	1981	OOU	
11	[121 122]	2w-2-2-2wRE	(GB	420461/11	1981		
			(HE	9110	1981	OOU	
12	[123 124]	2w-2-2-2wRE	(GB	420461/12	1981		
			(HE	9111	1981	OOU	
13	[125 126]	2w-2-2-2wRE	(GB	420461/13	1981		
			(HE	9112	1981	OOU	
14	[127 128]	2w-2-2-2wRE	(GB	420461/14	1981		
			(HE	9113	1981	OOU	
15	[129 130]	2w-2-2-2wRE	(GB	420461/15	1981		
			(HE	9114	1981	OOU	
16	[131 132]	2w-2-2-2wRE	(GB	420461/16	1981		
			(HE	9115	1981	OOU	
17	[133 134]	2w-2-2-2wRE	(GB	420461/17	1981		
			(HE	9116	1981	OOU	
18	[135 136]	2w-2-2-2wRE	(GB	420461/18	1981		
			(HE	9117	1981	OOU	
19	[137 138]	2w-2-2-2wRE	(GB	420461/19	1981		
			(HE	9118	1981	OOU	
20	[139 140]	2w-2-2-2wRE	(GB	420461/20	1981		
			(HE	9119	1981	OOU	
23	[145 146]	2w-2-2-2wRE	(GB	420461/23	1981		
			(HE	9122	1981	OOU	
24	[147 148]	2w-2-2-2wRE	(GB	420461/24	1981		
			(HE	9123	1981	OOU	
25	[149 150]	2w-2-2-2wRE	(GB	420461/25	1981		
			(HE	9124	1981	OOU	
26	[151 152]	2w-2-2-2wRE	(GB	420461/26	1981		
			(HE	9125	1981	OOU	
27	[153 154]	2w-2-2-2wRE	(GB	420461/27	1982		
			(HE	9126	1982	OOU	
28	[155 156]	2w-2-2-2wRE	(GB	420461/28	1982		
			(HE	9127	1982	OOU	
29	[157 158]	2w-2-2-2wRE	(GB	420461/29	1982		
			(HE	9128	1982	OOU	
30	[159 160]	2w-2-2-2wRE	(GB	420461/30	1982		
			(HE	9129	1982	OOU	
31	[161 162]	2w-2-2-2wRE	(GB	420461/31	1982		
			(HE	9130	1982	OOU	
32	[163 164]	2w-2-2-2wRE	(GB	420461/32	1982		
			(HE	9131	1982	OOU	
33	GREAT EAST EXPRESS	2w-2-2-2wRE	(GB	420461/33	1982		
	[165 166]		(HE	9132	1982	OOU	
34	[167 168]	2w-2-2-2wRE	(GB	420461/34	1982		
			(HE	9133	1982	OOU	
752	[101 102]	2w-2-2-2wRE	EEDK	752	1930	OOU	c
–		2w-2-2-2wRE	EEDK	753	1930		d
35	[107 108]	2w-2-2-2wRE	EEDK	755	1930	OOU	

36	[105 106]	2w-2-2-2wRE	EEDK	756	1930	OOU	
759	[110 115]	2w-2-2-2wRE	EEDK	759	1930	OOU	c
39	[121 122]	2w-2-2-2wRE	EEDK	762	1930	OOU	
763	[123 124]	2w-2-2-2wRE	EEDK	763	1930	OOU	c
793	[183 184]	2w-2-2-2wRE	EEDK	793	1930	OOU	c
795	[187 188]	2w-2-2-2wRE	EEDK	795	1930	OOU	c
797	[191 192]	2w-2-2-2wRE	EEDK	797	1930	OOU	
799	[195 196]	2w-2-2-2wRE	EEDK	799	1930	OOU	c
802	[201 202]	2w-2-2-2wRE	EEDK	802	1930	OOU	c
804	[205 206]	2w-2-2-2wRE	EEDK	804	1930	OOU	c
41	[207 208]	2w-2-2-2wRE	EEDK	805	1930	OOU	
810	[217 218]	2w-2-2-2wRE	EEDK	810	1930	OOU	
43	[219 220]	2w-2-2-2wRE	EEDK	811	1930	OOU	
813	[223 224]	2w-2-2-2wRE	EEDK	813	1930	OOU	c
45	[226]	2-2wRE	EEDK	814	1930	OOU	e
46	[227 228]	2w-2-2-2wRE	EEDK	815	1930	OOU	
816	[229 230]	2w-2-2-2wRE	EEDK	816	1930	OOU	c
817	[231 232]	2w-2-2-2wRE	EEDK	817	1930	OOU	c
818	[233 234]	2w-2-2-2wRE	EEDK	818	1930	OOU	c
820	[237 238]	2w-2-2-2wRE	EEDK	820	1931	OOU	c
–	[239 240]	2w-2-2-2wRE	EEDK	821	1931	OOU	d
822	[241 242]	2w-2-2-2wRE	EEDK	822	1931	OOU	c
826	[249 250]	2w-2-2-2wRE	EEDK	826	1931	OOU	c
49	[251 252]	2w-2-2-2wRE	EEDK	827	1931	OOU	
830	[257 258]	2w-2-2-2wRE	EEDK	830	1931	OOU	c
925	[445 446]	2w-2-2-2wRE	EEDK	925	1936	OOU	c
51	[457 458]	2w-2-2-2wRE	EEDK	931	1936	OOU	
932	[459 460]	2w-2-2-2wRE	EEDK	932	1936	OOU	c
50	[235 451]	2w-2-2-2wRE	EE			OOU	f
55	[236 452]	2w-2-2-2wRE	EE			OOU	f

a also contains parts of EE 3334-EES 8313 / 1962
b supplied as spare power units for 01 to 34 (orig 501 to 534) but now forming 01
c stored in a disused tunnel at Rathbone Place
d converted to a passenger car
e one power unit only; other preserved at Mail Rail exhibition
f contains a power unit from EEDK 819/1930 and EEDK 928/1936

SIEMENS AG, SIEMENS RAIL SYSTEMS, STRAWBERRY HILL TRAINCARE FACILITY, off SHACKLEGATE LANE, TEDDINGTON — TW11 8SF

Gauge 4ft 8½in www.siemens.com **TQ 154720**

3417	62236 GORDON PETTITT	4-4wRER	York		1969

TRANSPORT for LONDON, DOCKLANDS LIGHT RAILWAY LTD,

(Operated by Keolis Amey Docklands Ltd)

Locomotives are kept at :
Castor Lane, Poplar E14 0DS TQ 376806
Armada Way, Beckton E6 7FB TQ 442812

Gauge 4ft 8½in www.tfl.gov.uk

(992)	92	4wDM	Wkm	11622	1986

993	KYLIE	4wBE/RE		RFSK	V339	1991
			reb	HE	9385	2014
994	KEVIN KEARNEY	0-4-0DH		GECT	5577	1979

TRANSPORT for LONDON, LONDON TRAMLINK, THERAPIA LANE DEPOT, COOMBER WAY, CROYDON CR0 4TQ

Gauge 4ft 8½in www.tfl.gov.uk TQ 300669

-	4wBE	R/R	Zephir	2331	2011

TRANSPORT for LONDON, LONDON UNDERGROUND LTD

London underground railways maintenance. www.tfl.gov.uk

Locomotives are kept at :-

AW	Acton Works, Bollo Lane W3 8BZ	TQ 196791
EC	Ealing Common Depot, Uxbridge Road W5 3PA	TQ 189802
GG	Golders Green Depot, Finchley Road NW11 7NU	TQ 253875
LB	Lillie Bridge Depot SW6 1TP	TQ 250782
N	Neasden Depot NW10 1PH	TQ 206858
NF	Northfields Depot, Northfields Avenue W5 4UB	TQ 167789
NP	Northumberland Park Depot N17 0XE	TQ 349907
U	Upminster Depot RM14 1XL	TQ 570871
WR	West Ruislip Depot HA4 6NS	TQ 094862

Locomotives may also be found between duties at :-

Cockfosters Depot N14 4UT	TQ 288962
Hainault Depot IG6 2UU	TQ 450918
Hammersmith Depot W6 7PY	TQ 234787
Highgate Depot (N10 3JN)	TQ 279886
London Road Depot SE1 6LW	TQ 315793
Morden Depot SM4 5PT	TQ 255680
Stonebridge Park Depot NW10 0RL	TQ 192845
Stratford Market Depot E15 2SP	TQ 388836

Gauge 4ft 8½in

1 (L1)	BRITTA LOTTA		4wDH	Schöma	5403	1996
2 (L2)	NIKKI		4wBE/RE	CE	B4607.02	2015
		a rebuild of	4wDH	Schöma	5404	1996
3 (L3)	CLAIRE		4wDH	Schöma	5405	1996
4 (L4)	PAM		4wBE/RE	CE	B4607.04	2015
		a rebuild of	4wDH	Schöma	5406	1996
5 (L5)	SOPHIE		4wBE/RE	CE	B4607.05	2015
		a rebuild of	4wDH	Schöma	5407	1996
6 (L6)	DENISE		4wBE/RE	CE	B4607.06	2015
		a rebuild of	4wDH	Schöma	5408	1996
7 (L7)	ANNEMARIE		4wBE/RE	CE	B4607.07	2015
		a rebuild of	4wDH	Schöma	5409	1996
8 (L8)	EMMA		4wBE/RE	CE	B4607.08	2015
		a rebuild of	4wDH	Schöma	5410	1996
9 (L9)	DEBORA		4wDH	Schöma	5411	1996
10 (L10)	CLEMENTINE		4wBE/RE	CE	B4607.10	2015
		a rebuild of	4wDH	Schöma	5412	1996
11 (L11)	JOAN		4wBE/RE	CE	B4607.11	2015
		a rebuild of	4wDH	Schöma	5413	1996
12 (L12)	MELANIE		4wDH	Schöma	5414	1996
13 (L13)	MICHELE		4wBE/RE	CE	B4607.13	2015
		a rebuild of	4wDH	Schöma	5415	1996

14 (L14)	CAROL	4wBE/RE	CE	B4607.14	2015
	a rebuild of	4wDH	Schöma	5416	1996
L 15	69015 97715	4w-4wBE/RE	MetCam		1970
		reb	Acton		2018
L 16	69016 (97716)	4w-4wBE/RE	MetCam		1970
		reb	Acton		2017
L 17	(97717)	4w-4wBE/RE	MetCam		1970
L 18	69018 (97718)	4w-4wBE/RE	MetCam		1970
		reb	Acton		2013
L 19	69019 (97719)	4w-4wBE/RE	MetCam		1970
		reb	Acton		2016
L 20	64020 97720	4w-4wBE/RE	MetCam		1964
		reb	Acton		2017
L 21	64021 (97721)	4w-4wBE/RE	MetCam		1964
		reb	Acton		2018
L 22	(97722)	4w-4wBE/RE	MetCam		1965
		reb	Acton		2014
L 23	(97723)	4w-4wBE/RE	MetCam		1965
		reb	Acton		2012
L 24	64024 (97724)	4w-4wBE/RE	MetCam		1965
		reb	Acton		2009
L 25	64025 (97725)	4w-4wBE/RE	MetCam		1965
		reb	Acton		2010
L 26	64026 (97726)	4w-4wBE/RE	MetCam		1965
		reb	Acton		2012
L 27	64027 (97727)	4w-4wBE/RE	MetCam		1965
		reb	Acton		2011
L 28	64028 (97728)	4w-4wBE/RE	MetCam		1965
		reb	Acton		2013
L 29	64029 (97729)	4w-4wBE/RE	MetCam		1965
L 30	64030 (97730)	4w-4wBE/RE	MetCam		1965
		reb	Acton		2012
L 31	(97731)	4w-4wBE/RE	MetCam		1965
		reb	Acton		2012
L 32	64032 97732	4w-4wBE/RE	MetCam		1965
		reb	Acton		2014
L 44	73044	4w-4wBE/RE	Don	L44	1973
		reb	Acton		2015
L 45	73045	4w-4wBE/RE	Don	L45	1974
		reb	Acton		2017
L 46	73046	4w-4wBE/RE	Don	L46	1974
		reb	Acton		2016
L 47	73047 97747	4w-4wBE/RE	Don	L47	1974
		reb	Acton		2018
L 48	73048	4w-4wBE/RE	Don	L48	1974
		reb	Acton		2017
L 49	73049	4w-4wBE/RE	Don	L49	1974
		reb	Acton		2014
L 50	73050 97750	4w-4wBE/RE	Don	L50	1974
		reb	Acton		2016
L 51	97751	4w-4wBE/RE	Don	L51	1974
		reb	Acton		2014
L 52	73052 97752	4w-4wBE/RE	Don	L52	1974
		reb	Acton		2017
L 53	73053 97753	4w-4wBE/RE	Don	L53	1974
		reb	Acton		2016

L 54		4w-4wBE/RE		Don	L54	1974	
			reb	Acton		2017	
L 132	(3901)	4w-4wRE		Cravens		1960	a
			reb	Derby		1987	
L 133	(3905)	4w-4wRE		Cravens		1960	a
			reb	Derby		1987	
PC2	6724	4wDH		Schöma	6724	2014	b
PC1	6725	4wDH		Schöma	6725	2014	b
–		4wBE		Harmill	1	2013	N
–		4wBE	R/R	Harmill	2	2013	N
–		4wBE		Harmill	(3?)	c2014	EC
–		4wBE		Niteq	B184	2001	NP
–		4wBE		Niteq	B222	2005	NF
–		4wBE		Niteq	B285	2009	N
–		4wBE		Niteq	B379	2016	U
–		4wBE		SET		2003	GG
–		4wBE	R/R	Zephir	2556	2015	AW
1406		2-2w-2w-2RER		MetCam		1962	c
1407		2-2w-2w-2RER		MetCam		1962	c
1441		2-2w-2w-2RER		MetCam		1962	c
1570		2-2w-2w-2RER		MetCam		1963	c
1681		2-2w-2w-2RER		MetCam		1963	c
1682		2-2w-2w-2RER		MetCam		1963	c
1690		2-2w-2w-2RER		MetCam		1963	OOU
1691		2-2w-2w-2RER		MetCam		1963	
3007		2-2w-2w-2RER		MetCam		1967	d
3022		2-2w-2w-2RER		MetCam		1968	d
3079		2-2w-2w-2RER		MetCam		1969	e
3107		2-2w-2w-2RER		MetCam		1967	d
3122		2-2w-2w-2RER		MetCam		1968	d
3179		2-2w-2w-2RER		MetCam		1969	e
3213		2-2w-2w-2RER		MetCam		1972	e
3229		2-2w-2w-2RER		MetCam		1973	f
3313		2-2w-2w-2RER		MetCam		1972	e
3329		2-2w-2w-2RER		MetCam		1973	f
9125		2-2w-2w-2RER		MetCam		1961	c
9441		2-2w-2w-2RER		MetCam		1962	c
9459		2-2w-2w-2RER		MetCam		1962	c
9577		2-2w-2w-2RER		MetCam		1963	c
9691		2-2w-2w-2RER		MetCam		1963	c
92442		4w-4wRER		BRE(D)		1994	g
93442		4w-4wRER		BRE(D)		1994	g
AC795709		4wBER		Bance	068/98	1998	
AC795701		2w-2BER		Bance	070/98	1998	
AC795704		2w-2BER		Bance	071/98	1998	
AC795703		2w-2BER		Bance	072/98	1998	
AC795705		2w-2BER		Bance	073/98	1998	
AC795706		2w-2BER		Bance	074/98	1998	
AC795702		2w-2BER		Bance	075/98	1998	
AC795710		2w-2BER		Bance	076/99	1998	
–		2w-2BER		Bance	077/99	1999	
AC795707		2w-2BER		Bance	078/99	1999	

AC795708	2w-2BER	Bance	079/99	1999
AC790309	2w-2BER	Bance	107/03	2003
AC790310	2w-2BER	Bance	108/03	2003
–	2w-2BER	Bance	263	2013
AC795713	2w-2BER	Consillia	05/001	2005
MTP106002	2w-2BER	Consillia	05/002	2005
AC795711	2w-2BER	Consillia	05/003	2005
AC795712	2w-2BER	Consillia	05/004	2005
MTP106003	2w-2BER	Consillia	05/005	2005
MTP106005	2w-2BER	Consillia	05/007	2005
MTP106001	2w-2BER	Consillia	05/008	2005
AC795724	2w-2BER	Consillia	06/009	2006
AC795723	2w-2BER	Consillia	06/010	2006
AC795722	2w-2BER	Consillia	06/011	2006
AC795721	2w-2BER	Consillia	06/013	2006
AC795716	2w-2BER	Consillia	06/014	2006
AC795717	2w-2BER	Consillia	06/017	2006
AC795714	2w-2BER	Consillia	06/018	2006
AC795720	2w-2BER	Consillia	06/019	2006
AC795726	2w-2BER	Consillia	13/0046	2013
AC795725	2w-2BER	Consillia	13/0047	2013
MEC3016003	2w-2BER	Consillia	14/0049	2014
MEC3016002	2w-2BER	Consillia	14/0050	2014
MEC3016001	2w-2BER	Consillia	14/0051	2014
AC304101	2w-2BER	Consillia	15/0055	2015

a track recording pilot motor cars
b rail grinding train power cars
c Sandite Car Unit
d set aside for conversion to replace TCC1/TCC5
e asset inspection train
f training and filming unit (Aldwych Branch)
g Sandite / ATO Test Unit

TRANSPORT for LONDON, FLEET MANAGEMENT, ACTON WORKS, 130 BOLLO LANE, ACTON — W3 8BZ

Gauge 4ft 8½in — **TQ 196791**

C622 EWT L85	4wDM	R/R	_(Unimog	126262	1986
			(Zweiweg	1130	1986

PRESERVATION SITES

C. BLUMSOM, BLUMSOM GROUP, BLUMSOM TIMBER CENTRE, MAPLE WHARF, 36-38 RIVER ROAD, BARKING — IG11 0DN

Gauge 1524mm www.blumsomtimbercentre.com **TQ 455824**

794	0-6-0T	OC	TK	350	1925

COOPERS LANE PRIMARY SCHOOL, GROVE PARK, LEWISHAM — SE12 OLF

Gauge 4ft 8½in www.cooperslane.lewisham.sch.uk **TQ 405728**

7027	2-2w-2w-2RER	MetCam		1980	Dsm

CRAVENS HERITAGE TRAINS LTD

c/o London Underground Ltd, Northfields Depot, London — W5 4UB

Gauge 4ft 8½in www.cravensheritagetrains.co.uk **TQ 167789**

3906	4w-4wRER	Cravens		1960
3907	4w-4wRER	Cravens		1960

c/o London Underground Ltd, Hainault Depot — IG6 2UU

Gauge 4ft 8½in www.cravensheritagetrains.co.uk **TQ 450918**

1506	2-2w-2w-2RER	MetCam		1962
1507	2-2w-2w-2RER	MetCam		1962
9507	2-2w-2w-2RER	MetCam		1962

CROSSNESS ENGINE TRUST, THE CROSSNESS PUMPING STATION, THE OLD WORKS, CROSSNESS SEWAGE TREATMENT WORKS, BELVEDERE ROAD, ABBEY WOOD — SE2 9AQ

Gauge 2ft 0in www.crossness.org.uk **TQ 485810**

"BUSY BASIL"	4wDH	s/o	SL		1986	

Gauge 1ft 6in

No. 1 WOOLWICH	0-4-0T	OC	AE	1748	1916	

EUROSTORAGE LTD, TAVISTOCK ROAD, WEST DRAYTON — UB7 7QT

Gauge 4ft 8½in www.eurostorage.co.uk **TQ 054802**

21147	4w-4wRER	MetCam		1949	OOU

THE FRIENDS OF THE PUMP HOUSE, WALTHAMSTOW PUMP HOUSE MUSEUM, 10 SOUTH ACCESS ROAD, LOW HALL LANE, WALTHAMSTOW — E17 8AX

Gauge 4ft 8½in www.e17pumphouse.org.uk **TQ 363882**

3186 (3122)	2-2w-2w-2RER	MetCam		1968

GREAT ORMOND STREET CHILDRENS HOSPITAL — WC1N 3JH

Gauge 4ft 8½in www.gosh.nhs.uk **TQ 305821**

3634	4w-4wRER	MetCam		1988	Dsm

Reduced in length to 32ft and converted to hospital radio studio.

G.W.R. PRESERVATION GROUP LTD, SOUTHALL RAILWAY CENTRE, GLADE LANE, SOUTHALL — UB2 4SE

Gauge 4ft 8½in www.gwrpg.co.uk **TQ 133798**

9682		0-6-0PT	IC	Sdn		1949	
A.E.C. No.1		4wDM		AEC		1938	
ARMY 251	FRANCIS BAILY OF THATCHAM	0-4-0DM		RH	390772	1956	
–		4wDM		RH	416568	1957	
–		4wDMR		BD	3706	1975	
–		2w-2PMR		DC	1895	1950	DsmT

THE HAMPTON & KEMPTON WATERWORKS RAILWAY LTD, c/o KEMPTON STEAM ENGINES TRUST, KEMPTON PARK WATER TREATMENT WORKS, SNAKEY LANE, HANWORTH — TW13 6XH

Private site, operates on advertised public opening dates only

Gauge 2ft 0in www.hamptonkemptonrailway.org.uk **TQ 109709**

3	DARENT		0-4-0T	OC	AB	984	1903
		rebuilt as	0-4-0ST	OC	Provan Group		2003
	HOUNSLOW		4wPH		SPL	No.1	2008
–			4wDH		HE	9338	1994
	SPELTHORNE		4wDH		HE	9357	1994
–			4wDM		MR	4023	1926

KINGS COLLEGE, STRAND — WC2R 2LS

Gauge 1ft 3in www.kcl.ac.uk **TQ 308808**

PEARL	2-2-2WT	IC	B(C)		1860

LONDON MUSEUM OF WATER & STEAM, KEW BRIDGE WATERWORKS RAILWAY, GREEN DRAGON LANE, BRENTFORD — TW8 0EN

Gauge 2ft 0in www.waterandsteam.org.uk **TQ 188780**

	THOMAS WICKSTEED	0-4-0ST	OC	_(Kew		1995	
				(HE	3906	2009	a
2	ALISTER	4wDM		L	44052	1958	

a locomotive part built at Kew and completed by HE

NETWORK RAIL, STEWARTS LANE DEPOT, DICKENS STREET, NINE ELMS — SW8 3EP

Privately preserved locomotives and multiple units based here.

Gauge 4ft 8½in **TQ 288766**

3053	CAR No.92	4w-4wRER	MetCam	1932
3053	CAR No.93	4w-4wRER	MetCam	1932

NEWHAM BOROUGH COUNCIL,
MERIDIAN SQUARE, STRATFORD STATION, STRATFORD E15 1XD

Gauge 4ft 8½in **TQ 386843**

ROBERT	0-6-0ST	OC	AE	2068	1933	

POSTAL HERITAGE TRUST, THE POSTAL MUSEUM and MAIL RAIL,
FREELING HOUSE, PHOENIX PLACE, MOUNT PLEASANT, LONDON WC1X ODA

Gauge 2ft 0in www.postalmuseum.org **TQ 309823**

–		4w Atmospheric Car			1861	
–		4wRE	EEDK	601	1926	
3		4wBE	EEDK	704	1926	
45	[225]	2w-2RE	EEDK	814	1930	a
824	[245 246]	2w-2-2-2wRE	EEDK	824	1931	
(21)	141 142	2w-2-2-2wRE	(GB	420461/21	1981	
			(HE	9120	1981	

a single power unit only

Gauge 2ft 0in - Mail Rail experience

–	4-4w-4-4BER	SL	21625/21626	2016
–	4-4w-4-4BER	SL	21630/21631	2016

THE ROYAL GREENWICH TRUST SCHOOL,
FERRANTI CLOSE, GREENWICH, LONDON SE7 8LJ

Gauge 4ft 8½in www.rgts17.rgtrustchool.net **TQ 417790**

5701	4w-4wRER	MetCam	1977

SCIENCE MUSEUM, EXHIBITION ROAD, SOUTH KENSINGTON SW7 2DD

Gauge 5ft 0in www.sciencemuseum.org.uk **TQ 268793**

"PUFFING BILLY"	4w	VCG	Hedley		1827-1832	a

a incorporates parts of locomotive of same name built c1814

Gauge 4ft 8½in

1868	2-2-2	OC	Crewe	20	1845

TRANSPORT for LONDON, LONDON TRANSPORT MUSEUM
Covent Garden Piazza WC2E 7BB

Gauge 4ft 8½in www.ltmuseum.co.uk

23		4-4-0T	OC	BP	710	1866	
5	JOHN HAMPDEN	Bo-BoRE		VL		1922	
4248		4w-4RER		GRC&W	O/5026	1924	
11182		2-2w-2w-2RER		MetCam		1939	
No.13		4wRE		(BP		1890	
				(M&P		1890	a

a carried number plate from No.1 until probable identity established c1990

The Depot, Museum Depot, Gunnersbury Lane, Acton Town **W3 9BQ**
Gauge 4ft 8½in www.ltmuseum.co.uk **TQ 193798**

4416		2w-2-2-2wRE		BRCW		1939	a
ESL107		4w-4-4-4wRE		_(BRCW		1903	
				‾(MetAmal		1903	
			reb	Acton		1939	
5034		4w-4RER		Cravens		1962	
61		4w-4wRER		EEDK	1151	1940	
L35		4w-4wBE/RE		GRC&W	O/7903	1938	
4184		4w-4RER		GRC&W	O/5026	1924	a
4417	(L127)	2w-2-2-2wRE		GRC&W	O/8060	1939	
(3370)	L134	4w-4RE		MetC&W		1927	
			reb	Acton		1967	
3327		4w-4RER		MetCam		1929	
(3693)	L131	4w-4RE		MetCam		1934	
			reb	Acton		1967	
12048		4w-4wRER		MetCam		1938	
10012		4w-4wRER		MetCam		1938	
11012		4w-4wRER		MetCam		1938	
22679		2w-2-2-2wRER		MetCam		1952	
16		4w-4wRER		MetCam		1986	
3052		4w-4wRER		MetCam		1968	
3530		4w-4wRER		MetCam		1972	
3734		4w-4wRER		MetCam		1988	
5721		4w-4wRER		MetCam		1978	

a currently under restoration at London Underground Ltd, Acton Works

VILLAGE UNDERGROUND,
54 HOLYWELL LANE, GREAT EASTERN STREET, SHOREDITCH **EC2A 3PQ**
Gauge 4ft 8½in www.villageunderground.co.uk **TQ 334823**

3662	4w-4wRER	MetCam		1988
3733	4w-4wRER	MetCam		1988

GREATER MANCHESTER

INDUSTRIAL SITES

ALSTOM TRANSPORT,
Manchester Traincare Centre, Kirkmanshulme Lane, Longsight, Manchester
(part of Alstom Holdings S.A.) **M12 4HR**
Gauge 4ft 8½in www.alstom.com **SJ 868960**

(D3566)	08451	LONGSIGHT T.M.D.	0-6-0DE		Derby		1958

Longsight Wheel Lathe Depot, New Bank Street, Longsight **M12 4HW**
Gauge 4ft 8½in **SJ 865962**

–	4wDH	R/R	Unilok	4009	2002

ASHTON PACKET BOAT CO LTD,
ASHTON BOATYARD, HANOVER STREET NORTH, GUIDE BRIDGE

Gauge 2ft 0in **SJ 920978**

–	4wDM		FH	2325	1941	a
–	4wDM		HE	2820	1943	
–	4wDM		HE	6012	1960	
–	4wDM		RH	200761	1941	
6	4wBE		WR	7967	1978	

a stored on behalf of owners.

BIFFA WASTE SERVICES LTD, OLDHAM ROAD, MANCHESTER M40 8RR

Gauge 4ft 8½in www.biffa.co.uk **SJ 859999**

See Appendix 1 for details

D.C.T. CIVIL ENGINEERING, PLANT DEPOT, ASHCROFT HOUSE, BREDBURY PARKWAY, BREDBURY PARK INDUSTRIAL ESTATE, BREDBURY, STOCKPORT (ceased trading) SK6 2SN

Locomotives may be present in this yard between use on contracts.

Gauge 2ft 0in

03	4wBE		CE	B1524/2	1977	
–	4wBE		CE			

V.J. DONEGHAN (PLANT) LTD, TUNNELLING & CIVIL ENGINEERING CONTRACTORS, BIRD HALL INDUSTRIAL PARK, BIRD HALL LANE, STOCKPORT SK3 0WT

Locomotives are present in this yard between use on contracts.

Gauge 1ft 6in www.doneghan.co.uk **SJ 879886**

PL 158	4wBE		CE	5640	1969	
2	4wBE		CE			
–	4wBE		CE			
4	4wBE		CE			

Three of the four battery boxes used by the above locomotives carry CE plates:- 5841/68, 5628/69 and 5866/71 respectively.

FINETURRET LTD,
UNIT 1, CROFT INDUSTRIAL ESTATE, CROFT LANE, BURY BL9 8QG

Locomotives are present in this yard between use on contracts

Gauge 2ft 0in www.fineturret.co.uk **SD 812084**

–	4wBE		CE	5590	1969	
		reb	CE	B4174B	1996	
–	4wBE		CE	5590/6	1969	
		reb	CE	B4174A	1996	
4	4wBE		CE	(5949)	1972	a

a one of CE 5949A or CE 5949B

JOHN MARROW, BRYN ENGINEERING SERVICES, UNIT C, SCOT LANE INDUSTRIAL ESTATE, SCOT LANE, BLACKROD, BOLTON BL6 5SL

Private Site. Other locomotives occasionally present for overhaul, repairs, etc. **SD 624090**

Gauge 4ft 8½in

45		0-4-0ST	OC	AB	2352	1954
–		0-6-0ST	OC	AE	1600	1912
WD 71499		0-6-0ST	IC	HC	1776	1944
S112	REVENGE	0-6-0ST	IC	HE	2414	1941
	SIR ROBERT PEEL	0-6-0ST	IC	HE	3776	1952
35	NORMAN	0-6-0ST	IC	RSHN	7086	1943

MARSH TRACKWORKS - TRACK RECOVERY SERVICES, PARK ROAD, TRAFFORD PARK, STRETFORD M32 8RA

Gauge 4ft 8½in **SJ 790955**

9036		2w-2PMR		Wkm	8196	1958

RILEY & SON (E) LTD, BUCKLEY WELLS LOCOMOTIVE WORKS, BURY BL9 0TY

Locomotives usually present for maintenance, repair and restoration.

Gauge 4ft 8½in www.rileysuk.com **SD 799101**

34067	TANGMERE	4-6-2	3C	Bton		1947
(35009)	(SHAW SAVILL)	4-6-2	3C	Elh		1942

RILEY & SON (E) LTD, PREMIER LOCOMOTIVE WORKS, SEFTON STREET, HEYWOOD OL10 2JF

Locomotives usually present for maintenance, repair and restoration.

Gauge 4ft 8½in www.rileysuk.com **SD 863101**

45212		4-6-0	OC	AW	1253	1935
47298		0-6-0T	IC	HE	1463	1924

TRANSPORT for GREATER MANCHESTER, METROLINK

(Operated by Keolis Amey Metrolink Ltd)

Ayres Road Depot, Old Trafford, Manchester M16 0PX

Gauge 4ft 8½in www.tfgm.com **SJ 813957**

–		4wBE		Zephir	2385	2012

Queens Road Depot, Cheetham Hill, Manchester M8 0RY

Gauge 4ft 8½in www.tfgm.com **SD 848005**

1027		4wDH		_(GECT	5862	1991
				(RFSK	V337	1991
	L253 HKK	4wDM	R/R	Multicar	000406	1994

TXM PLANT LTD, t/a TXM RAIL, HASLEMERE INDUSTRIAL ESTATE, PARK LANE, WIGAN ROAD, WIGAN WN4 0EL

Headquarters : The Grange, Harnet Drive, Wolverton Mill, Milton Keynes, Buckinghamshire MK12 5NE

Gauge 4ft 8½in www.txmrail.co.uk **SD 569017**

7146	99709 977048-9		4wDM	R/R	Unimog	072061	1981	a
7097	99709 977038-7		4wDM	R/R	Unimog	107598	c1983	a
7150	99709 977046-0		4wDM	R/R	Unimog	181336	1995	a
7147	99709 977044-5		4wDM	R/R	Unimog	184949	1999	a
7149	99709 977045-2		4wDM	R/R	Unimog	185657	c1999	a
5020	99709 979023-7	FA51 XOV	4wDM	R/R	Unimog	199725	2001	a

a based here but work at other locations as required

SIEMENS AG, SIEMENS RAIL SYSTEMS, ARDWICK TRAINCARE FACILITY, RONDIN ROAD, ARDWICK, MANCHESTER M12 6BF

Gauge 4ft 8½in www.siemens.com **SJ 864972**

(01551)	LANCELOT	0-4-0DH	EEV	D1122	1966
–		4wBE	CPM	1731.02UK	2005

A.E. YATES LTD, CRANFIELD ROAD, LOSTOCK INDUSTRIAL ESTATE, LOSTOCK, BOLTON BL6 4SB

Locomotives are present in this yard between contracts.

Gauge 2ft 0in www.aeyates.co.uk **SD 650098**

5137		4wBE		CE	5866A	1971
–		4wBE		CE		

Gauge 1ft 6in

2019	29	0-4-0BE		WR		
–		4wBE		CE	B2200B	1979
			reb	CE	B3990	1993

PRESERVATION SITES

EAST LANCASHIRE LIGHT RAILWAY CO LTD

Locomotives are kept at :-

Bolton Street Station, Bury BL9 0EY SD 802107
Buckley Wells Carriage Shops, Bury BL9 0TY SD 800103
Bury Diesel Depot BL9 0LN SD 799101
Castlecroft Road Goods Shed, Bury BL9 0LN SD 803109

Gauge 4ft 8½in www.eastlancsrailway.org.uk

3855		2-8-0	OC	Sdn		1942
7229		2-8-2T	OC	Sdn		1935
34092	CITY OF WELLS	4-6-2	3C	Bton		1949
(42765)	13065	2-6-0	OC	Crewe	5757	1927
44871		4-6-0	OC	Crewe		1945
45407	THE LANCASHIRE FUSILIER	4-6-0	OC	AW	1462	1937
(46428)		2-6-0	OC	Crewe		1948
47324		0-6-0T	IC	NBH	23403	1926

80097		2-6-4T	OC	Bton		1954	
No.1		0-4-0ST	OC	AB	1927	1927	
752		0-6-0ST	IC	BP	1989	1881	
(32)	1 (GOTHENBURG)	0-6-0T	IC	HC	680	1903	
WD 132	SAPPER	0-6-0ST	IC	HE	3163	1944	
			reb	HE	3885	1964	
–		0-4-0ST	OC	P	1370	1915	a
(D99)	45135 3rd CARABINIER	1Co-Co1DE		Crewe		1961	
(D120)	45108	1Co-Co1DE		Crewe		1961	
(D306)	40106	1Co-Co1DE		(EE	2726	1960	
	ATLANTIC CONVEYOR			(RSHD	8136	1960	
(D335)	40135	1Co-Co1DE		(EE	3081	1961	
				(VF	D631	1961	
(D)345	(40145)	1Co-Co1DE		(EE	3091	1961	
				(VF	D641	1961	
(D415)	50015 VALIANT	Co-CoDE		(EE	3785	1967	
				(EEV	D1156	1967	
D832	ONSLAUGHT	B-BDH		Sdn		1960	
D1041	WESTERN PRINCE	C-CDH		Crewe		1962	
D1501	(47402) (GATESHEAD)	Co-CoDE		BT	343	1962	
(D1643	47059 47631) 47765	Co-CoDE		Crewe		1964	
D2062	(03062)	0-6-0DM		Don		1959	
(D2956	01003)	0-4-0DM		AB	398	1956	
(D2997)	07013	0-6-0DE		RH	480698	1962	
(D3232)	08164 PRUDENCE	0-6-0DE		Dar		1956	
(D3594	08479) 13594	0-6-0DE		Hor		1958	
(D4112)	09024	0-6-0DE		Hor		1961	
(D4174)	08944	0-6-0DE		Dar		1962	
D5054	(24054) PHIL SOUTHERN	Bo-BoDE		Crewe		1959	
(D5533	31115) 31466	A1A-A1ADE		BT	132	1959	
D5705	(TDB 968006) FRANK	Co-BoDE		MV		1958	
(D6525)	33109 CAPTAIN BILL SMITH R.N.R.						
		Bo-BoDE		BRCW	DEL117	1960	
(D)6536	(33117) ETHEL 4	Bo-BoDE		BRCW	DEL128	1960	
(D6564	33046)	Bo-BoDE		BRCW	DEL156	1961	
(D6809)	37109	Co-CoDE		(EE	3238	1963	
				(VF	D763	1963	
(D6823	37123) 37679	Co-CoDE		(EE	3268	1963	
				(EES	8383	1963	
D7076		B-BDH		BPH	7980	1963	
D7629	(25279)	Bo-BoDE		BP	8039	1965	
(D8087)	20087 SALTLEY L.I.P.	Bo-BoDE		(EE	2993	1961	
	'HERCULES'			(RSHD	8245	1961	b
D8110	(20110)	Bo-BoDE		(EE	3016	1962	
				(RSHD	8268	1962	b
D8233	(ADB 968001)	Bo-BoDE		BTH	1131	1959	c
D9502		0-6-0DH		Sdn		1964	
D9531	ERNEST	0-6-0DH		Sdn		1965	
(E6001)	73001 (73901)	Bo-BoDE/RE		Elh		1962	
56006		Co-CoDE		Electro		1976	
	(BENZOLE)	4wDM		FH	3438	1950	
	ARUNDEL CASTLE	0-6-0DE		HC	D1076	1959	
–		4wDM		MR	9009	1948	
(M50437)	53437	2-2w-2w-2DMR		BRCW		1957	

M50455	(53455)	2-2w-2w-2DMR	BRCW		1957
(M50494)	53494	2-2w-2w-2DMR	BRCW		1957
M50517	(53517)	2-2w-2w-2DMR	BRCW		1957
(50556)	53556 N682	2-2w-2w-2DMR	BRCW		1958
W51339		2-2w-2w-2DMR	PSteel		1960
W51382		2-2w-2w-2DMR	PSteel		1960
Sc51485		2-2w-2w-2DMR	Cravens		1959
E 51813		2-2w-2w-2DMR	BRCW		1961
E 51842		2-2w-2w-2DMR	BRCW		1961
W55001	L101 (TDB 975023)	2-2w-2w-2DMR	GRC&W		1958
S60130	1305	4-4wDER	Afd/Elh		1962
M 65451		4w-4wRER	Wolverton		1959
8099		4w-4wDER	DerbyC&W		1978
105639	R135 NAU 99709 976014-9	4wDMR R/R	(Landrover		1998
			(Perm		1998

a currently at Beamish Museum, Stanley, Co. Durham
b property of Harry Needle Railroad Co Ltd, Barrow Hill, Derbyshire
c carries plate 1118/1959

INCEPTION HOLDINGS SarL, JUNGLE EXPRESS, AMAZONIA, BOLTON MARKET PLACE SHOPPING CENTRE, BOLTON — BL1 2AL

Gauge 600mm www.boltonmarketplace.co.uk **SD 716094**

THE SPINNING MULE	4w-4wBE s/o	Sartori		2016

LANCASHIRE MINING MUSEUM - AT ASTLEY GREEN, RED ROSE STEAM SOCIETY, ASTLEY GREEN COLLIERY MUSEUM, HIGHER GREEN LANE, ASTLEY GREEN, TYLDESLEY — M29 7JB

Gauge 4ft 8½in www.lancashireminingmuseum.org **SJ 705998**

–		4wDM	RH	244580	1946

Gauge 3ft 0in

No.17		0-6-0DMF	HC	DM781	1953
No.11		0-6-0DMF	HC	DM1058	1957
No.20	BEM 401	0-6-0DMF	HC	DM1120	1958
No.18	9306/103 1	0-6-0DMF	HC	DM1270	1961
No.15	20/105/624	0-6-0DMF	(HC	DM1439	1978
			(HE	8821	1978
–		0-6-0DMF	HE	4816	1955
No.9	20/125/101 M.T.R.	4w-4wDHF	HE	9227	1986

Gauge 2ft 6in

–		4wDHF	(AB	621	1981
			(HE	8567	1981
–		4wDHF	(AB	622	1981
			(HE	8568	1981
3		4wBEF	(EE	2417	1957
			(RSHN	7936	1957
4	1/44/17	0-6-0DMF	HC	DM1173	1959
5	1/44/18	0-6-0DMF	HC	DM1352	1967
6	PLT No.1/44/19	0-6-0DMF	HC	DM1413	1970
7	1/44/20	0-6-0DMF	HC	DM1414	1970

–		0-4-0DMF		HE	3411	1947	
	1-44-170	0-6-0DMF		HE	8575	1978	
	1-44-174	0-6-0DMF		HE	8577	1978	

Gauge 2ft 4in

(No.6)		0-6-0DMF		HC	DM970	1957	

Gauge 2ft 1in

2	KESTREL	0-6-0DMF		HC	DM647	1954	

Gauge 2ft 0in

	(STACY)	0-6-0DMF		HC	DM804	1951	
T1		0-6-0DMF		HC	DM840	1954	
14	GEORGE	0-6-0DMF		HC	DM929	1955	
No.16	WARRIOR	0-6-0DMF		HC	DM933	1956	
–		0-4-0DMF		HC	DM1164	1959	
R4		0-6-0DMF		_(HC	DM1443	1980	
				(HE	8843	1980	
3		0-4-0DMF		HE	6048	1961	
75	ROGER BOWEN	0-4-0DMF		HE	7375	1973	
09	CALVERTON	4wDHF		HE	7519	1977	
	MOLE	4wDMF		HE	8834	1978	
03	LIONHEART	4wDHF		HE	8909	1979	
	NEWTON	4wDH		HE	8975	1979	
	SANDY	4wDM		MR	11218	1962	
	POINT OF AYR	4wDMF		RH	497547	1963	
A19		4wBEF		MV/BP	989	1955	

THE MUSEUM OF SCIENCE & INDUSTRY IN MANCHESTER, LIVERPOOL ROAD, CASTLEFIELD, MANCHESTER — M3 4FP

Gauge 5ft 6in www.mosi.org.uk **SJ 831978**

No.3157		4-4-0	IC	VF	2759	1911	a

Gauge 4ft 8½in

No.1	ROCKET	0-2-2	OC	RS	19	1829	
	PLANET	2-2-0	IC	MSI		1992	
	"NOVELTY"	2-2-0VBWT	VC	ScienceMus		1929	b
	AGECROFT No.1	0-4-0ST	OC	RSHN	7416	1948	
(27001)	ARIADNE 1505	Co-CoWE		Gorton	1066	1954	
–		4wBE		_(EE	1378	1944	
				(Bg	3217	1944	

Gauge 3ft 6in

2352		4-8-2+2-8-4T	4C	BP	6639	1930	

Gauge 3ft 0in

No.3	PENDER	2-4-0T	OC	BP	1255	1873	c

a carries works plates VF 3064/1911
b a replica of original locomotive built 1829 by Braithwaite & Ericsson
c locomotive is sectioned to show moving parts

PGC DEMOLITION, junction of BURY OLD ROAD and MOSS HALL ROAD, THE MOUNT, HEAP BROW, near BURY **BL9 7JW**
Gauge 4ft 8½in **SD 828102**

LOT 26 YARD No.766	0-4-0DM		RH	414300	1957	

PRIVATE SITE, MANCHESTER
Gauge 2ft 0in

–	4wDM		JF	18892	1931	

SALFORD CITY COUNCIL, CADISHEAD BY - PASS, CADISHEAD WAY, IRLAM **M44 5AL**
Gauge 4ft 8½in **SJ 716922**

–	0-4-0F	OC	P	2155	1955	

WIGAN COUNCIL, WIGAN LEISURE & CULTURE TRUST LTD, HAIGH RAILWAY, HAIGH WOODLAND PARK, COPPERAS LANE, HAIGH **WN2 1PE**
Gauge 1ft 3in www.haighwoodlandpark.uk **SD 601082**

	HELEN	0-6-2DH	s/o	AK	41	1992	
	"THE CUB"	4-4wDH		Minirail		1954	
		2w-2-2-2w+4-4DMR		MossDW		2017	a
KD1		4-4w-4-4w-4DER		RRS		1983	b Dsm
	"RACHEL"	2-8-2DH	s/o	SL	7217	1972	
	rebuilt			CCLR	1492	1992	
	rebuilt as	2-8-0DH		MossDW		2017	

a constructed using part of RRS/1983 railcar
b currently off site for restoration

MERSEYSIDE

INDUSTRIAL SITES

ALSTOM TRANSPORT, LIVERPOOL TRAINCARE CENTRE, (EDGE HILL), PICTON ROAD, WAVERTREE, LIVERPOOL (part of Alstom Holdings S.A.) **L15 4LD**
Gauge 4ft 8½in www.alstom.com **SJ 380898**

(D3958) 08790	0-6-0DE		Derby		1960	

ARRIVA UK TRAINS LTD - ARRIVA TRAINS NORTH LTD ALLERTON TRAINCARE DEPOT, KNAVESMIRE WAY, off WOOLTON ROAD, ALLERTON, LIVERPOOL **L19 5NA**
Gauge 4ft 8½in www.northernrailway.co.uk **SJ 414849**

See Section 7 for details

EMERGENCY SERVICES TRAINING CENTRE LTD, EAST STREET, SEACOMBE, WIRRAL — CH41 1BY

Gauge 4ft 8½in www.emergencyservices-training.com **SJ 325905**

64649	508201 (508101)	4w-4wRER	York(BRE)	1980
64712	508209 (508121)	4w-4wRER	York(BRE)	1980

GB RAILFREIGHT LTD, c/o FORD MOTOR CO LTD, SPEKE ROAD, SPEKE, LIVERPOOL — L19 2JR

Gauge 4ft 8½in **SJ 399844**

(D3657)	08502	0-6-0DE	Don	1958	a
(D3986)	08818 MOLLY	0-6-0DE	Derby	1960	a

a on hire from Harry Needle Railroad Co Ltd, Derbyshire

MERSEYRAIL ELECTRICS 2002 LTD t/a MERSEYRAIL, KIRKDALE DEPOT, off MARSH STREET, KIRKDALE (an Abellio-SERCO joint venture) — L20 2BL

Gauge 4ft 8½in www.merseyrail.org **SJ 346938**

See Section 7 for details

MERSEYSIDE FIRE & RESCUE SERVICE, KIRKDALE COMMUNITY FIRE STATION, STUDHOLME STREET, BANK HALL, BOOTLE — L20 8ET

Gauge 4ft 8½in www.merseyfire.gov.uk **SJ 341935**

ERV 1	99709 901016-4	2w-2wBE	Bance ERV2-186/07	2007	a
ERV 2	99709 901017-2	2w-2wBE	Bance ERV2-204/08	2008	a
ERV 3	99709 901018-0	2w-2wBE	Bance ERV2-205/08	2008	a

a parts from each vehicle are interchangeable

One vehicle is based at Network Rail, James Street Station, Liverpool

PRESERVATION SITES

G. FAIRHURST, PRIVATE LOCATION

Gauge 2ft 0in

–	4wPM	LancTan	1958	Dsm

KNOWSLEY SAFARI PARK, KNOWSLEY HALL, near PRESCOT — L34 4AN

Gauge 1ft 3in www.knowsleysafariexperience.co.uk **SJ 460936**

–	2-6-0DH s/o	SL	343.2.91	1991	

LAKESIDE MINIATURE RAILWAY, MARINE LAKE, SOUTHPORT — PR8 1RX

Gauge 1ft 3in — **SD 331174**

	DUKE OF EDINBURGH	4-6-2DE	s/o	Barlow		1950	
	GOLDEN JUBILEE 1911-1961	4-6wDE	s/o	Barlow		1963	
No.3	JENNY	2-6-2DM	s/o	MossAJ		2006	
	PRINCESS ANNE	6-6wDH		SL		1971	

A.E. LEWIS, MEADOW VIEW, 170 HALL LANE, SIMONSWOOD

Gauge 1ft 3in

No.1	GORAM	2w-2-4BER	s/o	Hayne		1977	
–		4w-4BE		Hayne		1977	Dsm a

a unpowered and in use as a wagon

MERSEYSIDE DEVELOPMENT CORPORATION, junction of DERBY ROAD and BANKFIELD STREET, BANK HALL, BOOTLE, LIVERPOOL — L20 8EA

Gauge 4ft 8½in — **SJ 339937**

THE ATLANTIC AVENUE 1998	4wDM		RH	224347	1945	

NATIONAL MUSEUMS LIVERPOOL, Museum Of Liverpool, Pier Head, Liverpool — L3 1DG

Gauge 4ft 8½in www.liverpoolmuseums.org.uk/mol — **SJ 340899**

	LION	0-4-2	IC	TK&L		1838	
No.3		4w-4wRER		BM		1892	

Vehicle Workshop, Juniper Street, Bootle — L20 8QJ

Gauge 4ft 8½in — **SJ 344936**

M.D.& H.B. No.1	M.D.E. 1	0-6-0ST	OC	AE	1465	1904
	(CECIL RAIKES)	0-6-4T	IC	BP	2605	1885

Locomotives not on public display. Visitors by appointment only.

NEWTON-LE-WILLOWS PRIMARY SCHOOL, SANDERLING ROAD, NEWTON-LE-WILLOWS — WA12 9UF

Gauge 2ft 6in www.newton.st-helens.sch.uk — **SJ 584954**

PARKSIDE	4wDH		HE	8825	1978

SOUTHPORT PIER RAILWAY, SOUTHPORT — PR8 1QX

Gauge 3ft 6in **Closed** — **SD 335176**

–	2-2w-2w-2+2-2w-2w-2BER	SL/UK LOCO	2005	OOU a

a stored off site at Sefton Council yard, Southport

NORFOLK

INDUSTRIAL SITES

ABELLIO EAST ANGLIA LTD, CREMORNE LANE, NORWICH — NR1 1TZ

Gauge 4ft 8½in www.abelliogreateranglia.co.uk **TG 246078**

See Section 7 for details

GREAT EASTERN TRACTION, HARDINGHAM STATION, HARDINGHAM, near WYMONDHAM

Gauge 4ft 8½in www.greateasterntraction.co.uk **TG 050055**

11		0-6-0DE	BT	804	1978	
No.9	HARDINGHAM	0-4-0DE	RH	512842	1965	
8	DL 82	0-6-0DH	RR	10272	1967	
2		0-4-0DH	(RSHD	8368	1962	
			(WB	3213	1962	
	99709 901004-0	2w-2PMR	Lesmac	LMS014	2006	
	99709 901009-9	2w-2PMR	Lesmac	LMS019	2006	
	99709 901010-7	2w-2PMR	Lesmac	LMS020	2006	

Gauge 2ft 6in

05 581		4wDH	AK	18	1985	
	RACHEL	4wDH	Byers		1998	a

a incorporates frame of AB 562 / 1971

Gauge 2ft 0in

L2	05/580 JEFFREY	4wDH	AK	19	1985

NORTH NORFOLK RAILWAY ENGINEERING, WEYBOURNE — NR25 7HN

www.nnr-engineering.co.uk **TG 118419**

Locomotives under restoration usually present

OLYMPIC AQUATIC ENGINEERS, UNIT 10, OAKTREE BUSINESS PARK, BASEY ROAD, RACKHEATH INDUSTRIAL ESTATE, NORWICH — NR13 6PZ

Gauge monorail www.olympicaquaticengineers.co.uk **TG 282140**

002	IVOR	2a-2DH	AK	M002	1989
–		2a-2DH	AK	M003	1989
–		2w-2wDH	OAE		

Locomotives present in yard between contracts

PRESERVATION SITES

BRESSINGHAM STEAM PRESERVATION COMPANY, BRESSINGHAM LIVE STEAM MUSEUM, LOW ROAD, near DISS IP22 2AA

Gauge 5ft 0in www.bressingham.co.uk **TM 080806**

(1144)		2-8-0	OC	TK	571	1948	OOU

Gauge 4ft 8½in

(30102)	GRANVILLE	0-4-0T	OC	9E		1893
32662	(662)	0-6-0T	IC	Bton		1875
(41966)	80 THUNDERSLEY	4-4-2T	OC	RS	3367	1909
(62785)	No.490	2-4-0	IC	Str	836	1894
(68633)	No.87	0-6-0T	IC	Str	1249	1904
	(ROBERT KETT)	0-4-0F	OC	AB	1472	1916
6841	WILLIAM FRANCIS	0-4-0+0-4-0T	4C	BP	6841	1937
3193	NORFOLK REGIMENT	0-6-0ST	IC	HE	3193	1944
			reb	HE	3887	1964
No.25		0-4-0ST	OC	N	5087	1896
377	7	2-6-0	OC	Nohab	1164	1919
	MILLFIELD	0-4-0CT	OC	RSHN	7070	1942
5865	PEER GYNT	2-10-0	OC	Schichau	4216	1944
1	COUNTY SCHOOL	0-4-0DH		RH	497753	1963
11103 // 11104		0-4-0DM		_(VF	D297	1956
				(DC	2583	1956

Gauge 2ft 0in (line laid to accommodate 1ft 11½in & 1ft 10¾in gauge locomotives)

No.2	BEVAN	0-4-0PT	OC	_(Braithwaite No.2		
				(Bressingham		2009
	GWYNEDD	0-4-0STT	OC	HE	316	1883
	GEORGE SHOLTO	0-4-0ST	OC	HE	994	1909
3	FERNILEE	0-4-0VBTT	VC	_(Wbton	(3)	2015
				(Brookside No.8		2015
7	"TOBY"	4wDM	s/o	MR	22210	1964
–		4wDHF		HE	8911	1980
	BOVIS	4wBEF		HE	9155	1991

Gauge 1ft 3in — Waveney Valley Railway

	ST.CHRISTOPHER	2-6-2T	OC	ESR	311	2001
No.1662	ROSENKAVALIER	4-6-2	OC	Krupp	1662	1937
1663	MÄNNERTREU	4-6-2	OC	Krupp	1663	1937
D6353	JOE BROWN ENGINEER	4w-4wDM		BrownJ		1998
	IVOR	4wDH		Frenze		1979
			reb	‡	No.1	1986

‡ rebuilt by Gray's Engineering, Diss

BURE VALLEY RAILWAY (1991) LTD, THE BROADLAND LINE, AYLSHAM NR11 6BW

Gauge 1ft 3in www.bvrw.co.uk **TG 197265**

1	WROXHAM BROAD	2-6-2DM	s/o	Guest		1964	a
	rebuilt as	2-6-4T	OC	Winson		1992	
6	BLICKLING HALL	2-6-2	OC	Winson	12	1994	
7	SPITFIRE	2-6-2	OC	Winson	14	1994	

8	JOHN OF GAUNT	2-6-2T	OC	Winson/BVR	16	1998	
9	MARK TIMOTHY	2-6-4T	OC	Winson	20	1999	
			reb	AK	69R	2003	
3	2nd AIR DIVISION USAAF	4w-4wDH		BVR		1989	
PW 7		2-2w-4BEF		GB	6132	1966	b
4	RUSTY	4wDM		HE	4556	1954	
	rebuilt as	4wDH		EAGIT		1994	
7		4wDM		LB	51989	1960	

a carries plates G&S 20/1964
b converted to a non-powered wagon

CAISTER CASTLE CAR COLLECTION, CAISTER CASTLE, CASTLE LANE, CAISTER-ON-SEA, near GREAT YARMOUTH — NR 30 5SN

Gauge 4ft 8½in www.caistercastle.co.uk **TG 504123**

42	RHONDDA	0-6-0ST	IC	MW	2010	1921

FENGATE AGRICULTURAL LIGHT RAILWAY, FENGATE FARM, WEETING, near BRANDON — IP27 0QF

Gauge 1524mm www.weetingrally.co.uk **TL 770885**

1060		2-8-2	OC	LO	172	1954	

Gauge 4ft 8½in

	LITTLE BARFORD	0-4-0ST	OC	AB	2069	1939	a

Gauge 3ft 0in

No.7	(TYNWALD)	2-4-0T	OC	BP	2038	1880	Dsm

a based here, but may visit other locations

Visiting locomotives occasionally present.

R.& A. JENKINS, FRANSHAM STATION, GREAT FRANSHAM, near SWAFFHAM

Gauge 4ft 8½in **TF 888135**

–	4wDM	RH	398611	1956

Gauge 2ft 0in

–	4wDM	RH	422573	1958

M.W. MAYES, MECHANICAL ENGINEER, STATION YARD, STATION ROAD, YAXHAM, near DEREHAM — NR19 1RD

Gauge 4ft 8½in **TG 003102**

No.2		0-4-0ST	OC	RSHN	7818	1954
	"GEORGE"	4wVBT	VCG	S	9596	1955

Gauge 1ft 11½in

No.1		0-4-0VBT	VCG	Potter		1970
No.2	RUSTY	4wDM		L	32801	1948

MID-NORFOLK RAILWAY SOCIETY, DEREHAM STATION, STATION ROAD, DEREHAM

NR19 1DF

Gauge 4ft 8½in www.mnr.org.uk **TF 995128**

(D419)	50019 RAMILLIES	Co-CoDE		_(EE (EEV	3789 D1160	1967 1967	
(D1886)	47367	Co-CoDE		BT	648	1965	
(D1933	47255) 47596 ALDBURGH FESTIVAL	Co-CoDE		BT	695	1966	
(D2197)	03197	0-6-0DM		Sdn		1961	a
D2334		0-6-0DM		_(RSHD (DC	8193 2715	1961 1961	
(D4015)	08847	0-6-0DE		Hor		1961	b
(D5683)	31255	A1A-A1A DE		BT	284	1961	
(D6905	37205) 37688	Co-CoDE		_(EE (EEV	3383 D849	1963 1963	
(D8069)	20069 DEREHAM NEATHERED HIGH SCHOOL 1912-2012	Bo-BoDE		_(EE (RSHD	2975 8227	1961 1961	
B.S.C.1		0-6-0DH		EEV	D1049	1965	
(51226)		2-2w-2w-2DMR		MetCam		1958	
(51434)	MATTHEW SMITH 1974-2002	2-2w-2w-2DMR		MetCam		1959	
(51499)		2-2w-2w-2DMR		MetCam		1959	
51503		2-2w-2w-2DMR		MetCam		1959	
51942	L233	2-2w-2w-2DMR		DerbyC&W		1961	
55009	109	2-2w-2w-2DMR		GRC&W		1958	
68004	(9004) 931 094	4w-4wRE/BER		Afd/Elh		1960	
	B445 WPO	4wDMR	R/R	Bruff	502	c1984	
960225	GEORGE T. RASEY	2w-2PMR		Wkm	1308	1933	

a on hire to Sonic Rail Services Ltd

b property of British American Railway Services, Stanhope, Co. Durham

NORTH NORFOLK RAILWAY plc

Locomotives are kept at :-

Holt	NR25 6AJ	TG 094398
Sheringham	NR26 8RA	TG 156430
Weybourne	NR25 7HN	TG 118419

Gauge 4ft 8½in www.nnrailway.co.uk

53809	(13809)	2-8-0	OC	RS	3895	1925
61306	(1306) MAYFLOWER	4-6-0	OC	NBQ	26207	1948
(61572)	No.8572	4-6-0	IC reb	BP MaLoWa	6488	1928 1994
(65462	7564) 564	0-6-0	IC	Str		1912
76084		2-6-0	OC	Hor		1957
90775	THE ROYAL NORFOLK REGIMENT	2-10-0	OC	NBH	25438	1943
92203	BLACK PRINCE	2-10-0	OC	Sdn		1959
	WISSINGTON	0-6-0ST	IC	HC	1700	1938
1982	RING HAW	0-6-0ST	IC	HE	1982	1940
(D2051)		0-6-0DM		Don		1959
D2063	(03063)	0-6-0DM		Don		1959
D3935	(08767)	0-6-0DE		Hor		1961
D3940	(08772)	0-6-0DE		Derby		1960
(D5207)	25057	Bo-BoDE		Derby		1963

D5631	(31207)		A1A-A1A DE		BT	231	1960
D6732	(37032) MIRAGE		Co-CoDE		_(EE (VF	2895 D611	1961 1961
12131			0-6-0DE		Dar		1952
M51188			2-2w-2w-2DMR		MetCam		1958
M51192			2-2w-2w-2DMR		MetCam		1958
E51228			2-2w-2w-2DMR		MetCam		1958
98801	DAVID PINKERTON RAILWAY ENGINEER		4wDHR		Kershaw	45.121A	1992

ALAN POWELL, NORWICH

Gauge 1ft 3in

–		2-4wPM	s/o	Powell		1997

FRED ROUT, BROOMHILL FARM, LEYS LANE, BUCKENHAM, near ATTLEBOROUGH — NR17 1NS

Gauge 4ft 8½in

No.21		0-6-0DM		HC	D707	1950

STRUMPSHAW STEAM MUSEUM, STRUMPSHAW HALL, near ACLE — NR13 4HR

Gauge 1ft 11½in www.strumpshawsteammuseum.co.uk **TG 345065**

No.6	GINETTE MARIE	0-4-0WT	OC	Jung	7509	1937	OOU
5	JIMMY	4wDM	s/o	MR	7192	1937	

Gauge 1ft 3in

2	CAGNEY	4-4-0	OC	McGarigle		1902	

THURSFORD COLLECTION, THURSFORD GREEN, THURSFORD, near FAKENHAM — NR21 0AS

Gauge 1ft 10¾in **R.T.C.** www.thursford.com **TF 980345**

	CACKLER	0-4-0ST	OC	HE	671	1898	OOU

WHITWELL & REEPHAM STATION, WHITWELL ROAD, REEPHAM — NR10 4GA

Gauge 4ft 8½in www.whitwellstation.com **TG 092216**

	ANNIE	0-4-0ST	OC	AB	945	1904
	VICTORY	0-4-0ST	OC	AB	2199	1945
	AGECROFT No.3	0-4-0ST	OC	RSHN	7681	1951
1	"GEORGIE"	4wDH		BD	3733	1977
D1171	"WESTERN PRIDE"	0-6-0DM		HC	D1171	1959
D2700		0-4-0DH		NBQ	27426	1955
	"TIPOCKITY"	4wDM		RH	466629	1962
	SWANWORTH	4wDM		RH	518494	1967
	Q454 JTT	4wDM	R/R	Unimog	027869	1999
W51370		2-2w-2w-2DMR		PSteel		1960

W51412		2-2w-2w-2DMR		PSteel		1960	
(DX 68051 TR 6)		2w-2PMR		Wkm	6901	1954	a
	rebuilt as	2w-2DHR					
9023		2w-2PMR		Wkm	8087	1958	
–		2w-2PMR		Wkm		1932	Dsm b

a currently off site for restoration
b carries plate Wkm 529/1932

YAXHAM LIGHT RAILWAY,
OLD STATION HOUSE, STATION ROAD, YAXHAM, near DEREHAM **NR19 1RD**
Gauge 1ft 11½in **TG 003102**

	KIDBROOKE	0-4-0ST	OC	WB	2043	1917
No.3	PEST	4wDM		L	40011	1954
–		4wDM		LB	54684	1965
No.4	GOOFY	4wDM		OK	(6501?)	c1936
No.6	LOD/758097 COLONEL	4wDM		RH	202967	1940
No.7		4wDM		RH	170369	1934
No.10	OUSEL	4wDM		MR	7153	1937
No.13		4wDM		MR	7474	1940
No.14	LOD 758375 ARMY 25	4wDM		RH	222100	1943
No.18		4wDM		FH	3982	1962
No.19	PENLEE	4wDM		HE	2666	1942
44		4wDM		Moës		c1955

UNKNOWN OWNER c/o M.A.T. WASTE DISPOSAL LTD,
LANGHORNS LANE, OUTWELL, WISBECH
Gauge 4ft 8½in **TF 522042**

D1995	5 HEATHER	0-4-0DM	(VF	D293	1955
			(DC	2566	1955

NORTHAMPTONSHIRE

INDUSTRIAL SITES

GEISMAR (UK) LTD, SALTHOUSE ROAD, BRACKMILLS INDUSTRIAL ESTATE,
NORTHAMPTON **NN4 7EX**
Railcars under construction or repair are usually present. www.geismar.com **SP 781858**

PROLOGIS R.F.I., D.I.R.F.T.,
DAVENTRY INTERNATIONAL RAILFREIGHT TERMINAL,
RAILPORT APPROACH, CRICK **NN6 7ES**
Gauge 4ft 8½in www.dirft.com **SP 566725, 573721**

(D3560)	08445	0-6-0DE	Derby	1958	a

(D3738) 08571	0-6-0DE		Crewe		1959	b
LILY IZABELLA	6wDH		TH	V325	1987	a

a property of LH Group Services Ltd, Barton-under-Needwood, Staffordshire
b property of Wabtec Rail Ltd, Doncaster, South Yorkshire

SIEMENS AG, SIEMENS RAIL SYSTEMS, KINGS HEATH TRAINCARE FACILITY, HEATHFIELD WAY, KINGS HEATH, NORTHAMPTON NN5 7QP

Gauge 4ft 8½in www.siemens.com **SP 745612**

–	4wBE		CPM	1731.01	2005

PRESERVATION SITES

BILLING AQUADROME LTD, WILLOW LAKE MINIATURE TRAIN, CROW LANE, GREATER BILLING, near NORTHAMPTON NN3 9DA

(part of the Pure Leisure Group Ltd)

Gauge 1ft 3in www.billingaquadrome.com **SP 808615**

362	2-8-0gasH	s/o	SL	15.5.78	1978
converted to	2-8-0DH				

CORBY BOROUGH COUNCIL, LEISURE & CULTURE, EAST CARLTON COUNTRY PARK, EAST CARLTON, near CORBY LE16 8YF

Gauge 4ft 8½in www.corby.gov.uk **SP 834893**

–	0-6-0ST	OC	HL	3827	1934

J.EVANS, PRIVATE LOCATION, NORTHAMPTON

Gauge 2ft 0in

–	2w-2PMR	EvansJ		2009	a

a built using parts from Wkm 2449

HIGHAM FERRERS LOCOMOTIVES

Private Site, near Rushden

Gauge 1000mm

ND 3647	4wDM	MR	22144	1962

Private Site, near Wellingborough

Gauge 3ft 0in

18	4wBE	GB	1570	1938

IRCHESTER NARROW GAUGE RAILWAY TRUST, IRCHESTER COUNTRY PARK, GIPSY LANE, LITTLE IRCHESTER — NN29 7DL

Gauge 1000mm www.irchesterrailwaymuseum.co.uk **SP 906660**

	(4)	CAMBRAI	0-6-0T	OC	Corpet	493	1888	
No.85	1		0-6-0ST	OC	P	1870	1934	
(No.86	2)		0-6-0ST	OC	P	1871	1934	
(No.87	3	89-12) 8315/87	0-6-0ST	OC	P	2029	1942	
	9	THE ROCK	0-4-0DM		HE	2419	1941	
ND 3645	10	MILFORD	4wDM		RH	211679	1941	
(ED 10)	11	EDWARD CHARLES HAMPTON	4wDM		RH	411322	1958	

Gauge 3ft 0in

LR 3084	14		4wDM		MR	3797	1926	
			a rebuild of		MR	1363	1918	
H 048	BERTIE		0-6-0DM		RH	281290	1949	a
	5		4wDMF		RH	338439	1953	

a on loan to British American Railway Services Ltd, Stanhope, Co. Durham

NORTHAMPTON STEAM RAILWAY, PITSFORD & BRAMPTON STATION, CHAPEL BRAMPTON, NORTHAMPTON

Locomotives are kept at : Pitsford & Brampton NN6 8BA SP 734667, 737662
Brampton Lane NN6 8AA SP 737656

Gauge 4ft 8½in www.nlr.org.uk

3862		2-8-0	OC	Sdn		1942	
5967	BICKMARSH HALL	4-6-0	OC	Sdn		1937	
(RRM 23)	"FIREFLY"	0-4-0ST	OC	AB	776	1899	
	WALESWOOD	0-4-0ST	OC	HC	750	1906	
	WESTMINSTER	0-6-0ST	OC	P	1378	1914	
–		0-4-0ST	OC	P	2104	1950	a
No.7		0-4-0ST	OC	P	2130	1951	
(D1855)	47205 47395	Co-CoDE		Crewe		1965	
(D5821)	31289 PHOENIX	A1A-A1ADE		BT	322	1961	
No.21		0-4-0DM		JF	4210094	1954	
	rebuilt as	0-4-0DH		JF		1966	
(No.1	MERRY TOM)	4wDM		RH	275886	1949	
(764	SIR GILES ISHAM)	0-4-0DM		RH	319286	1953	
No.53	SIR ALFRED WOOD	0-6-0DM		RH	319294	1953	
–		0-4-0DH		TH	146C	1964	
	a rebuild of	0-4-0DM		JF	4210018	1950	b
–		2w-2BER		Bance		c1990	
	(99709 901013-1)	2w-2PMR		Lesmac	LMS008	2006	

a actually built 1948 but plates are dated as shown
b TH 146C incorporates frame and wheels (only) of the JF loco

NORTHAMPTONSHIRE IRONSTONE RAILWAY TRUST LTD, HUNSBURY HILL INDUSTRIAL MUSEUM, NORTHAMPTON — NN4 9UW

Gauge 4ft 8½in www.nir.org.uk **SP 735584**

9365	89-17	BELVEDERE	4wVBT	VCG	S	9365	1946

	(MUSKETEER)	4wVBT	VCG	S	9369	1946	
–		4wDH		FH	3967	1961	
–		0-4-0DM		HC	D697	1950	
–		0-4-0DM		HE	2087	1940	Dsm
	CHARLES WAKE	0-4-0DM		JF	4220001	1959	
–		4wDM		RH	321734	1952	a Dsm
4002	S 13004 S	4w-4wRER		Elh		1949	
5176	S14351	4w-4wRER		Lancing/Elh		1955	
5176	S14352	4w-4wRER		Lancing/Elh		1955	
960236		2w-2PMR		Wkm	1519	1934	

a frame in use as a wagon

PRIVATE OWNER, near NORTHAMPTON

Gauge 3ft 0in

–		4wDM		RH	418764	1957

RUSHDEN HISTORICAL TRANSPORT SOCIETY, THE OLD STATION, STATION APPROACH, RECTORY ROAD, RUSHDEN NN10 0AW

Gauge 4ft 8½in www.rhts.co.uk **SP 957672**

	EDMUNDSONS	0-4-0ST	OC	AB	2168	1943
–		0-4-0ST	OC	AB	2323	1952
	THE BLUE CIRCLE	2-2-0WT	G	AP	9449	1926
	"CHERWELL"	0-6-0ST	OC	WB	2654	1942
(D2179)	03179 CLIVE	0-6-0DM		Sdn		1961
	WD 70048	0-4-0DM		AB	363	1942
	LES FORSTER	4wDH		S	10159	1963
	YARD No.10433	0-4-0DM		(VF	5258	1945
				(DC	2177	1945
			reb	YEC	L121	1993
(55029)	977968	2-2w-2w-2DMR		PSteel		1960
059	(53.0551-1)	2w-2DMR		Robel 54.13-RW82		1975
	99709 901041-2	2w-2PMR		Geismar ST/01/34		2001

WICKSTEED PARK LAKESIDE RAILWAY, WICKSTEED LEISURE PARK, KETTERING NN15 6NJ

Gauge 2ft 0in www.wicksteedpark.org **SP 883770**

	MERLIN	0-4-0DH	s/o	AK	86	2010
2042	LADY OF THE LAKE	0-4-0DM	s/o	Bg	2042	1931
2043	KING ARTHUR	0-4-0DH	s/o	Bg	2043	1931
	CHEYENNE	4wDM	s/o	MR	22224	1966

J. WOOLMER, 15 BAKERS LANE, WOODFORD, KETTERING

Gauge 2ft 0in **SP 969769**

89-18	4wPM		L	14006	1940
–	4wDM		L	36743	1951

NORTHUMBERLAND

OUR DEFINITION OF "NORTHUMBERLAND" INCLUDES THOSE LOCATIONS WHICH WERE SITUATED WITHIN THAT PART OF THE FORMER COUNTY OF TYNE & WEAR WHICH IS NORTH OF THE RIVER TYNE AND NOW CONSISTS OF A NUMBER OF UNITARY AUTHORITIES.

INDUSTRIAL SITES

DEFENCE INFRASTRUCTURE ORGANISATION, DEFENCE TRAINING ESTATES - NORTH, REDESDALE RANGES MILITARY TARGET RAILWAY, REDESDALE RANGES, BUSHMAN'S ROAD, SILLS, ROCHESTER, NORTHUMBERLAND

(operated by Landmarc Support Services Ltd)

Gauge 2ft 6in **R.T.C.** **NT 826999**

–		2w-2PM	Wkm	3245	1943	OOU
6	SILLS 1	2w-2PM	Wkm	11686	1990	OOU
7	SILLS 2	2w-2PM	Wkm	11687	1990	OOU
(8)	SILLS 4	2w-2PM	Wkm	11688	1990	OOU
9	SILLS 3	2w-2PM	Wkm	11689	1990	OOU

Gauge 1000mm – White Spot Target Railway (elevated railway) **NT 862045**

2337	4wBE	SAAB	108	1986	OOU a
2338	4wBE	SAAB	(133?	19xx)	OOU a

a stored at Otterburn Camp

LYNEMOUTH POWER LTD, LYNEMOUTH POWER STATION, LYNEMOUTH, ASHINGTON NE63 9NW

(part of EP UK Investments Ltd, a subsidiary of EPH Energetický a Průmslový holdings)

Gauge 4ft 8½in www.lynemouthpower.com **NZ 302903**

TNS 106	2-2wDMR	Robel	56.27-10-AG37	1982

NEXUS TYNE WEAR METRO, SOUTH GOSFORTH CAR SHEDS, NEWCASTLE-UPON-TYNE NE3 1XD / NE3 5DG

Gauge 4ft 8½in www.nexus.org.uk **NZ 250686**

BL1	97901	4wBE/WE		HE	9174	1989	
BL2	97902	4wBE/WE		HE	9175	1989	
BL3	97903	4wBE/WE		HE	9176	1989	
YF59 YHK	99709 976055-2	4wDMR	R/R	_(Ford (Aquarius	833760	2009 2010	
YR10 FDP		4wDMR	R/R	_(Ford (Aquarius	833517	2010 2011	
T597 OKK		4wDM	R/R	SRS		1999	
F171 DUA		4wDM	R/R	_(Unimog (Zweiweg	140729 1188	1988 1988	OOU
NA52 JNN		4wDM	R/R	Unimog	200083	2003	
NK12 CWZ	99709 979101-1	4wDM	R/R	Unimog	228343	2012	
NK16 GUX	99709 942161-9	4wDM	R/R	_(Unimog (LHGroup	239027	2016 2016	

NK16 GUW	99709 942162-7	4wDM	R/R	_(Unimog	239031	2016	
				(LHGroup		2016	
AE66 CPU	99709 977041-1	4wDM	R/R	_(Unimog	242230	2016	
				(LHGroup		2016	
AE66 CPV	99709 977042-9	4wDM	R/R	_(Unimog	242499	2016	
				(LHGroup		2016	
–		2w-2BER		Bance	094	2000	
HCT 035		4wDH		Perm	035	1993	OOU

NORTHUMBRIA RAIL LTD, BEDLINGTON

Adminstration address only.

The following is a FLEET LIST of locomotives owned by (or in the care of) this contractor

Gauge 4ft 8½in

801		Bo-BoDE	Alco	77120	1950	a
323.674-2	SIMONSIDE / SPLUTTER	4wDH	Gmd	4991	1957	a
323-539-7	CHEVIOT	4wDH	Gmd	4861	1955	b
424839		0-4-0DE	RH	424839	1959	c
1212	HELGA	4w-4DMR	EK		1958	a
RB 004		4wDMR	Leyland/DerbyC&W			
				RB004	1984	d
MPP 10188		2w-2BER	Bance	098/00	2000	d
HCT 022		4wDH	Perm	022	1988	e

a currently at Nene Valley Railway, Cambridgeshire
b currently at Arlington Fleet Group Ltd, Eastleigh, Hampshire
c currently at M. Fairnington, Wooler, Northumberland
d currently at Waverley Route Heritage Association, Whitrope, Borders
e currently at Scottish Railway Preservation Society, Bo'ness, West Lothian

RIO TINTO ALCAN PRIMARY METALS EUROPE LTD, SHIP UNLOADING FACILITY, NORTH BLYTH — NE24 1SE

Gauge 4ft 8½in www.riotinto.com **NZ 315821**

1046	4wDH	R/R	Unilok	4020	2011	

IAN STOREY ENGINEERING LTD, HEPSCOTT, near MORPETH — NE61 6LN

Gauge 4ft 8½in **NZ 223844**

(DB 965053)	2w-2PMR	Wkm	7576	1956	Dsm

PRESERVATION SITES

ALLENHEADS HERITAGE CENTRE, ALLENHEADS — NE47 9HN

Gauge 2ft 0in www.northumberlandlife.org/allenheadstrust **NY 859453**

–	0-4-0BE	WR

ALN VALLEY RAILWAY,
LLOYD'S FIELD, LIONHEART ENTERPRISE PARK, ALNWICK — NE66 2PQ

Gauge 4ft 8½in www.alnvalleyrailway.co.uk **NU 202122**

No.9	RICHBORO	0-6-0T	OC	HC	1243	1917	
48	9103/48	0-6-0ST	IC	HE	2864	1943	
No.60		0-6-0ST	IC	HE	3686	1948	
	PENICUIK	0-4-0ST	OC	HL	3799	1935	
(01564)	12088	0-6-0DE		Derby		1951	
	20/110/711	0-6-0DH		AB	615	1977	
			reb	AB	6719	1987	
	"DRAX"	0-6-0DM		EES	8199	1963	
–		4wDM		RH	265617	1948	
–		0-4-0DE		RH	312989	1952	
(DX 68805)	BRIAN	4wDMR		Perm	006	1986	
(DX 68806)		4wDMR		Perm	007	1986	
(DB 965952)	BUZZ	2w-2PMR		Wkm	10648	1972	
(PWM 2807	B170W)	2w-2PMR		Wkm	4985	1949	Dsm a

a converted to non-powered passenger trailer

CHAINBRIDGE HONEY FARM and TRANSPORT MUSEUM,
HORNCLIFFE, BERWICK-UPON-TWEED — TD15 2XT

Gauge 4ft 8½in www.chainbridgehoney.com **NT 936507**

2234	(45010)	4+0-4-0VBTR	Derby	1904	Dsm a

a converted to hauled coaching stock

DANIEL FARM SHOP & TEA ROOMS, SLED LANE, WYLAM — NE41 8JH

Gauge 2ft 6in www.danielfarm.co.uk **NT 933385**

CANNONBALL	4wDM	s/o	RH	175403	1935

HEATHERSLAW LIGHT RAILWAY CO LTD,
HEATHERSLAW MILL, CORNHILL-ON-TWEED — TD12 4TJ

Gauge 1ft 3in www.heatherslawlightrailway.co.uk **NZ 123637**

BUNTY	2-6-0T	OC	(SmithN		c2005	
			(AK	85R	2010	a
THE LADY AUGUSTA	0-4-2	OC	TaylorB		1989	
BINKY	0-6-0DH		Heatherslaw		2014	
rebuild of	6wDH		SmithN		1989	b
CLIVE	4w-4wDH		SmithN		2000	

a Locomotive part built by SmithN and completed by AK
b uses parts supplied by AK

H. MEERS, BEAL STATION GOODS YARD, BEAL — TD15 2PB

Gauge 4ft 8½in **NU 063426**

1611	0-4-0ST	OC	P	1611	1923

NORTH TYNESIDE COUNCIL,
TYNE & WEAR ARCHIVES & MUSEUMS, STEPHENSON RAILWAY MUSEUM,
MIDDLE ENGINE LANE, Near CHIRTON **NE29 8DX**

Museum locomotives work services for the North Tyneside Steam Railway

Gauge 4ft 8½in www.stephensonrailwaymuseum.org.uk **NZ 323693**

A.No.5		0-6-0PT	IC	K	2509	1883	
ASHINGTON No.5	JACKIE MILBURN 1924-1988	0-6-0ST	OC	P	1970	1939	
	"BILLY"	0-4-0	VC	GS(K)		1816	
No.1	TED GARRETT JP, DL, MP	0-6-0T	OC	RSHN	7683	1951	
401	THOMAS BURT M.P. 1837-1922	0-6-0ST	OC	WB	2994	1950	
(D2078	03078) No.D2078	0-6-0DM		Don		1959	
(D4145)	08915	0-6-0DE		Hor		1962	
3267	(DE 900730)	4w-4wRER		York		1904	
10		0-6-0DM		Consett		1958	
E4		Bo-BoWE/BE		Siemens	457	1909	

M. FAIRNINGTON, AGRICULTURAL ENGINEERS,
UNIT A, BERWICK ROAD, WOOLER **NE71 6AH**

Gauge 4ft 8½in **NT 995287**

424839	NORTHUMBRIA	0-4-0DE		RH	424839	1959	a

a property of Northumbria Rail Ltd, Bedlington, Northumberland

locomotives for restoration occasionally present

WOODHORN NARROW GAUGE RAILWAY,
WOODHORN COLLIERY MUSEUM, ASHINGTON **NE63 9YF**

Gauge 2ft 0in www.woodhornnarrowgaugerailway.weebly.com **NZ 287884**

No.1	BLACK DIAMOND / HUNSLET 6348 / SEAHAM VANE TEMPEST 1993	4wDM	HE	6348	1975
3	RIO GEN	4wDH	HE	9353	1995
No.2	EDWARD STANTON	4wDH	Schöma	5240	1991

NOTTINGHAMSHIRE

INDUSTRIAL SITES

BALFOUR BEATTY RAIL (UK) LTD,
OLD STATION YARD, DERBY ROAD, SANDIACRE **NG10 5AG**

Gauge 4ft 8½in www.balfourbeatty.com **SK 483360**

XTU 2104	H867 ARB	4wDM	R/R	Unimog	166648	1991

BLYTH PLANT & TOOL HIRE,
REDBRIDGE HOUSE, WORKSOP ROAD, BLYTH **S81 0TX**
Gauge 4ft 8½in **SK 622857**

F511 LRR	4wDM	R/R	_(Multicar		1989
			(Perm		1989

BODEN RAIL ENGINEERING LTD,
EASTCROFT DEPOT, QUEENS ROAD, off LONDON ROAD, NOTTINGHAM NG2 3AH
Gauge 4ft 8½in **SK 582391**

NCB No.11 BIRCH COPPICE	4wDH		TH	134C	1964
a rebuild of	4wVBT	VCG	S	9578	1954

mainline and preserved locomotives for maintenance/overhaul usually present

EDF ENERGY (UK) LTD, COTTAM POWER STATION,
OUTGANG LANE, COTTAM, RETFORD **DN22 0NP**
Gauge 4ft 8½in www.edfenergy.com **SK 815796**

TNS 105	2w-2DMR	Robel		
			56.27-10-AG36	1982

HARSCO RAIL LTD, UNIT 1, CHEWTON STREET, EASTWOOD,
NOTTINGHAM (part of the Harsco Corporation) www.harscorail.com **NG16 3HB**
Road/Rail vehicles under construction, repair, or for hire, are usually present. **SK 475462**

NOTTINGHAM TRAMS LTD, NOTTINGHAM EXPRESS TRANSIT,
WILKINSON STREET DEPOT, NOTTINGHAM **NG7 7NW**
Gauge 4ft 8½in www.thetram.net **SK 553423**

–	4wBE		ACI	20699	2013
–	4wBE		Niteq	B183	2003
FG52 WCC	4wDM	R/R	_(Unimog	199635	2002
			(Zagro	3139	2002
GK03 ACX	4wDM	R/R	_(Unimog	201624	2003
			(Zweiweg		2003

SIMS METAL MANAGEMENT LTD, HARRIMANS LANE, DUNKIRK,
NOTTINGHAM (member of the Sims Group Ltd) **NG7 2SD**
Locomotives for scrap are occasionally present
Gauge 4ft 8½in www.simsmm.co.uk **SK 549337**

–	0-6-0DH	S	10140	1962	OOU

PRESERVATION SITES

BILSTHORPE HERITAGE MUSEUM, rear of VILLAGE HALL, CROSS STREET, BILSTHORPE — NG22 8QY

Gauge 3ft 0in — **SK 646608**

No.	Name	Type	Cyl	Builder	Works No.	Year	Notes
(E2)		4wBEF		CE	B1504A	1977	
			reb	CE	B3156	1984	

K. & J. BOWNES & SONS, 1 CORNMILL FARM BUNGALOW, off OWDAY LANE, WALLINGWELLS, WORKSOP — S81 8DA

Gauge 4ft 8½in www.kjbownes.co.uk — **SK 574838**

No.	Name	Type	Cyl	Builder	Works No.	Year	Notes
–		0-6-0F	OC	WB	3019	1952	

Visitors by appointment only.

GREAT CENTRAL RAILWAY (NOTTINGHAM) LTD, NOTTINGHAM TRANSPORT HERITAGE CENTRE, MERE WAY, RUDDINGTON, NOTTINGHAM — NG11 6JS

Gauge 4ft 8½in www.gcrn.co.uk — **SK 575322**

No.	Name	Type	Cyl	Builder	Works No.	Year	Notes
(411.388)		2-8-0	OC	Alco	70284	1942	Dsm
–		2-8-0	OC	Alco	(70610	1943?)	Dsm
(411.09)		2-8-0	OC	BLW	69621	1943	Dsm
2890		0-6-0	IC	Gartell		2001	
	a rebuild of	0-6-0ST	IC	HE	2890	1943	
			reb	HE	3882	1962	
	"JULIA"	0-6-0ST	IC	HC	1682	1937	
	(DOLOBRAN)	0-6-0ST	IC	MW	1762	1910	
No.15	(8310/41) (RHYL)	0-6-0ST	IC	MW	2009	1921	
5	(ABERNANT)	0-6-0ST	IC	MW	2015	1921	
(45160	8476) 8274	2-8-0	OC	NBH	24648	1941	
56		0-6-0ST	IC	RSHN	7667	1950	
63	CORBY	0-6-0ST	IC	RSHN	7761	1954	
(D147)	46010	1Co-Co1DE		Derby		1961	
(D1994)	47292	Co-CoDE		Crewe		1966	
(D2118)	03118	0-6-0DM		Sdn		1959	
(D3180	08114) 13180	0-6-0DE		Derby		1955	
(D3952)	08784	0-6-0DE		Derby		1960	
(D4152)	08922	0-6-0DE		Hor		1962	
(D5580)	31162	A1A-A1ADE		BT	180	1960	
(D6709)	37009 (37340)	Co-CoDE		(EE	2872	1960	
				(VF	D588	1960	
D7629	(25279)	Bo-BoDE		BP	8039	1965	
(D8154)	20154	Bo-BoDE		(EE	3625	1966	
				(EEV	D1024	1966	a
41001	(43000) 252001	Bo-BoDE		Crewe		1972	
56097		Co-CoDE		Don		1981	
D2959		0-4-0DE		RH	449754	1961	
(50645)	M53(645) (T061)	2-2w-2w-2DMR		DerbyC&W		1958	
50926	(W53926 977814)	2-2w-2w-2DMR		DerbyC&W		1959	

W51138	T003	2-2w-2w-2DMR		DerbyC&W		1958	
W51151	T004	2-2w-2w-2DMR		DerbyC&W		1958	
XLR 8104	R649 JFE	4wDM	R/R	_(Landrover (Raynesway		1997 1997	
DX68807		4wDMR		Perm	008	1986	

a currently at Electro-Motive Services International Ltd, Longport, Staffordshire

SHERWOOD FOREST RAILWAY, SHERWOOD FOREST FARM PARK, LAMB PENS LANE, EDWINSTOWE, MANSFIELD NG21 9HL

Gauge 1ft 3in www.sherwoodforestrailway.com **SK 586655**

No.1	SMOKEY JOE	0-4-0STT	OC	HardyK	01	1991	
2	PET	0-4-0ST	OC	HardyK	02	1998	
3	ANNE	4wBE		HardyK	E3	1993	
1		2-2wPM		Rosewall		1947	
			reb	Vanstone		1980	OOU
4	"GEORGE"	4wDH		SherwoodF		2005	
	"MILLIE"	4wDH		SherwoodF		2015	
	a rebuild of	2w-2DM		Longleat?			

SUNDOWN ADVENTURELAND, TRESWELL ROAD, RAMPTON, near RETFORD DN22 0HX

Gauge 1ft 6in www.sundownadventureland.co.uk **SK 793792**

–	2-4-0RE	s/o	SL	546.4.93	1993	

NEIL WHITE, PRIVATE LOCATION

Gauge 4ft 8½in

1677	2-2w-2w-2RER	MetCam		c1960

OXFORDSHIRE

INDUSTRIAL SITES

MINISTRY OF DEFENCE, DEFENCE STORAGE & DISTRIBUTION CENTRE, BICESTER MILITARY RAILWAY

See Section 6 for details.

SCIENTIFIC RESEARCH LABORATORY, PRIVATE LOCATION

Gauge 5ft 6in

RLJ 7	2w-2CE	R/R	HU		1958
		reb	CE	B4611	2015

PRESERVATION SITES

PETER ALEXANDER, OXFORD AREA

Gauge 4ft 8½in

B55W	PWM 4316	2w-2PMR	Wkm	7519	1956

THE BEECHES LIGHT RAILWAY

Private railway, not open to the public.

Gauge 2ft 0in

19		0-4-0ST+WT OC		SS	3518	1888
37		2w-2-2-2wRE		EEDK	760	1930
44	[221 222]	2w-2-2-2wRE		EEDK	812	1930
	COL. FREDERICK WYLIE	4wDH		HE	9349	1994
			reb	AK		2004
–		2-2wPMR		HE	9901	2008
–		4wDM		HU	38384	1930
	MAJOR GERALD SCOTT	4wDM		MR	21619	1957

BLENHEIM PALACE RAILWAY, PLEASURE GARDENS, WOODSTOCK — OX20 1PP

Gauge 1ft 3in www.blenheimpalace.com **SP 444163**

SIR WINSTON CHURCHILL	0-6-2DH	s/o	AK	39	1992
WINSTON	0-6-0DH	s/o	AK	94	2014

CHINNOR & PRINCES RISBOROUGH RAILWAY ASSOCIATION, STATION APPROACH, STATION ROAD, CHINNOR — OX39 4ER

Gauge 4ft 8½in www.chinnorrailway.co.uk **SP 756002**

D3018	(13018 08011) HAVERSHAM	0-6-0DE	Derby		1953
(D3993	08825)	0-6-0DE	Derby		1960
(D5581	31163) 97205	A1A-A1ADE	BT	181	1960
(D6927)	37227	Co-CoDE	_(EE	3413	1963
			(EEV	D871	1963
D8059	(20059)	Bo-BoDE	_(EE	2965	1961
			(RSHD	8217	1961
D8568		Bo-BoDE	CE	4365U/69	1963
	IRIS	0-6-0DH	RH	459515	1961
(51375)	977992 (960301)	2-2w-2w-2DMR	PSteel		1960
W55023		2-2w-2w-2DMR	PSteel		1960
W55024	(977858 960010)	2-2w-2w-2DMR	PSteel		1960
(61736)	1198	4w-4wRER	Elh		1960
(61737)	1198	4w-4wRER	Elh		1960
–		2w-2PMR	Bance	AL2025	1995
WD 9024		2w-2PMR	Wkm	7090	1955

CHOLSEY & WALLINGFORD RAILWAY, HITHERCROFT ROAD, WALLINGFORD — OX10 9DQ

Gauge 4ft 8½in www.cholsey-wallingford-railway.com **SU 600891**

(D3030	08022) LION	0-6-0DE		Derby		1953
(D3074	08060) 060 UNICORN	0-6-0DE		Dar		1953
(D3190	08123) "GEORGE MASON"	0-6-0DE		Derby		1955
	CARPENTER	0-4-0DM		FH	3270	1948
(T003)	CEPS 68097 105099	4wDHR		Perm	T003	1988
No.1	WILLY SKUNK	2w-2PMR		Wkm	8774	1960

COTSWOLD WILD LIFE PARK, BRADWELL GROVE, BURFORD — OX18 4JP

Gauge 2ft 0in www.cotswoldwildlifepark.co.uk **SP 237084**

No.3		0-4-0DH	s/o	AK	17	1985
No.4	BELLA	0-4-0DH	s/o	AK	68	2003

COULSDON OLD VEHICLE & ENGINEERING SOCIETY, and NETWORK SOUTHEAST RAILWAY SOCIETY, FINMERE STATION YARD, NEWTON PURCELL, near BICESTER — MK18 4AU

Gauge 4ft 8½in www.nsers.org **SP 629313**

(E6037)	73130	Bo-BoDE/RE	_(EE ¯(EEV	3709 E369	1966 1966
No.46	HEATHER	4wDM	RH	242868	1946
D3058		4wDM	RH	458961	1962
(M61183)	977349 (936003) No.3	4w-4wRER	Afd/Elh		1957
62043	1753 CHRIS GREEN	4w-4wRER	York		1965
(65302)	5703 (930 204 977874)	4w-4RER	Afd/Elh		1954
(65304)	5705 (930 204 977875)	4w-4RER	Afd/Elh		1954
(65379)	930 206 977925	4w-4RER	Afd/Elh		1954
(65382)	930 206 977924	4w-4RER	Afd/Elh		1954
1018		4w-4wRER	MetCam		1959
1304		2w-2-2-2wRER	MetCam		1961
9305		4w-4wRER	MetCam		1961
RTU 6647A	99709 901033-9	2w-2PMR	Geismar	ST/01/05	2001
RTU 50097	99709 901042-0	2w-2PMR	Geismar	ST/03/38	2003
RTU 13734	99709 901123-8	2w-2BER	Geismar	5E/0007	2006
MPP 13736	99709 901024-8	2w-2BER	Geismar	5E/0008	2006

R. DIXON, PRIVATE SITE, BICESTER

Gauge 2ft 6in

(10)		4wDHF		HE	9053	1981	Dsm

Gauge 2ft 0in

No.3	CDC	0-4-2PT	OC	HE	3758	1952
–		4wDM		LB	54181	1964
	DORIS	4wDM		RH	462365	1960
EL 9		4wBE		CE	5961C	1972
MBS 236		4wBE		LMM	1066	1950

GREAT WESTERN SOCIETY, DIDCOT RAILWAY CENTRE (OX11 7NJ)

Gauge 7ft 0¼in www.didcotrailwaycentre.org.uk **SU 524906**

	FIRE FLY	2-2-2	IC	FFP		1989	
	IRON DUKE	4-2-2	IC	Resco		1985	a

Gauge 4ft 8½in

1014	COUNTY OF GLAMORGAN	4-6-0	OC	_(Llangollen		2006	
				(GWS		2006	
	a rebuild of 7927	WILLINGTON HALL		Sdn		1950	
1338		0-4-0ST	OC	K	3799	1898	
1340	TROJAN	0-4-0ST	OC	AE	1386	1897	b
1363		0-6-0ST	OC	Sdn	2377	1910	
1466	(4866)	0-4-2T	IC	Sdn		1936	
3650		0-6-0PT	IC	Sdn		1939	
3738		0-6-0PT	IC	Sdn		1937	
3822		2-8-0	OC	Sdn		1940	
4079	PENDENNIS CASTLE	4-6-0	4C	Sdn		1924	
4144		2-6-2T	OC	Sdn		1946	
2999	LADY OF LEGEND	4-6-0	OC	GWS/Riley		2006	
	a rebuild of 4942	MAINDY HALL		Sdn		1929	
5051	EARL BATHURST	4-6-0	4C	Sdn		1936	
5227		2-8-0T	OC	Sdn		1924	
5322		2-6-0	OC	Sdn		1917	
5572		2-6-2T	OC	Sdn		1929	
5900	HINDERTON HALL	4-6-0	OC	Sdn		1931	
6023	KING EDWARD II	4-6-0	4C	Sdn		1930	
6106		2-6-2T	OC	Sdn		1931	
6697		0-6-2T	IC	AW	985	1928	
6998	BURTON AGNES HALL	4-6-0	OC	Sdn		1949	
7202		2-8-2T	OC	Sdn		1934	
7808	COOKHAM MANOR	4-6-0	OC	Sdn		1938	
No.5		0-4-0WT	OC	GE		1857	
	KING GEORGE	0-6-0ST	IC	HE	2409	1942	
No.1	BONNIE PRINCE CHARLIE	0-4-0ST	OC	RSHN	7544	1949	
93	4 +	0-4-0VBTR	OC	Sdn		1908	
			reb	_(TyseleyLW		2007	
				(Llangollen		2009	
18000		A1A-A1A GTE		_(BBC(S)	4559	1949	
				(SLM	3977	1949	
(D3771	08604) 604 PHANTOM	0-6-0DE		Derby		1959	
D9516		0-6-0DH		Sdn		1964	
(W22W)	No.22	Bo-BoDMR		Sdn		1940	
(5)		0-4-0PM		GWS		c1985	
	a rebuild of	0-4-0WE		DerbyC&W		1960	
DL26		0-6-0DM		HE	5238	1962	
(A 21 W)		2w-2PMR		Wkm	6892	1954	DsmT
(B 37 W	PWM 3963)	2w-2PMR		Wkm	6948	1955	DsmT
68007	PWM 4303	2w-2PMR		Wkm	7506	1956	

a incorporates parts of RSHN 7135 / 1944

b currently at Locomotive Maintenance Services Ltd, Loughborough, Leicestershire

P. ROGERS, PRIVATE LOCATION
Gauge 4ft 8½in

(A144	PWM 2176)	2w-2PMR		Wkm	4153	1946	OOU

JIM SHACKELL, WITNEY
Gauge 1ft 3in

No.103	JOHN	4-4-2	OC	Barnes	103	1921	
5751	PRINCE WILLIAM	4-6-2	OC	G&S		1949	b
–		2-6-2	OC	LemonB		1967	a
5127		4-4-0	OC	McGarigle		c1924	
712		0-4-0	OC	Morse		1951	
–		4-4-0PM		Morse		1939	

a not yet completed
b carries plate G&S 9/1946

RUTLAND

INDUSTRIAL SITES

HEIDELBERG CEMENT GROUP, t/a HANSON CEMENT, KETTON WORKS, KETTON (part of the Hanson Heidelberg Cement Group) PE9 3SX
Gauge 4ft 8½in www.heidelbergcement.com **SK 987057**

(D3460)	08375 21	0-6-0DE		Dar		1957	a
(D3789)	08622 H 028 19	0-6-0DE		Derby		1959	a

a on hire from British American Railway Services Ltd, Stanhope, Co. Durham

PRESERVATION SITES

RUTLAND RAILWAY MUSEUM LTD, t/a ROCKS BY RAIL : THE LIVING IRONSTONE MUSEUM, COTTESMORE IRON ORE SIDINGS, ASHWELL ROAD, COTTESMORE, near OAKHAM LE15 7BX
Gauge 4ft 8½in www.rocks-by-rail.org **SK 887137**

(RRM 2)		0-4-0ST	OC	AB	1931	1927
(RRM 6)	SIR THOMAS ROYDEN	0-4-0ST	OC	AB	2088	1940
(RRM196)	BELVOIR	0-6-0ST	OC	AB	2350	1954
	(STAMFORD)	0-6-0ST	OC	AE	1972	1927
	“RHOS”	0-6-0ST	OC	HC	1308	1918
(89-94)	(VIGILANT)	0-4-0ST	OC	HE	287	1883
(RRM 192)		0-6-0ST	OC	HL	3138	1915
(RRM 5)	SINGAPORE	0-4-0ST	OC	HL	3865	1936
RRM 1	UPPINGHAM	0-4-0ST	OC	P	1257	1912
RRM 4	(ELIZABETH)	0-4-0ST	OC	P	1759	1928

(RRM 183)	8310/11 (CRANFORD No.2)	0-6-0ST	OC	WB	2668	1942	
9	"EXTON PARK"	0-6-0ST	OC	YE	2521	1952	
	LUDWIG MOND	6wDE		GECT	5578	1980	c
(RRM 175)		0-4-0DH		JF	4220007	1960	
RRM 12		4wDM		RH	306092	1950	
(RRM 176)	No.198 ELIZABETH	0-4-0DE		RH	421436	1958	
RRM 150		0-4-0DE		RH	544997	1969	
RRM 21	(BETTY)	0-4-0DH		RR	10201	1964	
	JEAN	0-4-0DH		RR	10204	1965	
(RRM 186)	61 689/171 GRAHAM	0-4-0DH		RR	10207	1965	
No.8		4wDH		TH	178V	1967	
(RRM 177)	No.4 MR.D	4wDH		TH	186V	1967	
(RRM 16)	DE 5	0-6-0DE		YE	2791	1962	
68/038		2w-2DMR		Wkm	1947	1935	
	rebuilt as	2w-2DMR/BER		DonM		1983	DsmT b
(DB 965072)		2w-2PMR		Wkm	7587	1957	
–		2w-2PMR		Wkm		1956	DsmT a

a one of Wkm 7445/1956 or Wkm 7581/1956
b currently stored elsewhere
c property of Andrew Briddon

SHROPSHIRE

PRESERVATION SITES

CAMBRIAN HERITAGE RAILWAYS
Cambrian Railways Society, Oswestry

Locomotives can be found at : Blodwell Junction, Llanyblodwel SY10 8LZ SJ 252230
Oswestry Cycle & Railway Museum, Station Yard, Oswald Road, Oswestry SY11 1RE SJ 294297

Gauge 4ft 8½in www.cambrianrailways.com

–		0-6-0ST	OC	AB	885	1900	
–		0-4-0ST	OC	AB	2261	1949	
	NORMA	0-6-0ST	IC	HE	3770	1952	
	ADAM	0-4-0ST	OC	P	1430	1916	
	OLIVER VELTOM	0-4-0ST	OC	P	2131	1951	
7	HO37	0-6-0DH		EEV	D1201	1967	
	CYRIL	4wDM		FH	3541	1952	
	ALPHA	4wDM		FH	3953	1960	
–		0-4-0DM		HC	D843	1954	
	SCOTTIE	4wDM		RH	412427	1957	
	TELEMON	0-4-0DM		(VF (DC	D295 2568	1955 1955	
	PPM 30	2w-2F/BER		PPM	11	1995	
98205		4wDMR		Plasser	52760A	1985	

Cambrian Railway Trust, Llynclys Goods Yard, Llynclys **SY10 8BX**
Gauge 4ft 8½in **SJ 285239**

D3019	(13019)	0-6-0DE		Derby		1953	
–		0-6-0DH		EEV	D1230	1969	

11517	ALUN EVANS	0-4-0DE		RH	458641	1963	
W51187		2-2w-2w-2DMR		MetCam		1958	
W51205		2-2w-2w-2DMR		MetCam		1958	
W51512		2-2w-2w-2DMR		MetCam		1959	
–		2w-2PMR		Geismar	98/28	1998	a

a stored at private site in Weston Rhyn

G.FAIRHURST, c/o ENGLISH NATURE, MANOR HOUSE WORKS, near WHIXALL

Gauge 2ft 0in **SJ 505366**

–		4wPM		MR	1934	1919	a
–		4wDM		RH	191679	1938	a

a locomotives stored at a private location

GLYN VALLEY TRAMWAY, PRIVATE SITE

Gauge 600mm

2		4wPM		Fordson			

HADLEY LEARNING COMMUNITY, CRESCENT ROAD, HADLEY, TELFORD **TF1 5JU**

Gauge 4ft 0in www.hadleylearningcommunity.org.uk

–		4wG		TU		1987	

IRONBRIDGE GORGE MUSEUM TRUST LTD - IRONBRIDGE BIRTHPLACE OF INDUSTRY

Blists Hill Victorian Town, Legges Way, Madeley, Telford **TF7 5UD**

Gauge 4ft 8½in www.ironbridge.org.uk **SJ 693033**

–		0-6-0ST	OC	AB	782	1896	

Gauge 3ft 0in

–		4wG	OC	CastleGKN		1990	a

a working replica of Trevithick loco, operates on plateway-type flanged rails.

Gauge 2ft 0in

	SIR PETER GADSDEN	4wBE		AK	84	2008	

Coalbrookdale Museum of Iron, Coach Road, Coalbrookdale **TF8 7QD**

Locomotives on static display.

Gauge 4ft 8½in **SJ 668047**

–		0-4-0VBT	VCG	S	6155	1925	
	a rebuild of	0-4-0ST	OC	MW	437	1873	
–		0-4-0VBT	VCG	S	6185	1925	
	a rebuild of (6)	0-4-0ST	OC	Coalbrookdale		c1865	

Enginuity, Coach Road, Coalbrookdale **TF8 7DX**

Gauge 4ft 8½in **SJ 667048**

5		0-4-0ST	OC	Coalbrookdale		c1865	

DAVID LEWIS, WHITCHURCH

Gauge 4ft 8½in

–		0-4-0DM		JF	4210074	1952	

PRIVATE OWNER, PRIVATE LOCATION

Gauge 4ft 8½in

68019	(DE 320501)	2w-2PMR		Wkm	7598	1957	

OSWESTRY & DISTRICT NARROW GAUGE GROUP, WESTON WHARF, MORDA

Gauge 3ft 0in **SJ 298276**

A		4wDMF		RH	433388	1959	Dsm

Gauge 2ft 6in

ND 3051		0-4-0DM		HE	2022	1939	Dsm

Gauge 2ft 0in

	IORWERTH	0-4-0VBT	VC	ACAB		2008	
	BUSTA	0-4-0PM		ACAB	1	2004	Dsm
8		4wPM		KC		c1926	
89		4wDM		MR	4565	1928	
SO 39		4wDM		MR	7190	1937	
D		4wDM		MR	7463	1939	
T3		4wDM		MR	8738	1942	
	rebuilt as	4wDH				c1998	
8860		4wDM		MR	8860	1944	Dsm
	(WEDHOLME)	4wDM		MR	8885	1944	Dsm
	MOLE	4wDM		MR	22031	1959	
–	"WARRIOR"	4wDM		RH	191680	1938	Dsm
496038	BARNY	4wDM		RH	496038	1963	
	LLANFFORDA	4wDM		RH	496039	1963	
7	VULCAN	4wDM		RH	7002/0967/6	1967	
JMLM19		4wBE		WR	6502	1962	
			reb	WR	10102	1983	
JMLM20		4wBE		WR	6505	1962	
			reb	WR	10105	1983	
JMLM21		4wBE		WR	6503	1962	
			reb	WR	10104	1983	
JMLM22		4wBE		WR	6504	1962	
			reb	WR	10106	1983	
	(RATTY)	2w-2PM		BarberA			Dsm

SEVERN VALLEY RAILWAY CO LTD

Locomotives are kept at :-

Bewdley, Worcestershire	DY12 1BG	SO 793753
Bridgnorth, Shropshire	WV16 5DT	SO 715926
Highley, Shropshire	(WV16 6NU)	SO 749831
Kidderminster, Worcestershire	DY10 1QX	SO 836757

Gauge 4ft 8½in www.svr.co.uk

813		0-6-0ST	IC	HC	555	1900	

1450		0-4-2T	IC	Sdn		1935	
1501		0-6-0PT	OC	Sdn		1949	
2857		2-8-0	OC	Sdn		1918	
4150		2-6-2T	OC	Sdn		1947	
4930	HAGLEY HALL	4-6-0	OC	Sdn		1929	
7714		0-6-0PT	IC	KS	4449	1930	
7802	BRADLEY MANOR	4-6-0	OC	Sdn		1938	
34027	TAW VALLEY	4-6-2	3C	Bton		1946	
34053	SIR KEITH PARK	4-6-2	3C	Bton		1947	
42968		2-6-0	OC	Crewe	136	1934	
43106		2-6-0	OC	Dar	2148	1951	
75069		4-6-0	OC	Sdn		1955	
82045		2-6-2T	OC	SVR		2015	
	WARWICKSHIRE	0-6-0ST	IC	MW	2047	1926	
	DUNROBIN	0-4-4T	IC	SS	4085	1895	
	CATCH ME WHO CAN	2-2-0	IC	SVR		2008	
(D407)	50007 HERCULES	Co-CoDE		_(EE	3777	1967	
				(EEV	D1148	1967	
(D431)	50031 HOOD	Co-CoDE		_(EE	3801	1968	
				(EEV	D1172	1968	
(D433)	50033 GLORIOUS	Co-CoDE		_(EE	3803	1968	
				(EEV	D1174	1968	
(D435)	50035 (50135)	Co-CoDE		_(EE	3805	1968	
	(ARK ROYAL)			(EEV	D1176	1968	
(D444)	50044 EXETER	Co-CoDE		_(EE	3814	1968	
				(EEV	D1185	1968	
(D449)	50049 (50149)	Co-CoDE		_(EE	3819	1968	
	DEFIANCE			(EEV	D1190	1968	
D821	GREYHOUND	B-BDH		Sdn		1960	
D1013	WESTERN RANGER	C-CDH		Sdn		1962	
D1015	(89416) WESTERN CHAMPION	C-CDH		Sdn		1962	
D1062	WESTERN COURIER	C-CDH		Crewe		1963	
D3022	(08015)	0-6-0DE		Derby		1953	
D3201	(08133)	0-6-0DE		Derby		1955	
D3586	(08471)	0-6-0DE		Crewe		1958	
(D3802)	08635	0-6-0DE		Derby		1959	
(D4013	08845) 09107	0-6-0DE		Hor		1961	
D4100	(09012) DICK HARDY	0-6-0DE		Hor		1961	
(D4126)	08896 STEVEN DENT	0-6-0DE		Hor		1962	
D5410	(27059)	Bo-BoDE		BRCW	DEL253	1962	a
D7029		B-BDH		BPH	7923	1962	
D8188	(20188)	Bo-BoDE		_(EE	3669	1966	
				(EEV	D1064	1966	
D9551		0-6-0DH		Sdn		1965	
12099		0-6-0DE		Derby		1952	
D2957		0-4-0DM		RH	319290	1953	
D2960	SILVER SPOON	0-4-0DM		RH	281269	1950	
D2961		0-4-0DE		RH	418596	1957	
M50933		2-2w-2w-2DMR		DerbyC&W		1960	
M51941		2-2w-2w-2DMR		DerbyC&W		1960	
E52064		2-2w-2w-2DMR		DerbyC&W		1961	
(LMS 017)		2w-2PMR		Lesmac	LMS017	2006	
(PT 2P)		2w-2PMR		Wkm	1580	1934	

PWM 3189		2w-2PMR		Wkm	5019 1948	DsmT
DB 965054		2w-2PMR		Wkm	7577 1957	DsmT
(PT 1P TP 49P)		2w-2PMR		Wkm	7690 1957	
8085	(9021)	2w-2PMR		Wkm	8085 1958	

a currently at East Somerset Railway, Shepton Mallet, Somerset

The Engine House, Highley **(WV16 6NU)**
Gauge 4ft 8½in **SO 748829**

4566		2-6-2T	OC	Sdn		1924
5764		0-6-0PT	IC	Sdn		1929
7325		2-6-0	OC	Sdn		1932
7819	HINTON MANOR	4-6-0	OC	Sdn		1939
45110		4-6-0	OC	VF	4653	1935
46443		2-6-0	OC	Crewe		1950
47383		0-6-0T	IC	VF	3954	1926
48773	(8233 WD 70307)	2-8-0	OC	NBH	24607	1940
80079		2-6-4T	OC	Bton		1954
WD 600	GORDON	2-10-0	OC	NBH	25437	1943
	THE LADY ARMAGHDALE	0-6-0T	IC	HE	686	1898

SHREWSBURY STEAM TRUST, c/o SHROPSHIRE COUNCIL, SHREWSBURY MUSEUMS SERVICE, COLEHAM PUMPING STATION, LONGDEN COLEHAM, SHREWSBURY **SY3 7DN**

Gauge 2ft 0in www.colehampumpingstation.co.uk **SJ 496121**

–	2w-2BE	TargettR		2017

SHROPSHIRE MINES TRUST, SNAILBEACH LEAD MINES

Locomotives are kept at another, private, site.

Gauge 1ft 11½in www.shropshiremines.org.uk **SJ 375022**

–		0-4-0BE	WR	S7950	1978
RED DWARF		0-4-0BE	WR	5655	1956
	rebuilt as	4wBE	ShepherdFG		1980

THE TANAT VALLEY LIGHT RAILWAY CO LTD, NANTMAWR VISITOR CENTRE, NANTMAWR **SY10 9HW**

Gauge 4ft 8½in www.nantmawrvisitorcentre.co.uk **SJ 254244**

(2145)		0-4-0DM		HE	2145	1940
EE8416		4wDM		RH	338416	1953
SC52005	(977832)	2-2w-2w-2DMR		DerbyC&W		1961
(51993	977834 S003 960933)	2-2w-2w-2DMR		DerbyC&W		1961
(52012	977835 S003 960933)	2-2w-2w-2DMR		DerbyC&W		1960
SC52031		2-2w-2w-2DMR		DerbyC&W		1961
61937	(309616 977963) 960101	4w-4wWER		York		1962
(68012	DB 965065)	2w-2PMR		Wkm	7580	1956

Gauge 2ft 0in

14005	STEAM TRAM	4wVBT	G			1969
	rebuilt from	4wDM		L	14005	1940

39005		4wPM	L	39005	1952	
(C37)		2w-2PMR	Locospoor	B7281E		Dsm
	RAIL TAXI	4-2-0PMR	MorrisRP		1967	
PWM 2788	6887 (A16W)	2w-2PMR	Wkm	6887	1954	
	rebuilt as	2w-2DMR	ENG/GEM		1994	

Gauge monorail – The Monorail Collection

1795	HULL	1w1PM	RM	1795	1952	
2910	HAZLEMERE	1w1PM	RM	2910	1953	
3250	BOLTON (HEAP CLOUGH)	1w1PM	RM	3250	1953	
3906	WANTAGE	1w1PM	RM	3906	1954	
3981	(BISHOPS SUTTON)	1w1PM	RM	3981	1955	
4114	BOSTON 'A' (FOLDHILL)	1w1PM	RM	4114	1955	
4810	(BINFIELD)	1w1PM	RM	4810	1955	
4904	SACRISTON	1w1PM	RM	4904	1956	
(4989)	1989 STEYNING	1w1PM	RM	4989	1956	
5013	STROUD	1w1PM	RM	5013	1956	a Dsm
5041	BODMIN	1w1PM	RM	5041	1956	
(5074)		1w1PM	RM	5074	1956	b
(6560)		1w1PM	RM	6560	1957	b
6572	FELBRIDGE	1w1PM	RM	6572	1957	
7050	COOKHAM	1w1PM	RM	7050	1957	
7182	ESHER 'A'	1w1PM	RM	7182	1958	
7481	EDENBRIDGE	1w1PM	RM	7481	1958	
7493	HENFIELD	1w1PM	RM	7493	1958	
7498	HASLINGDEN	1w1PM	RM	7498	1958	
8071	CHIPPING NORTON	1w1PM	RM	8071	1959	
	UCKFIELD 'A'	1w1PM	RM	8073	1959	Dsm
8118	(GODSTONE)	2wPH	RM	8118	1959	
8415	BINFIELD 'B'	1w1PM	RM	8415	1959	Dsm
–		2wDH	RM	8423	1959	a
BATTERY ELECTRIC MONORAIL TRAM LOCO		1w1PM	RM	8564	1959	
	rebuilt as	1w1BE	PilbeamA			Dsm
8583	SHENFIELD	2wPH	RM	8583	1959	
–		1w1PM	RM	8611	1959	Dsm
8633	ESHER 'B'	1w1PM	RM	8633	1959	
8638	SLEAFORD	1w1PM	RM	8638	1960	
8653	(CHERTSEY)	2wPH	RM	8653	1959	
8663	(LLANGEFNI)	2wPH	RM	8663	1960	
8862	"UCKFIELD 'B'"	2wPH	RM	8862	1960	
8992	FAVERSHAM	1w1PM	RM	8992	1960	
9391	(WANSTEAD)	2wPH	RM	9391	1960	
9392	(WOODFORD)	2wPH	RM	9392	1960	
9536	CREWKERNE	1w1PM	RM	9536	1961	
9537	(KIDLINGTON)	1w1PM	RM	9537	1961	
9795	(NAYLAND)	2wPH	RM	9795	1960	
9811	(ESHER 'C')	2wPH	RM	9811	1961	
9812	(ESHER 'D')	2wPH	RM	9812	1961	
9852	TADLEY	2wPH	RM	9852	1961	
9853	"INGRAVE 'A'"	2wPH	RM	9853	1961	
9869	UCKFIELD 'C'	2wPH	RM	9869	1961	
9925	LLANGOLLEN	2wPH	RM	9925	1961	

10051	GILLINGHAM	1w1PM	RM	10051	1961		
10067	CAMBERLEY	2wPH	RM	10067	1961		
10069	(LIVERPOOL 'A')	2wPH	RM	10069	1961		
–		2wPH	RM	10070	1961	a	Dsm
	(SPALDING 'A')	2wGasH	RM	10248	1961	c	
10258	HESWALL	2wPH	RM	10258	1961		
10260	"EARLEY"	2wPH	RM	10260	1961		
	OWSTON FERRY	1w1PM	RM	10268	1961		
10422	HALE	2wPH	RM	10422	1961		
10593	SPALDING 'B'	2wGasH	RM	10593	1962	c	
10744		1w1PH	RM	10744	1962		
10782	HARWELL	1w1PH	RM	10782	1962		
10809	STORRINGTON	1w1PH	RM	10809	1962		OOU
–	(HULL 'B')	1w1PH	RM	10894	1962		Dsm
10900	BOURNEMOUTH 'A'	1w1PH	RM	10900	1962		OOU
10952	(HAVERHILL)	1w1PH	RM	10952	1962		OOU
11097	(AYLESFORD)	1w1PH	RM	11097	1962		OOU
11204	OXFORD	1w1PH	RM	11204	1963		OOU
11206	(EYNSHAM)	1w1PH	RM	11206	1963		OOU
–		1w1PH	RM	11338	1963		Dsm
–		1w1PH	RM	11451	1963		Dsm
11457	(WEST CALDER)	1w1PH	RM	11457	1963		OOU
"11459"	(TANKERSLEY)	1w1PH	RM	11459	1963		OOU
"11460"	ASHINGTON	1w1PH	RM	11460	1963		OOU
11809	HECKINGTON	1w1PH	RM	11809	1963	d	
11836	(BOWBURN)	1w1PH	RM	11836	1963		OOU
12124	WICKHAM	1w1PH	RM	12124	1964		OOU
12126	CHELTENHAM	1w1PH	RM	12126	1964		
12432	(HOWDEN)	1w1PH	RM	12432	1964		
"12438"	(DURHAM)	1w1PH	RM	12438	1964		OOU
–		1w1DH	RM	12625	1964		Dsm
–		1w1DH	RM	12626	1964		Dsm
12634	(HULL 'C')	1w1DH	RM	12634	1964		OOU
12649	MOSTYN	1w1GasH	RM	12649	1964		OOU
12713	ABINGDON	1w1PM	RM	12713	1964		
12869	(SWINDON)	1w1PH	RM	12869	1964		OOU
	LONDON 'A'	2wPH	RM	13300	1965	d	
	LONDON 'B'	2wPH	RM	13301	1965	d	
13313	SAUNDERSFOOT	2wPH	RM	13313	1965	d	
13318	"LIVERPOOL 'B'"	2wPH	RM	13318	1965		OOU
"13626"	BOURNEMOUTH 'B'	2wPH	RM	13626	1965		OOU
13629	LANCHESTER	2wPH	RM	13629	1965		
"13911"	(ESHER 'E')	2wPH	RM	13911	1965		OOU
13913	SWANWICK	2wPH	RM	13913	1965		OOU
13929	(INGRAVE 'B')	2wPH	RM	13929	1965		OOU
14753	WITTON GILBERT	2wPH	RM	14753	1966		OOU
14766	(WHARNCLIFFE-SIDE)	2wPH	RM	14766	1966		
	BANBURY	2wDH	RM	14769	1966		Dsm
14789	LEESWOOD	2wPH	RM	14789	1966		
14806	TIPTREE	2wPH	RM	14806	1966		OOU
15152	"WESTBURY"	2wDH	RM	15152	1967		OOU

15773	PENTRE HALKYN	2wPH		RM	15773	1967	d
15778	(HOLME UPON SPALDING MOOR)	2wPH		RM	15778	1967	
15784	(CONINGSBY)	2wDH		RM	15784	1967	e
No.2	MEXBOROUGH	2wDH		Metalair	20002	1968	
3	LINCOLN 'A'	2wDH		Metalair	20003	1968	
4	(BRENTWOOD 'A')	2wPH		Metalair	20004	1968	
5	(BRENTWOOD 'B')	2wPH		Metalair	20005	1968	
20007	FARINGDON	1w1DH		Metalair	20007	1968	
17	(MONO TANKER 'A')	2wDH		Metalair	20017	1968	
18	MONO TANKER 'B'	2wDH		Metalair	20018	1968	
19	LINCOLN 'B'	2wDH		Metalair	20019	1968	
25	SKELLINGTHORPE	2wDH		Metalair	20025	1968	
No.39	BROUGHTON ASTLEY	2wPH		Metalair	20039	1968	Dsm
44	CHIRK	2wPH		Metalair	20044	1968	
65		2wDH		Metalair	20065	1968	f
	SHEPSHED	2wDH		Metalair	20067	1968	
69		2wDH		Metalair	20069	1968	g
20091	PARTINGTON	1w1DH		Metalair	20091	1968	
No.140	RAILCAR A	2wDH		Metalair	20140	1969	e
No.141	RAILCAR B	2wDH		Metalair	20141	1969	e
No.146	MONOLOCO	0-2-0ST	OC	CM	11003	1998	h
161	LLANGEFNI (WITHERNSEA)	2wPH		Metalair	20161	1969	d
No.212	POOLE	2wDH		Metalair	20212	1970	
222	TREFNANT	2wPH		Metalair	20222	1971	
20228		2wDH		Metalair	20228	1970	
20240	(SALISBURY)	2wDH		Metalair	20240	1971	
	DOWNTON	2wDH		Metalair	20242	1971	
265	BOSTON 'B'	2wDH		Metalair	20265	1973	i

a converted to unpowered wagon
b converted to permanent way manrider
c horticultural flat power wagon
d hydraulic tip power wagon
e power wagon
f converted to passenger carriage
g converted to barrel carrying wagon
h built on chassis of trailer wagon Metalair 20146/1969
i converted to railcar 2002

TELFORD HORSEHAY STEAM TRUST, TELFORD STEAM RAILWAY, THE OLD LOCO SHED, BRIDGE ROAD, HORSEHAY, TELFORD — TF4 2NF

Gauge 4ft 8½in www.telfordsteamrailway.co.uk **SJ 675073**

–		0-4-0F	OC	AB	1944	1927	OOU
"139	BEATTY"	0-4-0ST	OC	HL	3240	1917	
	ROCKET	0-4-0ST	OC	P	1722	1926	a b
3		0-4-0ST	OC	P	1990	1940	
(D3925)	08757 EAGLE	0-6-0DE		Hor		1961	
(D6963)	37263	Co-CoDE		(EE (EEV	3523 D952	1964 1964	
27414	TOM	0-4-0DH		NBQ	27414	1954	
183062	FOLLY	4wDM		RH	183062	1937	Dsm
D2971		0-4-0DM		RH	313394	1952	

D2959		4wDM	RH	382824	1955	
–		0-4-0DH	RH	525947	1968	
50479	(53479)	2-2w-2w-2DMR	BRCW		1958	
50531	(53531)	2-2w-2w-2DMR	BRCW		1958	
(W51950)		2-2w-2w-2DMR	DerbyC&W		1961	
(W52062)		2-2w-2w-2DMR	DerbyC&W		1961	

a property of Somerset & Dorset Locomotive Co Ltd
b currently at Tyseley Steam Locomotive Works Ltd, Birmingham

Gauge 2ft 0in - Telford Town Tramway **SJ 674072**

–		4wVBTram VCG	(Kierstead (AK		1979 1979	
–		4wDM	RH	222101	1943	Dsm

SOMERSET

INDUSTRIAL SITES

AGGREGATE INDUSTRIES UK LTD, MEREHEAD STONE TERMINAL, TORR WORKS, SHEPTON MALLET BA4 4RA

(part of Holcim Group; operated by Mendip Rail Ltd)

Gauge 4ft 8½in www.aggregate.com **ST 693426**

(D3810)	08643	0-6-0DE	Hor		1959	
(D4163)	08933	0-6-0DE	Dar		1962	
(D4177)	08947	0-6-0DE	Dar		1962	
277		6wDE	GECT	5475	1978	a
44	WESTERN YEOMAN II	Bo-BoDE	GM	798033-1	1980	

a property of Ed Murray & Sons Ltd, Hartlepool, Teesside

HANSON QUARRY PRODUCTS EUROPE LTD, t/a HANSON AGGREGATES, WHATLEY QUARRY, near FROME BA11 3LF

(part of the Heidelberg Cement Group; operated by Mendip Rail Ltd)

Gauge 4ft 8½in www.hanson.co.uk **ST 733479**

(D3817)	08650	0-6-0DE	Hor		1959
(D3819)	08652	0-6-0DE	Hor		1959
120	KENNETH JOHN WITCOMBE	4w-4wDE	GM	37903	1972
No.4		4wDH	TH	200V	1968

PRESERVATION SITES

BLATCHFORD LIGHT RAILWAY, EMBOROUGH QUARRY, EMBOROUGH, near WELLS

This a private railway, not open to the public.

Gauge 2ft 6in

9	YARD No.24	0-4-0DM	HE	2017	1939

2	YARD No. B6	0-4-0DM		HE	2266	1940
ND 3060		0-4-0DM		HE	2398	1941
	YARD No.1075	4wDH		HE	7447	1976
	YARD No.1074	4wDH		HE	7451	1976
10		4wDHF	RACK	HE	9057	1981
–		4wDM		RH	398101	1956

Gauge 750mm

–		4wDH?		GIA	55134	1966

Gauge 2ft 4in

	CORRIS	4wDM		RH	398102	1956

Mr. BOND, YEOVIL AREA

Gauge 3ft 0in

–		4wDM		JF	3930048	1951

Stored at unknown location

B. CLARKE, 11 PENN GARDENS, BATH

These locomotives are not on public display.

Gauge 2ft 0in

	CLARA	4wPM		Bonnymount		c1986
–		4wDM		L	38296	1952
	ADAM	4wDM		MR	9978	1954
–		4wDM		OK	7595	1937
–		4wDM		RH	213834	1942

CLEVEDON MINIATURE RAILWAY, SALTHOUSE FIELDS, CLEVEDON **BS21 7RH**

Gauge 1ft 3in **ST 397710**

5305		4-6-0BE	s/o	MossAJ		1999

EAST SOMERSET RAILWAY CO LTD, WEST CRANMORE RAILWAY STATION, SHEPTON MALLET **BA4 4QP**

Locomotives are kept at :- West Cranmore Station ST 664429
Merryfield Lane Station ST 654425

Gauge 4ft 8½in www.eastsomersetrailway.com

5239	GOLIATH	2-8-0T	OC	Sdn		1924
5637		0-6-2T	IC	Sdn		1925
46447		2-6-0	OC	Crewe		1950
1719	LADY NAN	0-4-0ST	OC	AB	1719	1920
1		0-6-0T	IC	RSHN	7609	1950
(D)6566	(33048)	Bo-BoDE		BRCW	DEL170	1961
51909	L231	2-2w-2w-2DMR		DerbyC&W		1960
DH16		4wDH		S	10175	1964
	CATTEWATER	4wDH		RR	10199	1964
39	"CABOT"	0-6-0DH		RR	10218	1965

–		0-6-0DH		RR	10221	1965	
2	JOAN	0-4-0DH		S	10165	1964	
–		2-2wBER		Cranmore		c2015	

GARTELL LIGHT RAILWAY, COMMON LANE, YENSTON, near TEMPLECOMBE — BA8 0NB

Gauge 2ft 0in www.newglr.weebly.com **ST 718218**

No.6	MR.G	0-4-2T	OC	NDLW	698	1998	
9	JEAN	0-4-0TT	OC	NDLW		2008	
No.5	ALISON	4wDH		AK	No.10	1983	
No.2	ANDREW	4wDH		BD	3699	1973	
No.1	AMANDA	Bo-BoDH		Gartell		2003	a
–		4wDM		RH	193984	1939	b

a incorporates parts from SL 7323 1973
b converted into a brake van

JJP HOLDINGS SOUTH WEST, UNIT 2, WESTLAND DISTRIBUTION PARK, WINTERSOKE ROAD, WESTON-SUPER-MARE — BS24 9AD

Gauge 4ft 8½in www.jjpsouthwest.co.uk ST 337594

7027	THORNBURY CASTLE	4-6-0	4C	Sdn		1949	

J. KESWICK, PRIVATE LOCATION, near BRIDGWATER

Gauge 2ft 0in

–		4wPM		FH	1830	1933	

PRIVATE OWNER, PRIVATE LOCATION, NORTH SOMERSET

Gauge 1ft 6in

12	GREENSBURG	4wBE		Greensburg	2368	1950	

PRIVATE OWNER, near RADSTOCK

Gauge 4ft 8½in

2		4wDH		TH	136C	1964	
	a rebuild of	4wVBT	VCG	S			

SANDFORD STATION RAILWAY HERITAGE CENTRE LTD, ST. MONICA TRUST RETIREMENT VILLAGE, STATION ROAD, SANDFORD, WINSCOMBE — BS25 5AA

Gauge 4ft 8½in www.sandfordstation.co.uk **ST 416595**

	MARJORIE	4wVBT	VCG	S	9387	1948	

SOMERSET & DORSET RAILWAY HERITAGE TRUST, MIDSOMER NORTON SOUTH STATION, SILVER STREET, MIDSOMER NORTON BA3 2EY

Gauge 4ft 8½in www.sdjr.co.uk **ST 664536**

	JOYCE	4wVBT	VCG	S	7109	1927	
D1120	DAVID J. COOK	0-6-0DH		EEV	D1120	1966	
D4095	(08881) 881	0-6-0DE		Hor		1961	
(PWM 4301	TR18)	2w-2PMR		Wkm	7504	1956	

SOMERSET & DORSET RAILWAY MUSEUM TRUST, WASHFORD STATION, WASHFORD TA23 OPP

Gauge 4ft 8½in www.sdrt.org.uk **ST 044412**

–		4wDM		RH	210479	1942	

Gauge 2ft 0in

–		4wDM		L	42319	1956	

SOUTH WEST MAIN LINE STEAM COMPANY plc YEOVIL RAILWAY CENTRE, YEOVIL JUNCTION STATION, YEOVIL BA22 9UU

Gauge 4ft 8½in www.yeovilrailway.freeservers.com **ST 570140**

	LORD FISHER	0-4-0ST	OC	AB	1398	1915	
	PECTIN	0-4-0ST	OC	P	1579	1921	
–		0-4-0DM		JF	22898	1940	Dsm
	SAM	0-4-0DM		JF	22900	1941	
44	COCKNEY REBEL	0-4-0DM		JF	4000007	1947	
DS1174	RIVER YEO	4wDM		RH	458959	1961	

WEST SOMERSET RAILWAY plc

Locomotives are kept at :-

Bishops Lydeard TA4 3RU ST 164290
Dunster TA24 6PJ SS 996447
Minehead TA24 5BG SS 975463
Williton Goods Yard TA4 4RQ ST 085416

Gauge 4ft 8½in www.west-somerset-railway.co.uk

4110		2-6-2T	OC	Sdn		1936	
4561		2-6-2T	OC	Sdn		1924	
4936	KINLET HALL	4-6-0	OC	Sdn		1929	a
6024	KING EDWARD I	4-6-0	4C	Sdn		1930	
6695		0-6-2T	IC	AW	983	1928	
6960	RAVENINGHAM HALL	4-6-0	OC	Sdn		1944	
7828	ODNEY MANOR	4-6-0	OC	Sdn		1950	
9351		2-6-0	OC	WSR		2004	
	rebuilt from 5193	2-6-2T	OC	Sdn		1934	
44422		0-6-0	IC	Derby		1927	
53808		2-8-0	OC	RS	3894	1925	
	FORESTER	0-4-0ST	OC	AB	1260	1911	
	CALEDONIA WORKS	0-4-0ST	OC	AB	1219	1910	
–		0-4-0F	OC	AB	1984	1930	
–		0-6-0T	OC	HC	1857	1952	

D1010	WESTERN CAMPAIGNER	C-CDH	Sdn		1962	
D1661	(47076 47613 47840) NORTH STAR					
		Co-CoDE	Crewe		1965	
D2133		0-6-0DM	Sdn		1960	
D4107	(09019)	0-6-0DE	Hor		1961	
(D)6566	(33048)	Bo-BoDE	BRCW	DEL170	1961	b
D6575	(33057)	Bo-BoDE	BRCW	DEL179	1961	
D7017		B-BDH	BPH	7911	1961	
D7018		B-BDH	BPH	7912	1961	
(D9518)	9312/95 No.7	0-6-0DH	Sdn		1964	
D9526		0-6-0DH	Sdn		1964	
–		0-4-0DH	AB	578	1972	
–		0-4-0DH	AB	579	1972	
200793	GOWER PRINCESS	4wDM	RH	200793	1940	
51663		2-2w-2w-2DMR	DerbyC&W		1960	c Dsm
51859	859	2-2w-2w-2DMR	DerbyC&W		1960	
51880	880	2-2w-2w-2DMR	DerbyC&W		1960	
(51887)		2-2w-2w-2DMR	DerbyC&W		1960	
4162		2w-2PMR	Geismar	ST/02/27	2002	
4163		2w-2PMR	Geismar	ST/02/28	2002	

a on loan from Tyseley Locomotive Works Ltd, Birmingham, West Midlands
b currently at East Somerset Railway, Cranmore
c frame used by PW Dept as a flat wagon

WESTONZOYLAND ENGINE TRUST, WESTONZOYLAND PUMPING STATION, MUSEUM OF STEAM POWER & LAND DRAINAGE, HOOPERS LANE, WESTONZOYLAND, near BRIDGWATER — TA7 OLS

Gauge 2ft 0in www.wzlet.org **ST 340328**

6299		4wPM	L	6299	1935
–		4wDM	L	34758	1949
024		4wDM	MR	40S310	1968

STAFFORDSHIRE

INDUSTRIAL SITES

BOMBARDIER TRANSPORTATION UK LTD, CENTRAL RIVERS DEPOT, BARTON-UNDER-NEEDWOOD — DE13 8ES

Gauge 4ft 8½in www.bombardier.com **SJ 203177**

(D4173)	08943	0-6-0DE	Dar		1962	a

a on hire from Harry Needle Railroad Co Ltd, Derbyshire

CLAYTON EQUIPMENT LTD,
SECOND AVENUE, CENTRUM 100, BURTON-UPON-TRENT DE14 2WF

New CE locomotives under construction and locomotives for repair usually present.

Gauge 2ft 0in www.claytonequipment.co.uk

	DIANE	4wBE		CE	RD001	2016	a

a demonstration locomotive

DB CARGO MAINTENANCE LTD, WHEILDON ROAD, STOKE-ON-TRENT ST4 4HP

Gauge 4ft 8½in uk.dbcargo.com **SJ 880439**

D3575	08460 SPIRIT OF THE OAK	0-6-0DE		Crewe		1958	a

a property of Railway Support Services Ltd, Wishaw, Warwickshire

ELECTRO-MOTIVE DIESEL LTD, LONGPORT GOODS YARD, BROOKSIDE INDUSTRIAL ESTATE, off STATION STREET, LONGPORT ST6 4NF

(Subsidiary of Progress Rail; A Caterpillar Company)

Gauge 4ft 8½in www.progressrail.com **SJ 855495**

(D3290)	08220	0-6-0DE		Derby		1956	a
(D8154)	20154	Bo-BoDE		(EE	3625	1966	
				(EEV	D1024	1966	
	(66048)	Co-CoDE		GMC	968702-48	1998	Dsm

a property of Railway Support Services Ltd, Wishaw, Warwickshire

F.M.B. / T.G.S., PRIVATE SITE

Locomotives under repair or restoration are usually present.

Gauge 2ft 6in **SU 769345**

(2	LENA)	0-4-2ST	OC	KS	1098	1910	
(6	LUCY)	0-4-2ST	OC	KS	1313	1916	
–		0-4-2ST	OC	KS	3025	1917	
–		4wDM		Plymouth		1939	
	a rebuild of	4wPM		Plymouth	513	1918	

LH GROUP SERVICES LTD (part of the Wabtec Rail Group), HUNSLET ENGINE COMPANY and LH ACCESS TECHNOLOGY LTD, GRAYCAR BUSINESS PARK, BARTON-UNDER-NEEDWOOD DE13 8EN

Locomotives under construction / repair / overhaul / for resale and hire usually present.

Gauge 4ft 8½in www.lh-group.com **SK 205183, 207186, 207190**

(D3516)	08401	0-6-0DE		Derby		1958	a
(D3560)	08445	0-6-0DE		Derby		1958	b
(D3782)	08615	0-6-0DE		Derby		1959	
(D3991)	08823	0-6-0DE		Derby		1960	
(D4041)	08873	0-6-0DE		Dar		1960	
	ALEX	0-6-0DH		AB	614	1977	
	SAM	0-6-0DH		AB	659	1982	c f
			reb	HAB	6768	1990	
			reb	HE		2004	

	EMILY	0-6-0DH		AB	660	1982	d g
			reb	HAB	6769	1990	
			reb	HE		2004	
	GILLIAN	0-6-0DH		EEV	D1137	1966	e
	BLUEBIRD	0-6-0DH		HE	8998	1981	
	REDWING	0-6-0DH		HE	8999	1981	
DH50-2		0-6-0DHF		TH	246V	1973	
			reb	HE	9377	2011	c
	EDWARD	4wDH		TH	267V	1976	h
DH50-1		0-6-0DH		TH	278V	1978	
			reb	HE	9376	2011	c
	LILY IZABELLA	6wDH		TH	V325	1987	b

a currently at ABP Hams Hall Railfreight Terminal, Coleshill, Warwickshire
b currently at Prologis RFI, DIRFT (Daventry), Crick, Northamptonshire
c currently at Tata Steel Europe, Trostre Works, Llanelli, South Wales
d currently at ICL UK (Cleveland) Ltd, Tees Dock, Teesside
e currently at European Metal Recycling Ltd, Sheffield, South Yorkshire
f carried ABG 660 plates from 2004 to 2007
g carried ABG 659 plates from 2004 to 2007
h currently at Puma Energy (UK) Ltd, Milford Haven, South Wales

MAGNOR PLANT, (part of Morgan Sindall Group plc), PLANT DEPOT, COLD MEECE, SWYNNERTON, STONE — ST15 OUD

Locomotives are present in this yard between use on contracts.

Gauge 2ft 0in www.morgansindall-plantdesk.co.uk **SJ 850325**

S191	263 023	4wBE		CE	B0131A	1973
S213	263 024	4wBE		CE	B0183A	1974
S237	263.008	4wBE		CE	B0471A	1975
S241	263 026	4wBE		CE	B0471E	1975
S242	263 027	4wBE		CE	B0471F	1975
			reb	CE	B3480/1A	1988
S261	263 028	4wBE		CE	B0941A	1976
S263	263 029	4wBE		CE	B0941C	1976
S264	263 030	4wBE		CE	B0948.1	1976
S265	263 031	4wBE		CE	B0948.2	1976
S275	263 033	4wBE		CE	B0957B	1976
			reb	CE	B3480/1B	1988
	263 077	4wBE		CE	B1559	1977
			reb	CE	B3214B	1985
1	263 053	4wBE		CE	5512/1	1968
5	263 054	4wBE		CE		
6	263 058	4wBE		CE		
7	263 032 (S271)	4wBE		CE	B0952.2	1976
			reb	CE	B3787	1991
8	263 059	4wBE		CE	5839B	1971
9	263 057	4wBE		CE	5839C	1971
1		4wDH		Schöma	6154	2007
2		4wDH		Schöma	6155	2007
3		4wDH		Schöma	6156	2007
4		4wDH		Schöma	6157	2007
5		4wDH		Schöma	6158	2007

Gauge 600mm

–	4wDH	Schöma	6299	2008
–	4wDH	Schöma	6300	2008

Gauge 1ft 6in

S147 263 016	4wBE	CE	5926/2	1972

D.J. MILNER HAULAGE LTD, BUTE STREET, LONGTON, STOKE-ON-TRENT — ST4 3PW

www.duncanmilnerhaulage.co.uk — SJ 904438

Locomotives stored for third parties occasionally present

NEMESIS RAIL LTD, BURTON RAIL DEPOT, DERBY ROAD, BURTON-UPON-TRENT — DE14 1RS

Locomotives for maintenance, overhaul and storage usually present

Gauge 4ft 8½in www.nemesisrail.com **SK 250244, SK 251244**

(D61) 45112 ROYAL ARMY ORDNANCE CORPS	1Co-Co1DE	Crewe		1962		
(D1713) 47488	Co-CoDE	BT	475	1964		
(D1927 47250 47600) 47744	Co-CoDE	BT	689	1966		
(D1932 47493) 47701	Co-CoDE	BT	694	1966		
(D2324) 2324	0-6-0DM	(RSHD (DC	8183 2705	1961 1961	a	
(D3236 08168) 13236	0-6-0DE	Dar		1956		
(D3577 08462) 08994	0-6-0DE	Crewe		1958	a	
(D3655) 08500	0-6-0DE	Don		1958	a	
(D3662) 08507	0-6-0DE	Don		1958		
(D3670) 09006	0-6-0DE	Dar		1959	a	
(D3742) 08575	0-6-0DE	Crewe		1959	b	OOU
(D3878) 08711	0-6-0DE	Crewe		1960	a	
(D4102) 09014	0-6-0DE	Hor		1961	a	
(D4121) 08891	0-6-0DE	Hor		1962	b	OOU
(D4148) 08918	0-6-0DE	Hor		1962	a	
D5217 (25067)	Bo-BoDE	Derby		1963		
(D5304) 26004	Bo-BoDE	BRCW	DEL49	1958		
(D5311) 26011	Bo-BoDE	BRCW	DEL56	1959		
(D5546) 31128 CHARYBDIS	A1A-A1ADE	BT	145	1959	c	
(D5547 31129) 31461	A1A-A1ADE	BT	146	1956		
(D5817) 31285	A1A-A1ADE	BT	318	1961	a	
(D6514) 33103 SWORDFISH	Bo-BoDE	BRCW	DEL106	1960	d	
(D6955) 37255	Co-CoDE	(EE (EEV	3512 D943	1964 1964		
(D7615) 25265	Bo-BoDE	Derby		1966		
(D8041 20041) 20904	Bo-BoDE	(EE (VF	2763 D488	1959 1959	a	
(D8083 20083) 20903	Bo-BoDE	(EE (RSHD	2989 8241	1961 1961	a	
X7202 KEMIRA 2 52	0-6-0DH	EEV	D1233	1968	a	
33	0-6-0DH	EEV-AEI	3998	1970	a	

–		0-6-0DH		EEV-AEI	5352	1971	a
X7215	KEMIRA 1 51	0-6-0DH		GECT	5380	1972	a
–		0-6-0DH		HE	7041	1971	a
28	3D 63/000/316	0-6-0DH		HE	7181	1970	a
	LAURA	0-6-0DH		HE	8805	1978	a
H 013		0-4-0DH		S	10137	1962	a
–		0-6-0DH		S	10186	1964	
			reb	HAB	6459	1989	a
9		0-6-0DH		TH	237V	1971	a
(W79976)		4wDMR		ACCars		1958	
	99709 909137-0	2w-2PMR		Geismar	ST/04/06	2004	

a property of Harry Needle Railroad Co Ltd, Derbyshire
b property of Freightliner Group Ltd
c based here for maintenance
d currently at Wyvernrail plc, Ecclesbourne Valley Railway, Derbyshire

REID FREIGHT SERVICES LTD, CINDERHILL INDUSTRIAL ESTATE, LONGTON, near STOKE-UPON-TRENT — ST3 5LB

Locomotives in transit may be present in yard. www.reidfreight.co.uk **SJ 925435**

STATFOLD ENGINEERING LTD, STATFOLD BARN FARM, ASHBY ROAD, TAMWORTH — B79 0BU

Locomotives under restoration usually present www.statfoldengineeringltd.co.uk **SK 240064**

J. WATSON & SONS LTD, SCRAP METAL MERCHANTS, COMMON ROAD, STAFFORD (Part of the Watson Group) — ST16 3DG

Locomotives for scrap or resale occasionally present. www.jameswatsonandsons.com **SJ 923249**

PRESERVATION SITES

CHASEWATER LIGHT RAILWAY AND MUSEUM COMPANY, CHASEWATER RAILWAY, CHASEWATER COUNTRY PARK, POOL ROAD, near BROWNHILLS — WS8 7NL

Gauge 4ft 8½in www.chasewaterrailway.co.uk **SK 034070**

No.3	(COLIN McANDREW)	0-4-0ST	OC	AB	1223	1911
431		0-6-0ST	OC	HC	431	1895
S.100		0-6-0T	OC	HC	1822	1949
	HOLLY BANK No.3	0-6-0ST	IC	HE	3783	1953
4	ASBESTOS	0-4-0ST	OC	HL	2780	1909
"No.11"	ALFRED PAGET	0-4-0ST	OC	N	2937	1882
917		0-4-0ST	OC	P	917	1902
	TEDDY	0-4-0ST	OC	P	2012	1941
5		4wVBT	VCG	S	9632	1957
	LINDA	0-4-0ST	OC	WB	2648	1941
"34"		0-4-0DE		BBT	3097	1956
No.5		0-4-0DM		Bg	3027	1939

MARSTON THOMPSON EVERSHED		0-4-0DM		Bg	3410	1955	
–		0-4-0DM		Bg	3590	1962	
	DERBYSHIRE STONE No.2	4wDM		FH	1891	1934	
D3429	(08359)	0-6-0DE		Crewe		1958	
306		0-6-0DH		GECT	5383	1973	a
251		6wDE		GECT	5414	1976	a
255		6wDE		GECT	5418	1976	a
(262)		6wDE		GECT	5431	1977	
267		6wDE		GECT	5464	1977	a
–		0-6-0DM		HC	D615	1938	
6678	4/33	0-4-0DH		HE	6678	1969	
			reb	HE		1982	
–		0-4-0DM		JF	4100013	1948	
–		0-4-0DH		JF	4220015	1962	
21		4wDM		KC	1612	1929	
15097	UBIQUE	4wPM		MR	1930	1919	
15099	MORRIS	4wDM		MR	2026	1920	b
RRM 106		0-4-0DH		NBQ	27656	1957	
D2911		0-4-0DH		NBQ	27876	1959	
–		4wDM		RH	305306	1952	
5300003	MYFANWY	0-4-0DH		_(RSHD	8366	1962	
				¯(WB	3211	1962	
–		4wDH		TH	111C	1961	
	a rebuild of	4wVBT	VCG	S			
–		0-6-0DH		TH	150C	1965	
	a rebuild of	0-6-0VBT	VCG	S	(9650	1957)	a
(01568)	HELEN	4wDH		TH	264V	1976	
	(HEM HEATH 3D)	0-6-0DM		WB	3119	1956	
–		4wDH		WB	3208	1961	
(AD 9118)		4wDMR		BD	3707	1975	
MPP 13731	99709 901028-9	2w-2BER		Geismar	5E/0006	2006	
900331		2w-2PMR		Wkm	496	1932	
(9033)		2w-2PMR		Wkm	6857	1954	

a property of Ed Murray & Sons Ltd, Hartlepool, Teesside
b plate reads MR 2028

Gauge 2ft 0in

8		4wDHF	HE	7385	1976	OOU
SO	30	4wDM	MR	5609	1931	
–		4wDM	RH	174535	1936	
	YD No.988	4wDM	RH	235729	1944	
	YKSMM 2004/3004	4wDMF	RH	441424	1960	
–		4wDMF	RH	480678	1961	Dsm

CHATTERLEY WHITFIELD MINING MUSEUM, TUNSTALL

Gauge 2ft 6in (Closed) www.chatterleywhitfieldfriends.org.uk **SJ 883531**

3	TOM	4wBEF	Bg	3555	1961
2	CW 1986 111 JERRY	4wBEF	Bg	3578	1961
–		4wBEF	_(EE	3223	1962
			¯(RSHD	8344	1962

CHURNET VALLEY RAILWAY (1992) plc

Locomotives are kept at :- Cheddleton (ST13 7EQ) SJ 983519
Oakamoor SK 046450

Gauge 4ft 8½in www.churnet-valley-railway.co.uk

No.	Name	Type	Cyl	Builder	Works No.	Year	Note
(48173	8173)	2-8-0	OC	Crewe		1943	
	KATIE	0-4-0ST	OC	AB	2226	1946	
6046	(411.144)	2-8-0	OC	BLW	72080	1945	
Tkh 2871		0-6-0T	OC	Chrz	2871	1951	
Tkh 2944	HOTSPUR	0-6-0T	OC	Chrz	2944	1952	
5197		2-8-0	OC	Lima	8856	1945	
D1107	47524	Co-CoDE		Crewe		1966	
D3800	(08633)	0-6-0DE		Derby		1959	
(D6513)	33102 SOPHIE	Bo-BoDE		BRCW	DEL105	1960	
(D6539)	33021 EASTLEIGH	Bo-BoDE		BRCW	DEL131	1961	
(D7672)	25322 (25912) TAMWORTH CASTLE	Bo-BoDE		Derby		1967	
(D8057)	20057	Bo-BoDE		(EE	2963	1961	
				(RSHD	8215	1961	
	BRIGHTSIDE	0-4-0DH		YE	2672	1959	
No.6	ROGER H BENNETT	0-6-0DE		YE	2748	1959	a
RT1		4w-4wDHR		Balfour Beatty		1993	b
(DX 68702)		4wDHR		Perm	007	1986	b
(68706	98706)	4wDHR		Perm	010	1986	b
(DX) 98710		4wDHR		Perm	011	1986	b
DX 98707		4wDHR		Perm	012	1986	b
(DX) 68801		4wDHR		Perm	002	1985	b
(DX) 68803	(RTU 8803)	4wDHR		Perm	004	1985	b
950021		2w-2PMR		Wkm	590	1932	
	C965 YOR	4wDMR	R/R	Bruff	523	1986	

a carries worksplate dated 1956
b property of B & R Track Services

DRAYTON MANOR PARK, FAZELEY, TAMWORTH B78 3TW

Gauge 2ft 0in www.draytonmanor.co.uk **SK 194016**

No.	Name	Type		Builder	Works No.	Year
1	THOMAS	4-4wDH	s/o	MetallbauE	081309	2008
6	PERCY	4-4wDH	s/o	MetallbauE		2008
	ROSIE	4-4wDH	s/o	MetallbauE		2009
	POLPERRO EXPRESS	4-4-0+4w-4wDH	s/o	SL	2151	2003

FOXFIELD LIGHT RAILWAY SOCIETY, FOXFIELD STEAM RAILWAY, BLYTHE BRIDGE, near STOKE-ON-TRENT

Locomotives are kept at :- Blythe Bridge (Caverswall Road) Station ST11 9BG SJ 957421
Site of former Foxfield Colliery, near Dilhorne SJ 976446

Gauge 4ft 8½in www.foxfieldrailway.co.uk

No.	Type	Cyl	Builder	Works No.	Year
No.2	0-6-2T	IC	Stoke		1923
–	0-4-0ST	OC	AE	1563	1908
–	0-4-0ST	OC	BP	1827	1879
4101	0-4-0CT	OC	D	4101	1901

No.	Name	Type	Cyl	Builder	Works No.	Year	Note
	BELLEROPHON	0-6-0WT	OC	Haydock	C	1874	a
	WHISTON	0-6-0ST	IC	HE	3694	1950	
7	WIMBLEBURY	0-6-0ST	IC	HE	3839	1956	
6		0-4-0ST	OC	Heath		1885	
			reb	CW		1934	
–		0-4-0ST	OC	HL	3581	1924	
	MOSS BAY	0-4-0ST	OC	KS	4167	1920	
–		0-4-0ST	OC	KS	4388	1926	
	“THE WELSHMAN”	0-6-0ST	IC	MW	1207	1890	
	HENRY CORT	0-4-0ST	OC	P	933	1903	
			reb	EV		1920	
			reb	EV		1933	
	“ACKTON HALL NO.3”	0-6-0ST	IC	P	1567	1920	
	"IRONBRIDGE"	0-4-0ST	OC	P	1803	1933	
No.11		0-4-0ST	OC	P	2081	1947	
–		4wVBT	VCG	S	9535	1952	
	LEWISHAM	0-6-0ST	OC	WB	2221	1927	
	HAWARDEN	0-4-0ST	OC	WB	2623	1940	
2	“KENT No.2”	0-4-0ST	OC	WB	2842	1946	
	FLORENCE No.2	0-6-0ST	OC	WB	3059	1953	
MEAFORD POWER STATION LOCOMOTIVE No.4							
	CLIVE	0-6-0DH		AB	486	1964	
	WD 820	0-4-0DM		(DC	2157	1940	
				(EEDK	1188	1940	
				(VF		1940	
–		4wBE/WE		EEDK	1130	1939	
RT1		0-6-0DM		JF	22497	1938	
–		6wDM		KS	4421	1929	
	HELEN	4wDM		MR	2262	1923	
	"HERCULES"	4wDM		RH	242915	1946	b
–		4wDM		RH	408496	1957	
–		0-4-0DE		RH	424841	1960	
	WOLSTANTON No.3	0-6-0DM		WB	3150	1959	
	BAGNALL	4wDH		WB	3207	1961	
	LUDSTONE	0-6-0DE		YE	2868	1962	
	–	2w-2BER		Bance	054	1998	
PWM 3764	(68065)	2w-2PMR		Wkm	6643	1953	

a carries works plate D
b rebuilt from 2ft 0in gauge

Gauge 1ft 6in

No.	Name	Type	Cyl	Builder	Works No.	Year	Note
	JMLM7	0-4-0BE		WR	M7548	1972	

M. HAMBLY, PRIVATE LOCATION, TAMWORTH

Gauge 4ft 8½in

No.	Name	Type	Builder	Works No.	Year	Note
–		2w-2PMR	Fairmont			a
–		2w-2PMR	Wkm	4146	1947	
(DX 68010	DB965987)	2w-2PMR	Wkm	7073	1955	b
DB 965071		2w-2PMR	Wkm	7586	1957	

a engine no.114866
b currently stored off site

LAWRENCE HODGKINSON, 17 BILLINGTON AVENUE, LITTLE HAYWOOD

These locomotives are kept at a private location.

Gauge 2ft 6in

–	4wDM	RH	441948	1959	
–	4wDM	RH	476112	1962	

Gauge 2ft 0in

–	4wPM	L	962	1928	

A. HODGSON

These locomotives are kept at a private location.

Gauge 2ft 6in

5	4wBE	BV	690	1974	
–	4wBE	WR	892	1935	

Gauge 2ft 0in

No.1 BESSIE	4wDM	RH	170374	1934	

LIME KILN WHARF INDUSTRIAL RAILWAY, LIME KILN BASIN, WHITEBRIDGE INDUSTRIAL ESTATE, WHITEBRIDGE LANE, STONE

Gauge 2ft 0in **SJ 894345**

LOD 758227	4wDM	MR	8813	1943	
LOD 758019	4wDM	MR	8820	1943	
–	4wDM	RH	171901	1934	a

a carries plate RH 191679

MILL MEECE PUMPING STATION PRESERVATION TRUST LTD, MILL MEECE PUMPING STATION, COTES HEATH, near ECCLESHALL ST21 6QU

Gauge 2ft 0in www.millmeecepumpingstation.co.uk **SJ 831339**

–	4wDM	L	39419	1953	

MOSELEY RAILWAY TRUST, APEDALE VALLEY LIGHT RAILWAY, APEDALE COMMUNITY COUNTRY PARK, LOOMER ROAD, CHESTERTON, NEWCASTLE-UNDER-LYME ST5 7LB

Gauge 4ft 8½in www.avlr.org.uk **SJ 823484**

"96" (74-Lb-00003)	0-4-0DM	BLW			
a rebuild of	0-4-0PM	BLW		1917	

Gauge 2ft 6in

OLD NICK	4wDH	AB	556	1970	
12 ELECTRA	4wBE	BV	565	1970	
"54" (YARD No.54 TO 235)	4wDMR	FH	2196	1940	
8	4wDH	HE	8830	1979	
11	4wDH	HE	8968	1980	
"70" "CRYSTAL"	4wBE	WR	K7070	1970	

Gauge 2ft 0in

104		0-6-0WT	OC	HC	1238	1916	
303		4-6-0T	OC	HE	1215	1916	
	STANHOPE	0-4-2ST	OC	KS	2395	1917	
–		0-6-0T	OC	KS	3014	1916	
	EDGAR	0-4-0T	OC	NBRES	NBR004	2017	a b
"1"	"BILLET"	4wBE		WR	C6717	1963	
"2"	CABLE MILL	4wBE		WR	C6716	1963	
3		4wDM		MR	8878	1944	
6	"10"	4wPM		MR	9104	1941	
7	(Z. & W. WADE)	4wDM		MR	8663	1941	
13	"THE PILK"	4wDM		MR	11142	1960	
14	"KNOTHOLE WORKER"	4wDM		MR	22045	1959	
18	L.C.W.W. 8103	4wDM		HE	6299	1964	
20		4wDM		MR	8748	1942	
21		4wDM		MR	8669	1941	
24		4wDM		HE	1974	1939	
"25"	ND 10448	4wDM		HE	6007	1963	
26	"TWUSK"	4wDM		HE	6018	1961	
27	ANNIE	4wDM		RH	198297	1939	
28	24 "DOROTHY"	4wDM		RH	198228	1940	
29	VANGUARD	4wDM		RH	195846	1939	
"30"	"FRIDEN"	4wDM		RH	237914	1946	a
"31"	"PLUTO"	4wDM		RH	189972	1938	
"32"	LISTER	4wDM		LB	52885	1962	
"33"		4wPM		MR	7033	1936	
34		4wDM		RH	164350	1933	
"35"	DX 68061 PWM 2214 (TR26)	2w-2DMR		Wkm	4131	1947	
37	7	4wDM		RH	260719	1948	
"No.38"	KENNETH	4wDM		RH	223749	1944	
"No.39"	LR 2832	4wPM		MR	1111	1918	
"40"	(739) "SLUDGE"	4wDM		SMH	40SD516	1979	
41		4wDM		MR	5821	1934	
No.42		4wDM		MR	7710	1939	
"43"		4wDM		SMH	104G063	1976	
44	CHAUMONT	4wDH		HU	LX1002	1968	
"45"	87008	4wDM		RH	179870	1936	
"47"	LR 3090	4wPM		MR	1369	1918	
"48"	R12 ND 6458	4wDM		RH	235725	1944	
50	DELTA	0-4-0DM		Dtz	10050	1931	
"52"		0-4-0PM	s/o	Bg	1695	1928	a
"53"		4wDM		FH	2306	1940	
"56"	13 CAT C	4wDH		HE	9082	1984	
"57"	3 TOE RAG	4wDH		HE	8827	1979	
"58"		4wDM		HC	D558	1930	
"59"	4588	4wPM		OK	4588	1932	
"60"	"LORD AUSTIN"	4wPM		MR	6035	1937	
"61"	(LR 3041)	4wDM		MR	1320	1918	
62	P396 81A03 "LBT"	4wDM		RH	497542	1963	
"64"		4wDMF		RH	256314	1949	
"65"	5 42005	4wDM		RH	223667	1943	

"66"	24.8 "PIKROSE"	4wBE		PWR	B0366	1993	
"67"	6 "AMENE"	4wBE		WR	D6912	1964	
(69)		4wBE		CE	B0475	1975	
71		4wBE		CE	5843	1971	
"72	LADY ANN"	4wBE		CE	B0922A	1975	
"74"		4wDM		OK	(3444	1930?)	
"80"		4wDM		LB	52610	1961	
"81"	LR 2573	4wDM		MR	2197	1923	
"82"	No.33 "DRACULA"	4wDM		FH	3582	1954	Dsm
"83"	"RHIWBACH"	2w-2PM		Rhiwbach		1935	Dsm
"84"		4wPM		H	984	1931	
"86"	LR 2638	4wDM		FH	2586	1941	
"89"	L2 GHOST	4wDM		R&R	84	1938	
"90"	DH 887 (ND 10393)	4wDH		BD	3756	1981	
6	"GENESIX"	4wPM		MR	7066	1938	
"92"		4wDM		RH	193974	1938	
–		4wDH		AK	46	1993	
	MERLIN	4wDH	s/o	HU	LX1001	1968	
–		4wBE		CE	B0495	1975	
7		4wBE		CE	B1854	1979	
"93"	"PROMETHEUS"	4wBE		CE	B4299	1998	
	ALD HAGUE	4wPM		FH	3465	1954	
–		4wDM		MR	7522	1948	
L5	T10	4wDM		MR	21520	1955	Dsm
–		4wDM		RH	191658	1938	
	"NOEL / THE CHAIR"	2w-2PMR		SharmanJ		2018	

a currently off site for restoration
b carries worksplate Decauville 684

Gauge 1ft 11½in

"91"	(RTT/767156)	2w-2PM		Wkm	3158	c1943

Apedale Heritage Centre, Loomer Road, Chesterton — ST5 7LB

Gauge 2ft 0in www.apedale.co.uk **SJ 822483**

"16"	6 "MARGARET"	4wDHF	HE	9056	1982
"78"	CITY OF GLOUCESTER	4wPM	MR	5038	1930
"36"	6 "COMMERCIAL" ORDER No. N/NS/N/146/149	4wDM	RH	280865	1949

THE NATIONAL BREWERY CENTRE, HORNINGLOW STREET, BURTON-ON-TRENT — DE14 1NG

(operated by Planning Solutions Ltd)

Gauge 4ft 8½in www.nationalbrewerycentre.co.uk **SK 248234**

No.9	0-4-0ST	OC	NR	5907	1901
No.20	4wDM		KC		1926

NORTH STAFFS & CHESHIRE TRACTION ENGINE CLUB, KLONDYKE MILL, DRAYCOTT-IN-THE-CLAY

Gauge 2ft 0in www.nsctec.co.uk **SK 156289**

	DOROTHY	0-4-0ST	OC	WB	1568	1899

Locomotive currently away for restoration

R. PHILLIPS

These locomotives are kept at a private location.

Gauge 2ft 0in

(4)	4wDM	RH	260716	1949
–	4wDM	MR	9778	1953
–	4wPM	SkinnerD		c1975

PRIVATE OWNER, STOKE-ON-TRENT

Gauge 4ft 8½in

	WALTER	4wBE	CE	B4427A	2006
	KITTY	4wBE	CE	B4427D	2006

Gauge 2ft 6in

	"HERCULES"	0-6-0DH	U23A	24376	1981

Gauge 2ft 0in

1	(LM 21)	4wDM	RH	243392	1946
MBS 387		4wBE	LMM	1053	1950

PRIVATE OWNER, UTTOXETER

Gauge 4ft 8½in

2957	4wDM	RH	512572	1965

THE RAIL TROLLEY TRUST, CHASEWATER HEATHS STATION, BURNTWOOD **WS7 3PG**

Gauge 4ft 8½in www.railtrolleytrust.co.uk **SK036087**

900312	YORK 21	2w-2PMR	Wkm		1931	a
(900332)		2w-2PMR	Wkm	497	1932	a
A159	PWM 2191	2w-2PMR	Wkm	4168	1948	a
TR 1	6872	2w-2PMR	Wkm	6872	1954	a
(1)		2w-2PMR	Wkm	6952	1955	a
DB 965079	68/016	2w-2PMR	Wkm	7594	1957	a

a stored off site

THE STAFFORDSHIRE NARROW GAUGE RAILWAY LTD, AMERTON RAILWAY, AMERTON FARM & CRAFT CENTRE, STOWE-BY-CHARTLEY, near STAFFORD

ST18 0LA

Gauge 3ft 0in www.amertonrailway.co.uk **SJ 993278**

No.1	(ED 10)	0-4-0ST	OC	WB	1889	1911	

Gauge 2ft 0in

–		0-8-0T	OC	Hen	14019	1916	
	JENNIE	0-4-0ST	OC	HE	3905	2007	a
No.56	(LORNA DOONE)	0-4-0ST	OC	KS	4250	1922	
	ISABEL	0-4-0ST	OC	WB	1491	1897	
A10	RNAD TRECWN	4wDH		BD	3782	1984	
GOLSPIE /	THE TRENTHAM EXPRESS	0-4-0DM	s/o	Bg	2085	1934	
	DREADNOUGHT	0-4-0DM	s/o	Bg	3024	1939	
–		0-4-0DM	s/o	Bg	3235	1947	
3		0-4-0DM		Dtz	19531	1937	
–		4wDM		FH	2025	1937	
	GORDON	4wDHF		HE	8561	1978	
	rebuilt as	4wDH				2004	
–		4wDM		Jung	5869	1934	
–		4wDM		MR	7471	1940	
87033		4wDM		MR	40SD501	1975	
ND 6507		4wDM		RH	221623	1943	
–		4wDM		RH	506491	1964	
P.W.No.1	DB 965082	2w-2DMR		Wkm	7597	1957	

a carries plate dated 2005

STATFOLD NARROW GAUGE MUSEUM TRUST LTD, STATFOLD BARN RAILWAY, near TAMWORTH

B79 0BU

Private site hosting regular ticket-only public open days; strictly no public access at any other time.

Gauge 4ft 8½in www.statfoldbarnrailway.co.uk **SK 240064, 242061**

	HODBARROW	0-4-0ST	OC	HE	299	1882	
–		0-4-0PM		Bg	680	1916	
"78"	LIBBIE	2w-2PMR		Bg/DC	1097	1920	
14/3	2705-A9	2w-2PMR		Fairmont	252319		
	CN 168-31	2w-2PMR		Fairmont			a
–		2w-2PMR		Wkm	4164	1948	Dsm

Gauge 3ft 6in

–		4wPMR		Wkm	5864	1951	Dsm

Gauge 3ft 0in

LM 110		4wDM		RH	379066	1954	
–		2w-2PMR		Wkm	4091	1946	Dsm

Gauge 2ft 6in

–		0-4-0T	OC	LaMeuse	3243	1926	b
50		0-6-0DH		HB	D1419	1971	
YARD No.26	(ND 6506) SAM "69"	0-4-0DM		HE	2019	1939	

(51)	TOM	0-6-0DH		_(HE (HC	8847 DM1447	1981 1981	
19	"83"	4wDHF	RACK	HE	9294	1990	

Gauge 750mm

5	TJEPPER	0-4-4-0T	4C	Jung	2279	1914	
1		0-4-0WT	OC	OK	614	1900	
5		0-4-4-0T	4C	OK	1473	1905	

Gauge 2ft 0in

	CEGIN	0-4-0WT	OC	AB	1991	1931	
"184"	MARCHLYN	0-4-0T	OC	AE	2067	1933	
–		4-6-0T	OC	BLW	44657	1916	
–		0-4-0	OC	Dav	1650	1918	
–		0-6-0T	OC	Decauville	1735	1919	
	FIJI	0-6-0	OC	HC	972	1912	
19		0-4-0ST	OC	HC	1056	1914	
	ALPHA	0-6-0T	OC	HC	1172	1924	
GP 39		0-6-0WT	OC	HC	1643	1930	
	CLOISTER	0-4-0ST	OC	HE	542	1891	
	SYBIL MARY	0-4-0ST	OC	HE	921	1906	
No.2	HOWARD	0-4-2ST	OC	HE	1842	1936	
	rebuilt as	0-4-2T	OC				
	rebuilt as	0-4-2ST	OC	Statfold		2014	
No.4	TRANGKIL No.4	0-4-2ST	OC	HE	3902	1971	c
	STATFOLD	0-4-0ST	OC	HE	3903	2005	
	JACK LANE	0-4-0ST	OC	HE	3904	2006	
	SID	0-4-0CA	IC	HE	9902	2009	
	SACCHARINE	0-4-2T	OC	JF	13355	1912	
9	"72" SF. DJATIBARANG	0-4-4-0T	4C	Jung	4878	1930	
No.1	SRAGI No.1 // S.S. TRAM BIMA	0-4-2T	OC	KraussS	4045	1899	
	DIANA	0-4-0T	OC	KS	1158	1917	
	ROGER	0-4-0ST	OC	KS	3128	1918	
	SRAGI 14 MAX	0-6-0WT	OC	OK	10750	1923	
No.1	HARROGATE	0-6-0ST	OC	P	2050	1944	
	ISIBUTU	4-4-0T	OC	WB	2820	1945	
No.6	HOWARD	0-4-0VB	VC	Wbton	2	2007	
	CHARLES	4wPM		Brookville	3526	1949	
–	(521166)	4wPM		Brookville	3746	1951	
	rebuilt as	4wDM		Statfold		2014	
–		4wPM		FH	(1776	1931 ?)	
1	D4	4wDM		Funkey	1001		
4	D5	4wDM		Funkey	1033		
	rebuilt as	4wDH					
–		0-6-0DMF		HC	DM803	1954	Dsm
2	ATLAS	4wDM		HE	2463	1944	
			reb	ALR	No.2	1983	
S.1985.0055		4wDM		HE	2959	1944	
–		4wDH		HE	8819	1979	
	STATFOLD WORKS	4wDH		HE	9332	1994	
–		4wDHF		HE	9351	1994	
W114H	WESTERN REEFS GOLD MINE	4wDH		HT	6720	1965	
D7	N13	4wDH		HT	7588	1968	

–		4wDM		HU	36863	1929	
8		4wPM		HU	39924	1924	
39581		4wDM		MR	8640	1941	DsmT
	CHARLIE	4wDM		MR	9976	1954	
	20777	0-4-0DM		OK	20777	1936	
2		4wPM		Plymouth	1891	1924	
	rebuilt as	4wDM					
No.7	TINY	4wDM		Plymouth	5800	1954	
No.8	TIM	4wDH		Plymouth	6137	1958	
	THE GOOSE	4-4wDMR		Statfold		2015	
No.3		4wPM		VIW	4049	1929	
5		4wPM		VIW	4196	1936	
	CONTEX 1	4wBE		CE	5940A	1972	Dsm
JMLM23	280537	4wBE		GB	420253	1970	
			reb	WR	1983		
			reb	CE	B4623	2016	
"82"		4wBE		WR	6092	1958	
(A155W	TR11 PWM 2187) "81"	4wDMR		CravenJ		1987	d
–		4-4wPMR		NemethJ		2009	e
	rebuilt as	4-2wPMR		HE	9903	2009	

Gauge 600mm

2	MINAS DE ALLER 2	0-6-0PT	OC	Corpet	439	1884	f
No.1	CDC	0-4-2PT	OC	HE	3756	1952	
(NG) 35		4wDH		HE	7010	1971	
			reb	HAB	6941	1988	

Gauge 1ft 11½in

	GERTRUDE	0-4-0ST	OC	HE	995	1909	g
	LIASSIC	0-6-0ST	OC	P	1632	1923	

Gauge 1ft 10¾in

	KING OF THE SCARLETS	0-4-0ST	OC	HE	492	1888	
	MICHAEL	0-4-0ST	OC	HE	1709	1932	

Gauge 1ft 6in

	JACK	0-4-0WT	OC	HE	684	1898	

a currently off site for overhaul
b carries La Meuse worksplate 3355 1929
c convertible to 2ft 6in gauge.
d built by J. Craven, Walesby, Nottinghamshire. Incorporates parts from Wkm 4164. Convertible to all gauges from 2ft 0in to 4ft 8½in.
e converted Landrover
f worksplate shows build date of 1884, but locomotive was completed in 1885
g locomotive is sectioned

JOHN STRIKE, PRIVATE LOCATION, near TAMWORTH

Gauge 2ft 0in

–		4wDM		RH	175127	1935	

SUFFOLK

INDUSTRIAL SITES

FELIXSTOWE DOCK & RAILWAY CO LTD, t/a PORT OF FELIXSTOWE, FELIXSTOWE (member of HPH Group) **IP11 3SY**
Gauge 4ft 8½in www.portoffelixstowe.co.uk **TM 285331**

Currently shunted using Freightliner locomotives - see Appendix 1

PRESERVATION SITES

EAST ANGLIA TRANSPORT MUSEUM SOCIETY, EAST SUFFOLK LIGHT RAILWAY, CHAPEL ROAD, CARLTON COLVILLE, LOWESTOFT **NR33 8RL**
Gauge 2ft 0in www.eatransportmuseum.org.uk **TM 505903**

–		4wDM	MR	5902	1932	a Dsm
2	ALDBURGH	4wDM	MR	5912	1934	
No.6	THORPNESS	4wDM	MR	22209	1964	
No.5	ORFORDNESS	4wDM	MR	22211	1964	
4	LEISTON	4wDM	RH	177604	1936	

a converted to a brake van

ANTHONY GOFF
Locomotive stored at private location.
Gauge 4ft 8½in

No. 1	0-4-0DM	JF	20337	1934

HALESWORTH TO SOUTHWOLD NARROW GAUGE RAILWAY SOCIETY, HALESWORTH
Gauge 900mm www.halesworthtosouthwoldrailway.co.uk

RS106	"HOLTON"	4wDH	RFSD	L106	1989

IPSWICH TRANSPORT MUSEUM, COBHAM ROAD, IPSWICH **IP3 9JD**
Gauge monorail www.ipswichtransportmuseum.co.uk **TM 193428**

–	1w1PH	RM	11345	1963	a

a not on public display

LONG SHOP MUSEUM, MAIN STREET, LEISTON **IP16 4ES**
Gauge 4ft 8½in www.longshopmuseum.co.uk **TM 443626**

"SIRAPITE"	4wT	G	AP	6158	1906

THE MID-SUFFOLK LIGHT RAILWAY COMPANY, BROCKFORD STATION, WETHERINGSETT, near STOWMARKET **IP14 5PW**

Gauge 4ft 8½in www.mslr.org.uk **TM 128659**

(68088)	985	0-4-0T	IC	Dar	(1205)	1923
–		0-4-0VBT	OC	Cockerill	2525	1907
(No.4)	"ALSTON"	0-6-0ST	OC	HC	1604	1928
–		0-4-0ST	OC	WB	2565	1936
–		4wDM		RH	294266	1951
(5)	ALSTON	0-4-0DM		RH	304470	1951
	(DE) 960220	2w-2PMR		Wkm	1949	1935

PLEASUREWOOD HILLS THEME PARK, LEISURE WAY, LOWESTOFT **NR32 4TZ**

Gauge 4ft 8½in www.pleasurewoodhills.com **TM 545965**

D.2069		4wDM		RH	305315	1952

Gauge 2ft 0in

167	7324 SUFFOLK PUNCH	4w-2-4wPH	s/o	Chance	79.50167.24	1979

S. SMITH, HERRINGFLEET HILLS, HERRINGFLEET

This is a private location.

Gauge 2ft 0in

–		4wDM	s/o	MR	9774	1953

SOUTHWOLD RAILWAY TRUST, STEAMWORKS, BLYTH ROAD, SOUTHWOLD **IP18 6AZ**

Gauge 3ft 0in www.southwoldrailway.co.uk TM 500765

	SCALDWELL	0-6-0ST	OC	P	1316	1913
"No.5"	MELLS	4wDH		MR	105H006	1969

STEAM TRACTION LTD, c/o WEBB TRUCK EQUIPMENT, MELFORD ROAD, ACTON, near SUDBURY **CO10 0BB**

Gauge 1524mm home.btconnect/extrareach **TL 883455**

1077		2-8-2	OC	Jung	11787	1953

STONHAM BARNS LEISURE & RETAIL VILLAGE, PETTAUGH ROAD, STONHAM ASPAL, near STOWMARKET **IP14 6AT**

Gauge 4ft 8½in www.stonhambarns.co.uk **TM 127472**

–		0-4-0F	OC	RSHN	7803	1954

SURREY

INDUSTRIAL SITES

R. BANCE & CO LTD, COCKROW HILL HOUSE, ST. MARY'S ROAD, SURBITON

Administration address only. Vehicles under construction, for repair or resale, at another, private, location.

Gauge 4ft 8½in www.bance.com

–		2w-2BER		Bance	064/98	1998
–		2w-2BER		Bance	095/00	2000
–		2w-2BER		Bance	255/11	2011

PRESERVATION SITES

M. HAYTER,
1 HEATHERVIEW COTTAGES, SHORTFIELD COMMON, FRENSHAM, FARNHAM

Gauge 2ft 0in

–	2w-2PM		Wkm	2981	1941	Dsm

DAVID JEFFCOT, HASLEMERE

Gauge 2ft 0in

–	4wDM		MR	22235	1965

MIZENS RAILWAY, BARRS HILL, KNAPHILL, WOKING GU21 2JW

Gauge 3ft 6in www.wokingminiaturerailwaysociety.com **SU 967593**

–	4-8-2T	OC	D	3819	1899

OLD KILN LIGHT RAILWAY, THE RURAL LIFE CENTRE, OLD KILN MUSEUM, THE REEDS, REEDS ROAD, TILFORD, near FARNHAM GU10 2DL

Gauge 2ft 0in www.oldkilnlightrailway.co.uk **SU 858434**

	PAMELA		0-4-0ST	OC	HE	920	1906	
	EMMET		0-4-0T		Moors Valley	20	1995	
		a rebuild of	0-4-0DM		OK	21160	1938	a
M.N.No.1	ELOUISE		0-6-0WT	OC	OK	9998	1922	
	ALTONIA		0-4-0DM	s/o	Bg	1769	1929	
		rebuilt as	0-4-0DH	s/o	BES		1992	
	EMILY		4wDM		FH	2528	1941	
HE 1944	STINKER		4wDM		HE	1944	1939	
(AD) 36	CHAMPION		4wDH		HE	7011	1971	
(NG) 37	WEY VALLEY		4wDH		HE	7012	1971	
				reb	HAB	6014	1988	
(NG) 38	WEYFARER		4wDH		HE	7013	1971	
	"WALTER"		4wDM		Moës		c1955	

	(FIDO)	4wPM		MR	5297	1931 Dsm
5713	(EAGLE)	4wDM		MR	5713	1936
(LOD/758039)	PHOEBE	4wDM		MR	8887	1944 Dsm
–		4wDM		MR	8981	1946
	SANDROCK	4wDM		RH	177639	1936
	RED DWARF	4wDM		RH	181820	1936
–		4wDM		#		
No.4	L12 (RTT 767093) LIZ	2w-2PM		Wkm	3031	1941
	rebuilt as	4wDM		Hayter		1973
	SUE	2w-2PM		Wkm	3287	1943
	rebuilt as	2w-2PMR		‡		1977

‡ rebuilt by E.J.Stephens, Wey Valley Light Railway, Farnham, Surrey
# replica Lister locomotive
a on loan from Moors Valley Railway, Dorset

The Rustics Timber Group, Woodland Tramway
Gauge 2ft 0in

–		2w-2PM		Wkm	3032	1941 Dsm

P. RAMPTON, PRIVATE SITE, HAMBLEDON
The following locomotives are stored at private locations.
The locomotives are not available for viewing or photography.

Gauge 2ft 0in

82		2-6-2+2-6-2T	4C	Hano	10634	1928

Gauge 600mm

	RENISHAW 4	0-4-4-0T	VCG	AE	2057	1931
7	SOTILLOS	0-6-2T	OC	Borsig	6022	1906
1	SABERO	0-6-0T	OC	Couillet	1140	1895
2	SAMELICES	0-6-0T	OC	Couillet	1209	1898
3	OLLEROS	0-6-0T	OC	Couillet	1318	1900
101		0-4-2T	OC	Hen	16073	1918
102		0-4-0T	OC	Hen	16043	1918
103		0-4-0T	OC	Hen	16045	1918
	RENISHAW 5	0-4-4-0T	4C	WB	2545	1936
	R.M.P. No.2	0-4-2T	OC	WB	2895	1948
No.18		4wDE		DK		c1918

Gauge 550mm

4	SAN JUSTO	0-4-2ST	OC	HC	639	1902
5	SANTA ANA	0-4-2ST	OC	HC	640	1902

FRANK SAXBY, GUILDFORD
Gauge 2ft 0in

	SUE	4wG	IC	Saxby	1943	1999

THORPE PLEASURE PARK, STAINES ROAD, CHERTSEY **KT16 8PN**
(a member of the R.M.C.Group)
Gauge 2ft 0in - Canada Creek Railway (Closed) www.thorpepark.com **TQ 035681**

C.C.R.89	No.1	4-4-0DH	s/o	SL	139/1.2.89	1989
C.C.R.89	No.2	4-4-0DH	s/o	SL	139/2.1.89	1989
C.C.R.94	No.3	4-4-0DH	s/o	SL	606.3.94	1994
	IVOR	4wDH	s/o	AK	11	1984

EAST SUSSEX

INDUSTRIAL SITES

ST. LEONARDS RAILWAY ENGINEERING LTD,
ST. LEONARDS WEST MARINA DEPOT,
BRIDGE WAY, ST. LEONARDS-ON-SEA **TN38 8AP**
Gauge 4ft 8½in www.hastingsdiesels.co.uk **TQ 775084**

D2995	(07011)	0-6-0DE		RH	480696	1962
			reb	Resco	L105	1978
S60000	1001 HASTINGS	4-4wDER		Afd/Elh		1957
S60001	1001	4-4wDER		Afd/Elh		1957
S60016	(S60226) 1012 MOUNTFIELD	4-4wDER		Afd/Elh		1957
S60019		4-4wDER		Afd/Elh		1957
S60118	1013 TUNBRIDGE WELLS	4-4wDER		Afd/Elh		1957
(S60145)	977939	4-4wDER		Elh		1962
(S60149)	977940	4-4wDER		Elh		1962

PRESERVATION SITES

BLUEBELL RAILWAY CO LTD

Locomotives are kept at :-

Horsted Keynes, West Sussex RH17 7BB TQ 372293
Kingscote RH19 4LD TQ 367355
Sheffield Park TN22 3QL TQ 403238

Gauge 4ft 8½in www.bluebell-railway.co.uk

(3217)	9017 EARL OF BERKELEY	4-4-0	IC	Sdn		1938	
(30064)	WD 1959	0-6-0T	OC	VIW	4432	1943	
(30096)	96 NORMANDY	0-4-0T	OC	9E		1893	
(30541)	541	0-6-0	IC	Elh		1939	
(30583)	No.488	4-4-2T	OC	N	3209	1885	
(30847)	847	4-6-0	OC	Elh		1936	
(30928)	No.928 STOWE	4-4-0	3C	Elh		1934	
(31027)	No.27 PRIMROSE	0-6-0T	IC	Afd		1910	Dsm
(31065)	No.65	0-6-0	IC	Afd		1896	
(31178	1178) 178	0-6-0T	IC	Afd		1910	
(31263)	No.263	0-4-4T	IC	Afd		1905	

(31323)	323 BLUEBELL	0-6-0T	IC	Afd		1910	
(31592)	No.592	0-6-0	IC	Longhedge		1901	
(31618)	No.1618	2-6-0	OC	Bton		1928	
(31638)	1638	2-6-0	OC	Afd		1931	
32424	BEACHY HEAD	4-4-2	OC	Bluebell		2007	
(32473)	473 BIRCH GROVE	0-6-2T	IC	Bton		1898	
(32636)	No.672 FENCHURCH	0-6-0T	IC	Bton		1872	
(32655)	55 STEPNEY	0-6-0T	IC	Bton		1875	
(34023)	21 C 123 BLACKMOOR VALE	4-6-2	3C	Bton		1946	
34059	SIR ARCHIBALD SINCLAIR	4-6-2	3C	Bton		1947	
(58850	27505)	0-6-0T	OC	Bow	181	1880	
73082	CAMELOT	4-6-0	OC	Derby		1955	
75027		4-6-0	OC	Sdn		1954	
78059		2-6-0	OC	Dar		1956	
80064		2-6-4T	OC	Bton		1953	
80100		2-6-4T	OC	Bton		1955	
80151		2-6-4T	OC	Bton		1957	
92240		2-10-0	OC	Crewe		1958	
No.3	CAPTAIN BAXTER	0-4-0T	OC	FJ	158	1877	
No.4	SHARPTHORN	0-6-0ST	IC	MW	641	1877	
D4106	(09018)	0-6-0DE		Hor		1961	
	BRITANNIA	4wPM		H	957	1926	a
–		4wDH		RR	10241	1966	
			reb	TH	247C	1973	
387	(2761 LT512)	4w-4wRER		Ashbury		1898	
			reb			1907	Dsm b
DXN 68001	01	2w-2DMR		Wkm	10708	1974	
			reb	Ashford		1993	

a runs on propane gas
b converted to hauled coaching stock

DRUSILLA'S ZOO PARK,
ALFRISTON ROAD, BERWICK, ALFRISTON, near EASTBOURNE — BN26 5QS

Gauge 2ft 0in www.drusillas.co.uk **TQ 524050**

–	0-4-2+4-4wDH	s/o	MetallbauE		2007	

THE CLAUDE JESSETT TRUST COMPANY, THE GREAT BUSH RAILWAY,
TINKERS PARK, MAIN ROAD, HADLOW DOWN, near UCKFIELD — TN22 4HS

Gauge 2ft 0in www.tinkerspark.com **TQ 538241**

(No.28)	SAO DOMINGOS	0-6-0WT	OC	OK	11784	1928	
(1	ANIMAL)	4wDM		MR		c1931	
			reb	‡			
(4)	MILD	4wDM		MR	8687	1941	
(No.36)	DRUSILLA	4wDM		MR	22236	1965	
–		4wDM		OK	5926	1935	
5	(ALPHA)	4wDM		RH	183744	1937	
(14)	(ALBANY)	4wDM		RH	213840	1941	
"No.24"	(No.29) R. J. BROWN	4wDM		RH	382820	1955	
21		4wBE		WR	5035	1954	Dsm

(No.22)	(LAMA)	4wBE		WR	5033	1953	
(23)		0-4-0BE		WR	M7534	1972	Dsm
(No.24)	TITCH	0-4-0BE		WR	M7535	1972	
(No.25)	WOLF	4wDM		MR	7469	1940	

‡ rebuilt by Ludlay Brick Co, Berwick, near Eastbourne, Sussex

LAVENDER LINE PRESERVATION SOCIETY, ISFIELD STATION, STATION ROAD, ISFIELD, near UCKFIELD — TN22 5XB

Gauge 4ft 8½in www.lavender-line.co.uk **TQ 452171**

–		0-4-0VBT	OC	Cockerill	2945	1920
D4113	09025	0-6-0DE		Hor		1961
221	QUEENBOROUGH	0-4-0DM		AB	354	1941
			reb	BIS		1957
15		4wDM		FH	3658	1953
			rep	Resco	L112	
01583	422 VALIANT	0-6-0DH		RH	459517	1961
830	WEM	0-4-0DM		(VF	5257	1945
				(DC	2176	1945
S60117	205018 1118	4-4wDER		Afd/Elh		1957
S60122	(205023) (1123)	4-4wDER		Afd/Elh		1959
60151	205033 (1133)	4-4wDER		Afd/Elh		1962
(61928)	309624 977966) 960102	4w-4wWER		York		1962
999507		4wDMR		Wkm	8025	1958
	99709 909130-5	2w-2DMR		Geismar	ST/03/56	2003
	ARMY 9035	2w-2PMR		Wkm	8195	1958

A. PRAGNELL, PRIVATE LOCATION, near HORSTED KEYNES

Gauge 4ft 8½in

(900855)	2w-2PMR	Wkm	6967	1954	a

a carries plate Wkm 7445/1956

ROTHER VALLEY RAILWAY LTD, ROBERTSBRIDGE JUNCTION STATION, STATION ROAD, ROBERTSBRIDGE — TN32 5DG

Gauge 4ft 8½in www.rvr.org.uk **TQ 734235**

D2112	(03112)	0-6-0DM		Don		1960	
(D99)	(11790)	0-4-0DM		(VF	D77	1947	
				(DC	2251	1947	
1	TITAN	0-4-0DM		(VF	D140	1951	
				(DC	2274	1951	
PX 205	(54187)	2w-2DH		Perm	205	1980	
			reb	RVR		c2005	a

a rebuilt from a track machine

TUNBRIDGE WELLS & ERIDGE RAILWAY PRESERVATION SOCIETY LTD, SPA VALLEY RAILWAY, GROOMBRIDGE STATION, NEWTON WILLOWS, GROOMBRIDGE, near TUNBRIDGE WELLS — TN3 9RD

Gauge 4ft 8½in www.spavalleyrailway.co.uk **TQ 532371**

For details of locomotives see under Kent entry.

VOLKS ELECTRIC RAILWAY, MADEIRA DRIVE, BRIGHTON — BN12 1EN

Gauge 2ft 8½in www.volkselectricrailway.co.uk **TQ 326035**

3	4wRER		VER		1892	Dsm
4	4wRER		VER		1892	
		reb	AK	99R	2017	
6	4wRER		VER		1901	
		reb	AK	101R	2018	
7	4wRER		VER		1901	
8	4wRER		VER		1901	
9	4wRER		VER		1910	
10	4wRER		VER		1926	
		reb	AK	102R	2018	
–	4wDM		AK	40SD530	1987	

WEST SUSSEX

PRESERVATION SITES

BLUEBELL RAILWAY CO LTD, EAST GRINSTEAD STATION

Gauge 4ft 8½in www.bluebell-railway.co.uk **TQ 385379**

3417	62236 GORDON PETTITT	4-4wRER		York		1969	a

a currently at Siemens Rail Systems, Strawberry Hill, Greater London

HOLLYCOMBE STEAM & WOODLAND GARDEN SOCIETY, HOLLYCOMBE STEAM COLLECTION, IRON HILL, LIPHOOK — GU30 7LP

Gauge 4ft 8½in www.hollycombe.co.uk **SU 852295**

50	COMMANDER B	0-4-0ST	OC	HL	2450	1899	

Gauge 3ft 0in

	"EXCELSIOR"	2-2-0WT	G	AP	1607	1880	

Gauge 2ft 0in Quarry Railway

70	CALEDONIA	0-4-0WT	OC	AB	1995	1931	
38	JERRY M	0-4-0ST	OC	HE	638	1895	
	FLOATER	4wDM		MR	60S318	1964	a
C.F.V.O.	No.25	4wDM		Plymouth	4848	1945	
16	JACK	4wDM		RH	203016	1940	

	LIZZIE	4wDH		Ruhrthaler	3909	1969
RTT 767165	(RTT/767149)	2w-2PM		Wkm	3238	1943

a brake column is stamped 11164 in error

ANDREW JOHNSTON, MIDHURST

Gauge 2ft 0in

	TOBY	4wPM		Carter		1986
–		4wDM		FH	2544	1942
–		4wDM	s/o	MR	8727	1941

SOUTHERN INDUSTRIAL HISTORY CENTRE TRUST, AMBERLEY MUSEUM, HOUGHTON BRIDGE, AMBERLEY, ARUNDEL **BN18 9LT**

Gauge 4ft 8½in www.amberleymuseum.co.uk **TQ 031122**

No.5	BURT	4wDM		MR	9019	1950

Gauge 3ft 2¼in

1	(TOWNSEND HOOK)	0-4-0T	OC	FJ	172L	1880
	MONTY	4wDM		OK	7269	1936

Gauge 2ft 11in

–		4wDM		MR	10161	1950

Gauge 2ft 6in

–		4wPM		HU	45913	1932

Gauge 2ft 0in

	BARBOUILLEUR	0-4-0WT	OC	Decauville	1126	1947
	POLAR BEAR	2-4-0T	OC	WB	1781	1905
	PETER	0-4-0ST	OC	WB	2067	1918
T0001	ND10261 00 NZ 26	4wDH		BD	3751	1980
808		2w-2-2-2wRE		EEDK	808	1931
	CCSW	4wDM		FH	1980	1936
–		4wPM		(FH	3627	1953 ?)
–		0-4-0DMF		HC	DM686	1948
	(THAKEHAM TILES No.3)	4wDM		HE	2208	1941
–		4wDM		HE	3097	1944
	(THAKEHAM TILES No.4)	4wDM		HE	3653	1948
12		4wDH		HE	8969	1980
–		4wDM		HE		
	"PELDON"	4wDM		JF	21295	1936
–		4wPM		L	33937	1949
–		4wPM		L	34521	1949
–		4wPM		MR	872	1918
			reb	MR	3720	1925
3101		4wPM		MR	1381	1918
27		4wDM		MR	5863	1934
–		4wDM		MR	11001	1956
	SONIA	4wDM		OK	4013	1930
6193	REDLAND	4wDM		OK	6193	1935
7741	THE MAJOR	4wDM		OK	7741	1937

No.2		4wDM		RH	166024	1933	
–		4wDM		RH	187081	1937	
–		4wDM		R&R	80	1937	
–		4wPM		Thakeham		c1946	
–		4wPM		Thakeham		c1950	
3		4wBE		BE	16303	c1917	a
–		4wBE		WR	4998	1953	
–		4wBE		WR	5031	1953	
–		4wBE		WR	(5034	1953?)	
–		0-4-0BE		WR	T8033	1979	
	RTT/767159	2w-2PM		Wkm	3161	c1943	
WD 904		2w-2PMR		Wkm	3403	1943	

a rebuilt using parts from BE 16306 c1917

Gauge 1ft 10in

23	0-4-0T	IC	Spence	23L	1921	a

Gauge monorail

–	2wPH	RM	9514	1960	

a worksplate dated 1920

TATES GARDEN CENTRES, SOUTH DOWNS HERITAGE CENTRE, BRIGHTON ROAD, HASSOCKS — BN6 9LY

Gauge 2ft 8½in www.southdownsheritagecentre.co.uk **TQ 299015**

9	4wRER		BE		1898
		reb	BE		c1911
		reb	BrightonTC		1950

TEESSIDE

OUR DEFINITION OF "TEESSIDE" INCORPORATES THE AREA COVERED BY THE FORMER COUNTY OF CLEVELAND, CREATED IN 1974, AND DISBANDED IN 1996 INTO FOUR SEPARATE COUNTY BOROUGHS OF HARTLEPOOL, MIDDLESBROUGH, REDCAR & CLEVELAND AND STOCKTON-ON-TEES.

INDUSTRIAL SITES

BRITISH STEEL LTD, SKINNINGROVE WORKS, CARLIN HOW, SALTBURN-BY-THE-SEA — TS13 4EE

(a Greybull Capital LLP company)

Gauge 4ft 8½in www.britishsteel.co.uk **NZ 708194**

	HARRY POTTER	0-6-0DH	RR	10220	1965	a	
(01560)		4wDH	RR	10229	1965	a	
(309)	'LOCO No.2815'	0-6-0DH	YE	2825	1961	a	OOU
(314)		0-6-0DH	YE	2832	1962	a	OOU

a property of Ed Murray & Sons Ltd, Hartlepool, Teesside

BRITISH STEEL LTD,
TEESSIDE BEAM MILL, LACKENBY WORKS, MIDDLESBROUGH

(a Greybull Capital LLP company)

Gauge 4ft 8½in www.britishsteel.co.uk NZ 558222, 568235

261		6wDE	GECT	5430	1977	a
(MURR 2)		4wDE	Moyse	1464	1979	a
01569	EMMA	4wDH	TH	281V	1978	a

a property of Ed Murray & Sons Ltd, Hartlepool, Teesside

COBRA MIDDLESBROUGH LTD,
MIDDLESBROUGH GOODS YARD, NORTH ROAD, MIDDLESBROUGH **TS2 1DQ**

Gauge 4ft 8½in www.cobra.ltd.uk **NZ 488209**

01567	ELIZABETH	4wDH	TH	276V	1977	

a property of Ed Murray & Sons Ltd, Hartlepool, Teesside

A.V. DAWSON LTD

Locomotives kept at
Automotive Coil Store, Depot Road, Middlesbrough TS2 1LE NZ 491213
Ayrton Store and Railhead, Forty Foot Road, Middlesbrough TS2 1HG NZ 493215
NSSB Wharf, Riverside Park Road, Middlesbrough TS2 1UT NZ 491216

Gauge 4ft 8½in www.av-dawson.com

(D3765)	08598	0-6-0DE	Derby		1959	
(D3767)	08600	0-6-0DE	Derby		1959	
(D3942)	08774 ARTHUR VERNON DAWSON	0-6-0DE	Derby		1960	
(D4142)	08912	0-6-0DE	Hor		1962	Dsm
	ELEANOR DAWSON	0-4-0DM	Bg/DC	2725	1963	Pvd
7900/58		0-4-0DM	RSHN	7900	1958	Pvd

EDF ENERGY plc, HARTLEPOOL POWER STATION, SEATON CAREW **TS25 2BZ**

Gauge 4ft 8½in www.edfenergy.com **NZ 532270**

H 058		4wDH	RR	10280	1968	a
–		0-6-0DH	EEV-AEI	4003	1971	b

a on hire from British American Railway Services Ltd, Stanhope, Co. Durham
b property of Ed Murray & Sons Ltd, Hartlepool, Teesside

ED MURRAY & SONS LTD, BRENDA ROAD, HARTLEPOOL **TS25 2BW**

The following is a FLEET LIST of the locomotives owned by this contractor.

Gauge 4ft 8½in www.edmurrays.com

Z.Z. 267		0-6-0DH	EEV-AEI	4003	1971	a
306		0-6-0DH	GECT	5383	1973	b
251		6wDE	GECT	5414	1976	b
252		6wDE	GECT	5415	1976	c
253		6wDE	GECT	5416	1976	c
255		6wDE	GECT	5418	1976	b
257		6wDE	GECT	5426	1977	c

261		6wDE		GECT	5430	1977	h	
265		6wDE		GECT	5462	1977	c	
266	SHERRIFFS	6wDE		GECT	5463	1977	c	
267		6wDE		GECT	5464	1977	b	
268		6wDE		GECT	5465	1977	d	
276	SPAWOOD	6wDE		GECT	5474	1978	c	
277		6wDE		GECT	5475	1978	o	
EM 1		6wDE		GECT		1977	e	
–		0-4-0DH		HE	7425	1981	f	
MURR 1		4wDE		Moyse	1364	1976	g	
(MURR 2)		4wDE		Moyse	1464	1979	h	
3	(2)	4wDE		Moyse	1365	1976	j	Dsm
	CHARLIE	0-6-0DH		S	10107	1963	k	
	YOGI	0-6-0DH		S	10166	1963	k	
	HARRY POTTER	0-6-0DH		RR	10220	1965	l	
	LADY POTTER	0-6-0DH		RR	10214	1964	l	
(01560)		4wDH		RR	10229	1965	l m	
–		0-6-0DH		RR	10255	1966	j	
–		0-6-0DH		TH	150C	1965		
	a rebuild of	0-6-0VBT	VCG	S	(9650	1957)	b	
(01567)	ELIZABETH	4wDH		TH	276V	1977	n	
(01569)	EMMA	4wDH		TH	281V	1978	h	
	LOCO No.9	4wDH		TH	287V	1980	j	
(01555)	JAMES	4wDH		TH	288V	1980	j	
(309)	'LOCO No.2815'	0-6-0DH		YE	2825	1961	l	OOU
(314)		0-6-0DH		YE	2832	1962	l	OOU

a currently at EDF energy plc, Hartlepool Power Station, Teesside
b currently at Chasewater Railway, Brownhills, Staffordshire
c currently at former Sahavirya Steel Industries UK Ltd, Lackenby Works, Teesside
d currently at ICL UK (Cleveland) Ltd, Tees Dock Terminal, Teesside
e currently at Tata Strip Pruducts UK, Llanwern Works, South Wales
f currently located at Ed Murray & Sons Ltd, Brenda Road, Hartlepool, Teesside
g currently at Liberty Steels, Hartlepool, Teesside
h currently at British Steel Ltd, Lackenby, Teesside
j currently at Ed Murray & Sons Ltd, Casebourne Road, Hartlepool (NZ 511313)
k currently at ICL UK (Cleveland) Ltd, Boulby Mine, Loftus, Teesside
l currently at British Steel Ltd, Skinningrove Works, Teesside
m spare locomotive – may be found at Hartlepool or Skinningrove
n currently at Cobra (Middlesbrough) Ltd, Middlesbrough, Teesside
o currently at Aggregate Industries UK Ltd, Merehead, Somerset

ICL UK (CLEVELAND) LTD

(part of the ICL Fertilizers Group)

Boulby Mine, Loftus **TS13 4UZ**

Gauge 4ft 8½in www.icl-uk.uk **NZ 763183**

LADY POTTER	0-6-0DH	RR	10214	1964	a
CHARLIE	0-6-0DH	S	10107	1963	a
YOGI	0-6-0DH	S	10166	1963	a

a property of Ed Murray & Sons Ltd, Hartlepool, Teesside

Gauge 1ft 6in — Underground **NZ 765189**

–	0-4-0BE	WR	7654	1974

Tees Dock Terminal, Grangetown (TS6 6UD)

Gauge 4ft 8½in **NZ 549235**

	EMILY	0-6-0DH		AB	660	1982	
			reb	HAB	6769	1990	
			reb	HE	2004	a	
268		6wDE		GECT	5465	1977	b

a property of LH Group Services Ltd, Barton-under-Needwood, Staffordshire
b property of Ed Murray & Sons Ltd, Hartlepool, Teesside

LIBERTY STEELS, LIBERTY PIPES DIVISION, LIBERTY STEEL HARTLEPOOL, SAW PIPE MILLS (42in and 84in Mills), BRENDA ROAD, HARTLEPOOL TS25 2EF

(part of the Liberty House Group; part of GFG Alliance)

Gauge 4ft 8½in www.libertyhousegroup.com **NZ 508288**

(MURR 1)	4wDE		Moyse	1364	1976	a

a property of Ed Murray & Sons Ltd, Hartlepool, Teesside

NORTHERN STEAM ENGINEERING LTD, LONSDALE HOUSE, ROSS ROAD, PORTRACK, STOCKTON-ON-TEES TS18 2NH

Gauge 4ft 8½in www.northernsteamengineering.co.uk **NZ 457189**

2253	2-8-0	OC	BLW	69496	1944

vehicles for overhaul / repair occasionally present

PD PORTS TEESPORT, TEES DOCK, GRANGETOWN TS6 6UD

Gauge 4ft 8½in www.pdports.co.uk **NZ 546232**

(D3538 08423) H 011 14	0-6-0DE	Derby		1958	a
(D3780) 08613 H 064	0-6-0DE	Derby		1959	a

a property of British American Railway Services Ltd, Stanhope, Co. Durham

SEMBCORP UTILITIES TEESSIDE LTD, WILTON INTERNATIONAL MANUFACTURING SITE, MIDDLESBROUGH (part of Sembcorp Industries Group)

Gauge 4ft 8½in R.T.C. www.sembcorp.co.uk **NZ 564218**

(D3911) 08743 BRYAN TURNER		0-6-0DE	Crewe		1960	
(D4133) 08903 JOHN W. ANTILL		0-6-0DE	Hor		1962	
RO.005 "WILTONIA"		2w-2DMR	Wkm	7591	1957	
	rebuilt as	2w-2DHR	YEC	L112	1992	
–		2w-2PMR	Wkm	7603	1957	DsmT

SAHAVIRIYA STEEL INDUSTRIES UK LTD (SSI UK), TEESSIDE WORKS

(Sahaviriya Steel Industries UK Ltd entered Administration on 2/10/2015)

Redcar Coke Ovens, Redcar

Gauge 4ft 8½in R.T.C. **NZ 562257**

1	4wWE	GB	420355/1	1976	OOU
2	4wWE	GB	420355/2	1976	OOU
3	4wWE	GB	420408	1977	OOU

South Bank Coke Ovens

Gauge 4ft 8½in **R.T.C.** **NZ 536214**

1		4wWE		HartlepoolBSC		1986	OOU
2		4wWE		HartlepoolBSC		1986	OOU

Lackenby Works, Middlesbrough

Gauge 4ft 8½in **R.T.C.** **NZ 563223**

252		6wDE		GECT	5415	1976	a	Dsm
253		6wDE		GECT	5416	1976	a	OOU
255		6wDE		GECT	5418	1976	a	OOU
257		6wDE		GECT	5426	1977	a	OOU
260		6wDE		GECT	5429	1977		OOU
265		6wDE		GECT	5462	1977	a	OOU
266	SHERRIFFS	6wDE		GECT	5463	1977	a	OOU
269		6wDE		GECT	5466	1977	a	OOU
276	SPAWOOD	6wDE		GECT	5474	1978	a	OOU
8.711		Bo-BoDE		MaK	1600.011	1996		Dsm
1		4wDM		Robel 54.12-107 AD184		1980		
			reb	Lackenby		2010	a	
2	REDCAR No.2	4wDM		Robel 54.12-107 AD183		1980	a	OOU

a property of Ed Murray & Sons Ltd, Hartlepool, Teesside

TATA STEEL EUROPE, TATA STEEL TUBES EUROPE, NORTH EAST PIPE MILLS, HARTLEPOOL 20in MILL, BRENDA ROAD, HARTLEPOOL TS25 2EF

(Part of the Tata Group)

Gauge 4ft 8½in www.tatasteelconstruction.com **NZ 505276, 505278**

425		6wDE	GECT	5425	1977	OOU
264	PORT MULGRAVE	6wDE	GECT	5461	1977	OOU

PRESERVATION SITES

CLEVELAND IRONSTONE MINING MUSEUM, OFF MILL LANE, DEEPDALE, SKINNINGROVE TS13 4AP

Gauge 2ft 0in www.ironstonemuseum.co.uk **NZ 712192**

No.9	4wDM	RH	182137	1936	Dsm

SALTBURN MINIATURE RAILWAY, VALLEY GARDENS, SALTBURN-BY-THE-SEA TS13 4SD

Gauge 1ft 3in www.saltburn-miniature-railway.org.uk **NZ 668214**

SALTBURN 150	4-6-2gas	s/o	_(Artisair	1972
			(RADev	1972
rebuilt as	4-6-2DH	s/o	SMR	2011
PRINCE CHARLES	4-6-2DE	s/o	Barlow	1953

BLACKLOCK R.	4-4-2	OC	SMR		2015	
	a rebuild of		MossAJ		2001	
GEORGE OUTHWAITE	0-4-0DH	s/o	_(SMR		1994	
			(WiltonICI		1994	

STOCKTON-ON-TEES BOROUGH COUNCIL, PRESTON PARK MUSEUM and GROUNDS, EAGLESCLIFFE, STOCKTON-ON-TEES TS18 3RH

Gauge 4ft 8½in www.prestonparkmuseum.co.uk **NZ 429158**

–	0-4-0VBT	VCG	HW	21	1870	

GRAEME WALTON-BINNS, MIDDLESBROUGH

Gauge 4ft 8½in

No.8	0-6-0T	OC	AB	1296	1912	

Gauge 2ft 0in

SINEMBE	4-4-0T	OC	WB	2287	1926	
A. BOULLE	4-4-0T	OC	WB	2627	1940	

WARWICKSHIRE

INDUSTRIAL SITES

ALLELYS HEAVY HAULAGE LTD, THE SLOUGH, STUDLEY B80 7EN

Locomotives in transit are also stored in the yard for short periods of time.

Gauge 4ft 8½in www.allelys.co.uk **SP 056636**

(D6723	37023)	Co-CoDE	_(EE	2886	1961	
			(VF	D601	1961	
(D6846)	37146	Co-CoDE	_(EE	3321	1963	
			(EEV	D820	1963	
	TEUCER	0-4-0DM	_(VF	D294	1955	
			(DC	2567	1955	a

a currently located at Haydon Way Farm, Studley (SP 077620)

ASSOCIATED BRITISH PORTS HOLDINGS LTD, ABP HAMS HALL RAIL FREIGHT TERMINAL, EDISON ROAD, COLESHILL, BIRMINGHAM B46 1DA

(owned by ABP (Jersey) Ltd)

Gauge 4ft 8½in www.hamshallrailterminal.co.uk **SP 198911**

(D3516)	08401	0-6-0DE	Derby	1958	a

a property of LH Group Services Ltd, Barton-under-Needwood, Staffordshire

CEVA FREIGHT UK LTD, BIRCH COPPICE BUSINESS PARK, DORDON, TAMWORTH — B78 1SE

Gauge 4ft 8½in www.cevalogistics.com **SP 251994**

9210		4wDM	R/R	Unimog	199146	2002	

COSTAIN GROUP plc, PLANT YARD, WATLING STREET, SHAWELL — LE17 6AR

Gauge 750mm www.costain.com **SP 538793**

B4536/1	CP20T.1	4wBE		CE	B4536.1	2011	
B4536/2	CP20T.2 1	4wBE		CE	B4536.2	2011	
B4536/3	CP20T.3 1	4wBE		CE	B4536.3	2012	
B4536/4	7	4wBE		CE	B4536.4	2012	
B4536/5		4wBE		CE	B4536.5	2012	

Gauge 610mm

B4539/01		4wBE		CE	B4539.1	2012	
B4539/02		4wBE		CE	B4539.2	2012	
			reb	CE	B4573	2012	
3		4wBE		CE	B4539.3	2012	
B4539/04		4wBE		CE	B4539.4	2012	
5		4wBE		CE	B4539.5	2012	
			reb	CE	B4632	2018	
6		4wBE		CE	B4539.6	2012	
7	DUKE OF YORK	4wBE		CE	B4540	2012	
SP 1176	2	4wBE		CE			a
SP 1177		4wBE		CE			a
SP 1178		4wBE		CE			a
SP 1179	LOCO 7	4wBE		CE			a
SP 1180	4	4wBE		CE			a
SP 1181	LOCO 8	4wBE		CE			a

a individual identities unknown - from CE 5590/1, CE 5590/2, CE 5590/8, CE (5949A or 5949B), CE 5949G and CE B0129

plant yard with locos present between contracts

EUROPEAN METAL RECYCLING LTD, METAL MERCHANTS AND PROCESSORS, TRINITY ROAD, KINGSBURY — B78 2LB

Locomotives for scrap are occasionally present at this location.

Gauge 4ft 8½in www.emrgroup.com **SP 219969**

(D3951)	08783	0-6-0DE		Derby		1960	
7		4wDH		S	10040	1960	

MINISTRY OF DEFENCE, DEFENCE MUNITIONS, KINETON

See Section 6 for details

MORGAN SINDALL, CONSTRUCTION, INFRASTRUCTURE & DESIGN DIVISION, PLANT DEPOT, WATLING STREET, CLIFTON-UPON-DUNSMORE, near RUGBY

(part of MorganSindall Group plc) **CV23 0AL**

Locomotives are occasionally present for repair www.morgansindall.com **SP 533788**

MOTORAIL UK LTD, QUINTON RAIL TECHNOLOGY CENTRE, STATION ROAD, LONG MARSTON, near STRATFORD-UPON-AVON **CV37 8PL**

Also mainline and former mainline locomotives and rolling stock stored for reuse/resale.

Gauge 4ft 8½in www.motorail.co.uk **SP 152472, 155476, 158474**

(D3745)	08578	0-6-0DE		Crewe		1959	a
(D3820)	08653 VERNON	0-6-0DE		Hor		1959	a
(D3868)	08701 TYNE 100	0-6-0DE		Hor		1960	a
(D8016)	20016	Bo-BoDE		_(EE ¯(VF	2363 D391	1957 1957	a
(D8081)	20081	Bo-BoDE		_(EE ¯(RSHD	2987 8239	1961 1961	a
(D8088	20088) 2017 37	Bo-BoDE		_(EE ¯(RSHD	2994 8246	1961 1961	a
(51352)		2-2w-2w-2DMR		PSteel		1960	
(51371)	977987 960301	2-2w-2w-2DMR		PSteel		1960	
51376		2-2w-2w-2DMR		PSteel		1960	
(51413)	977988 960301	2-2w-2w-2DMR		PSteel		1960	
(55025)	977859 960011 PANDORA	2-2w-2w-2DMR		PSteel		1960	
870		0-6-0DH		AB	509	1966	a
871		0-6-0DH		AB	510	1966	a
873	JOHN BOY	0-6-0DH		AB	512	1966	a
RRM 22		0-6-0DH		EEV	D1231	1967	a
–		0-6-0DH		GECT	5395	1974	a
No.503		0-6-0DH		HE	6614	1965	a
No.7	31	0-6-0DH		HE	6973	1969	a
	RACHAEL	0-6-0DH		HE	7003	1971	a
No.5		0-4-0DH		HE	7161	1970	a
–		0-6-0DH		HE	7276	1972	a
28		0-6-0DH		HE	7279	1972	a
	EMMA	0-6-0DH		HE	8902	1978	a
7		0-6-0DH		RR	10279	1968	a
H 014		0-4-0DH		S	10119	1962	a
01552		0-6-0DH		TH	167V	1966	a
01547	(266)	4wDH		TH	308V	1983	a
–		4wDH	R/R	Minilok	134	1986	a
–		8wDH	R/R	Minilok	160	1991	a
TNS 104		2w-2DMR		Robel	56.27-10-AG35	1983	a

a property of Harry Needle Railroad Co Ltd, Derbyshire

R.S.S. - RAILWAY SUPPORT SERVICES LTD,
c/o JOHN WATTS FARMS, RYE FARM, RYEFIELD LANE, WISHAW B76 9QA

Administration address : Montpellier House, Montpellier Drive, Cheltenham

Railway Support Services Ltd Hire Fleet

Gauge 4ft 8½in www.railwaysupportservices.co.uk **SP 180945**

(D3290)	08220		0-6-0DE	Derby		1956	a	
(D3520)	08405		0-6-0DE	Derby		1958	b	
(D3526)	08411		0-6-0DE	Derby		1958		OOU
(D3556)	08441		0-6-0DE	Derby		1958	c	
(D3575)	08460	SPIRIT OF THE OAK	0-6-0DE	Crewe		1958	d	
(D3595)	08480		0-6-0DE	Hor		1958	e	
(D3599)	08484	CAPTAIN NATHANIEL DARELL	0-6-0DE	Hor		1958	f	
(D3673)	08511		0-6-0DE	Dar		1958	g	
(D3700)	08536		0-6-0DE	Dar		1959		OOU
(D3735)	08568	ST ROLLOX	0-6-0DE	Crewe		1959		
(D3747)	08580		0-6-0DE	Crewe		1959	c	
(D3760)	08593		0-6-0DE	Crewe		1959		
(D3799)	08632		0-6-0DE	Derby		1959	b	
(D3837)	08670		0-6-0DE	Crewe		1960	c	
(D3850)	08683		0-6-0DE	Hor		1959	e	
(D3870)	08703		0-6-0DE	Hor		1960	h	
(D3876)	08709		0-6-0DE	Crewe		1960		
(D3906)	08738		0-6-0DE	Crewe		1960		
(D3920)	08752	LENNY	0-6-0DE	Crewe		1960		
(D4103)	09015	ROB	0-6-0DE	Hor		1961		
(D4151)	08921	PONGO	0-6-0DE	Hor		1962		
(D4157)	08927		0-6-0DE	Hor		1962		
(D4169)	08939		0-6-0DE	Dar		1962		
–			4wDH	TH	184V	1967	i	

a currently at Electromotive Diesels, Longport, Staffs
b currently at East Midland Trains Ltd, Neville Hill Depot, Leeds, West Yorkshire
c currently at London North Eastern Railway Ltd, Bounds Green, Gtr. London
d currently at Axiom Rail Ltd, Stoke-on-Trent, Staffordshire
e currently at Abellio Greater Anglia Ltd, Norwich Crown Point, Norwich
f currently at Hitachi Rail Europe, Newton Aycliffe, Co. Durham
g currently at Arriva Traincare, Cambridge Depot, Cambridgeshire
h currently at Network Rail, Springs Branch Depot, Wigan, Lancashire
i currently at D.T. Warehousing Ltd, Ely, Cambridgeshire

Vehicles in storage or transit

Gauge 4ft 8½in

45	COLWYN	0-6-0ST	IC	K	5470	1933		Dsm
(D)1662	47484	Co-CoDE		Crewe		1965		
(D3854	08687) 08995	0-6-0DE		Hor		1959		
(D3873)	08706	0-6-0DE		Crewe		1960	a	
(D3970)	08802	0-6-0DE		Derby		1960	a	
(D4014	08846) 003	0-6-0DE		Hor		1961		
	(TOBY)	4wDM	s/o	RH	235513	1945		
			reb	Shackerstone		2000		
	HAMBLE-LE-RICE	0-6-0DH		TH	294V	1981	a	
W51354		2-2w-2w-2DMR		PSteel		1960		

51396 L720	2-2w-2w-2DMR	PSteel		1960	
PWM 2189 (TR 13 DX 68053)	2w-2PMR	Wkm	4166	1948	

Locomotives in transit are also stored in the yard for short periods of time.

a property of Harry Needle Railroad Co Ltd, Barrow Hill, Derbyshire

PRISON SERVICE COLLEGE, NEWBOLD REVEL, STRETTON-UNDER-FOSSE, near RUGBY — CV23 0TH

Gauge 2ft 0in — **SP 455808**

–	4wDM	L	33651	1949	Pvd

SEVERN LAMB UK LTD, ARDEN INDUSTRIAL ESTATE, TYTHING ROAD, ALCESTER — B49 6ET

New miniature locomotives under construction, and locomotives in for repair, usually present.
www.severn-lamb.com **SP 097587**

PRESERVATION SITES

LA SERVICES LTD, THE ENGINEERS EMPORIUM, BRAMCOTE WORKS, BRAMCOTE FIELDS FARM, BRAMCOTE — CV11 6QL

Gauge 2ft 0in www.theengineersemporium.co.uk

–	4wDM	RH	452294	1960	Dsm

J.W. LEWIS, BEACON VIEW, BENTLEYS LANE, MAXSTOKE — B46 2QR

Gauge 4ft 8½in — **SP 244875**

AD 9124	4wDMR	BD	3713	1975
RS/140	4wDM	FH	3892	1958

JOHN PAYNE, PRIVATE LOCATION

Gauge 4ft 8½in

YD No.26653 TANGO	4wDH	BD	3730	1977	
00 NZ 66	4wDH	BD	3732	1977	
"11720" ND10490	0-4-0DH	NBQ	27648	1959	
–	4wDM	RH	210481	1941	Dsm
–	0-4-0DE	YE	2856	1961	

THE STRATFORD & BROADWAY RAILWAY SOCIETY, LONG MARSTON, near STRATFORD-UPON-AVON

Gauge 4ft 8½in **(Closed)** — **SP 153473**

WD 70047 MULBERRY	0-4-0DM	AB	362	1942

WEST MIDLANDS

INDUSTRIAL SITES

ALSTOM TRANSPORT, WOLVERHAMPTON (MIDLAND) TRAINCARE CENTRE, JONES ROAD, OXLEY, WOLVERHAMPTON (part of Alstom Holdings S.A.) WV10 6JQ

Gauge 4ft 8½in www.alstom.com **SJ 906010**

(D3784) 08617 STEVE PURSER	0-6-0DE	Derby		1959	

J.P.M. PARRY & ASSOCIATES LTD, t/a PARRY PEOPLE MOVERS LTD CORNGREAVES TRADING ESTATE, OVEREND ROAD, CRADLEY HEATH B64 7DD

New PPM railcars under construction or repair usually present

Gauge 4ft 8½in www.parrypeoplemovers.com **SO 948853**

PPM 35	2w-2F/BER	PPM	10	1997	a
(PPM 50 999900) 12	2w-2FER	PPM	12	2000	b
"EROS"	4wFER	SUSTRACO		2005	a c

Gauge 610mm / 760mm

–	2w-2FER	PPM	8	1994	b d Dsm

Gauge 2ft 0in

(6)	4wFER	PPM	6	1993	b
(7)	4wFER	PPM	7	1994	b
(9)	2w-2FER	PPM	9	1995	a e

a property of Sustainable Transport Co Ltd, Bristol
b property of Parry People Movers Ltd, Cradley Heath
c incomplete vehicle
d test-bed vehicle
e currently off site

SALFORD METALS LTD, 475 LICHFIELD ROAD, ASTON, BIRMINGHAM B6 7SP

Gauge 4ft 8½in www.salfordmetals.co.uk **SP 093899**

THE SIR HENRY TARONI EXPRESS	0-4-0DH		HE	7259	1971	
THE FLYING SCRAPMAN		reb	HE	9286	1987	

TATA STEEL EUROPE, TATA STEEL ROUND OAK RAIL, THE GATEWAY, BRIERLEY HILL (part of the Tata Group) DY5 1LJ

Gauge 4ft 8½in www.tatasteeleurope.com **SO 925879**

ROUND OAK ROSIE	0-6-0DH	AB	491	1964	OOU
293V	0-6-0DH	TH	293V	1980	Pvd

T.M.A. ENGINEERING LTD, TYBURN ROAD, ERDINGTON, BIRMINGHAM B24 8NQ

New locomotives under construction, and for repair, occasionally present. www.tmaeng.co.uk

WEST MIDLANDS TRAINS, t/a WEST MIDLANDS RAILWAY and LONDON NORTHWESTERN RAILWAY, TYSELEY DEPOT, WARWICK ROAD, TYSELEY, BIRMINGHAM **B11 2HL**

(an Abellio, JR East and Mitsui Joint Venture)

Gauge 4ft 8½in www.londonnorthwesternrailway.co.uk **SP 105840**

See Section 7 for details

WEST MIDLANDS TRAVEL LTD, t/a MIDLAND METRO, WEDNESBURY DEPOT, POTTERS LANE, WEDNESBURY (subsidiary of National Express Group plc) **WS10 0AR**

Gauge 4ft 8½in www.nxbus.co.uk/the-metro **SO 984945**

BX64 VMA	4wDM	R/R	Hako		c2015
Q179 VOH	4wDM	R/R	Unimog	166200	1991
ERNIE	4wBE		SET	WN1021/1	2007
(A513484)	2w-2PMR		Geismar	ST/08/03	2008

WILLENHALL COMMERCIALS LTD, ASHMORE LAKE WAY, WILLENHALL WV12 4LF

Gauge 610mm www.willenhall-commercials.co.uk **SP 965995**

2	VENTILLATA	0-6-0DH		OK	25996	1960

Gauge 600mm

4		0-4-0WT	OC	OK	9922	1922
5		0-4-0WT	OC	OK	10701	1924
3		0-4-0DM		Ruhrthaler	3526	1958

Private site. No casual visitors

PRESERVATION SITES

THE BLACK COUNTRY LIVING MUSEUM TRUST, THE BLACK COUNTRY LIVING MUSEUM, TIPTON ROAD, DUDLEY **DY1 4SQ**

Gauge 4ft 8½in www.bclm.co.uk **SO 950913**

2025	WINSTON CHURCHILL	0-6-0ST	IC	MW	2025	1923

CADBURY WORLD, LINDEN ROAD, BOURNVILLE, BIRMINGHAM **B30 1JR**

Gauge 4ft 8½in www.cadburyworld.co.uk SP 048813

No.14	CADBURY	0-4-0DM		HC	D1012	1956

COUNT LOUIS TRUST, c/o A. WALTON, BLOXWICH

Gauge 1ft 3in

	COUNT LOUIS	4-4-2	OC	BL	32	1923

L.C.P. PROPERTIES LTD, PENSNETT TRADING ESTATE, SHUT END — DY6 7NA

Gauge 4ft 8½in www.lcpproperties.co.uk **SO 900898**

–		4wDM		RH	215755	1942	

NATIONAL COLLEGE for HIGH SPEED RAIL, 2 LISTER STREET, ASTON, BIRMINGHAM — B7 4NG

Gauge 4ft 8½in www.nchsr.ac.uk **SP 078879**

3102	#BRUMSTAR	Bo-BoWE		GEC-Alsthom		1992	

J.E. SELWAY, near WEDNESBURY

Gauge 1ft 6in

–		4-4-2	OC	CurwenD		1951	a

a currently stored off site

THINKTANK – BIRMINGHAM SCIENCE MUSEUM, MILLENNIUM POINT, CURZON STREET, BIRMINGHAM — B4 7AP

Gauge 4ft 8½in www.birminghammuseums.org.uk/thinktank **SP 084875**

46235	CITY OF BIRMINGHAM	4-6-2	4C	Crewe		1939	

TYSELEY LOCOMOTIVE WORKS LTD and VINTAGE TRAINS LTD, THE STEAM DEPOT, WARWICK ROAD, TYSELEY, BIRMINGHAM — B11 2HL

Gauge 4ft 8½in www.tyseleylocoworks.co.uk **SP 105841**

2885		2-8-0	OC	Sdn		1938	
4121		2-6-2T	OC	Sdn		1937	
4588		2-6-2T	OC	Sdn		1927	
4936	KINLET HALL	4-6-0	OC	Sdn		1929	a
4965	ROOD ASHTON HALL	4-6-0	OC	Sdn		1929	
5043	EARL OF MOUNT EDGCUMBE	4-6-0	4C	Sdn		1936	
5080	DEFIANT	4-6-0	4C	Sdn		1939	
7029	CLUN CASTLE	4-6-0	4C	Sdn		1950	
(7752)	L94	0-6-0PT	IC	NBQ	24040	1930	
7760		0-6-0PT	IC	NBQ	24048	1930	
7812	ERLESTOKE MANOR	4-6-0	OC	Sdn		1939	
9600		0-6-0PT	IC	Sdn		1945	
34070	MANSTON	4-6-2	3C	Bton		1947	
(45593)	5593 KOLHAPUR	4-6-0	3C	NBQ	24151	1935	
71000	DUKE OF GLOUCESTER	4-6-2	3C	Crewe		1954	
–	(CADBURY)	0-4-0T	OC	AE	1977	1925	
3278		2-8-0	OC	Alco	71533	1944	
–		0-4-0VBT	VCG	Cockerill	3083	1924	
No.65		0-6-0T	OC			1994	
	a rebuild of	0-6-0ST	OC	HC	1631	1929	
–		0-4-0ST	OC	HL	2918	1912	
	F.D. & E. Co. No.3	0-4-0ST	OC	HL	3597	1926	
	ROCKET	0-4-0ST	OC	P	1722	1926	

	MERLIN	0-4-0ST	OC	P	1967	1939	
No.1		0-4-0ST	OC	P	2004	1941	
–		0-6-0ST	IC	RSHN	7289	1945	
No.670		2-2-2	IC	TyseleyLW		1989	b
(D318	97408) 40118	1Co-Co1DE		_(EE	2853	1961	
				(RSHD	8148	1961	
D1755	(47161 47541) 47773	Co-CoDE		BT	483	1964	
(D3029	08021) 13029	0-6-0DE		Derby		1953	
No.6	PRINCESS MARGARET	0-4-0DM		AB	376	1948	
–		0-4-0PM		Bg	800	1920	
–		4wDM		RH	299099	1950	
–		0-6-0DM		RH	347747	1957	

a currently at West Somerset Railway, Somerset
b not yet completed

WILTSHIRE

INDUSTRIAL SITES

HANSON QUARRY PRODUCTS EUROPE LTD, HARTHAM PARK QUARRY (PICKWICK QUARRY), PARK LANE, CORSHAM SN13 0QR

(part of the Heidelberg Cement Group)

Gauge 2ft 6in www.heidelbergcement.com **ST 855702**

–		4wDM	RH	359169	1953	OOU

Locomotive stored on surface

MINISTRY OF DEFENCE, DEFENCE COMMUNICATIONS SERVICES AGENCY, BASIL HILL SITE, PARK LANE, CORSHAM

Gauge 600mm - Underground **ST 851694**

WD No.1	M.W. 11025	4wDM	RH	179009	1936	OOU
28	M.W. 11026	4wDM	RH	(?)	1938	OOU

MINISTRY OF DEFENCE, LUDGERSHALL RAILHEAD

See Section 6 for details

PRESERVATION SITES

COLIN HATCH, HERITAGE & STEAM ENGINEERS, SWINDON

Gauge 1ft 3in www.hatchsteamengineers.co.uk

_		4-6-0	OC	SmithEL			a

a incomplete loco

LONGLEAT SAFARI PARK & ADVENTURE PARK, "JUNGLE EXPRESS", LONGLEAT LIGHT RAILWAY, LONGLEAT, WARMINSTER — BA12 7NW

Gauge 1ft 3in www.longleat.co.uk **ST 808432**

7	FLYNN		0-6-0DH		AK	79	2007
–			4w-4wDH		AK	104	2018
5	CEAWLIN		2-8-2DH	s/o	Longleat		1989
		a rebuild of	2-8-0DH	s/o	SL	75 356	1975
		rebuilt	0-6-2DH	s/o	Longleat		

JOHN PAYNE, PRIVATE LOCATION, near SALISBURY

Gauge 4ft 8½in

–			0-4-0DH		NBQ	27941	1961

SCIENCE MUSEUM GROUP AT WROUGHTON, RED BARN GATE, RED BARN, WROUGHTON, near SWINDON — SN4 9LT

Not on public display; visits by appointment only.

Gauge 1ft 11½in www.sciencemuseum.org.uk **SU 131790**

–			0-4-0DM		HE	4369	1951

PAUL STANFORD, CHIPPENHAM

Gauge 2ft 0in

3414			2w-2PM		Wkm	3414	1943
		rebuilt as	2w-2DM				1992

STEAM - MUSEUM OF THE GREAT WESTERN RAILWAY, SWINDON HERITAGE CENTRE, FIRE FLY AVENUE, SWINDON — SN2 2FY

Gauge 7ft 0¼in www.steam-museum.org.uk **SU 143849**

	NORTH STAR	2-2-2	IC	Sdn		1925	a

a replica, incorporating parts of the original, RS 150/1837

Gauge 4ft 8½in

2516		0-6-0	IC	Sdn	1557	1897
2818		2-8-0	OC	Sdn	2122	1905
(3440)	3717 CITY OF TRURO	4-4-0	IC	Sdn	2000	1903
4073	CAERPHILLY CASTLE	4-6-0	4C	Sdn		1923
4248		2-8-0T	OC	Sdn		1916
6000	KING GEORGE V	4-6-0	4C	Sdn		1927
9400		0-6-0PT	IC	Sdn		1947
A38W		2w-2PMR		Wkm	8505	1960

SWINDON & CRICKLADE RAILWAY SOCIETY, near SWINDON

Locomotives are kept at:– Blunsdon Station SN25 2DA SU 110897
Hayes Knoll SU 106907

Gauge 4ft 8½in www.swindon-cricklade-railway.org

5619		0-6-2T	IC	Sdn		1925	
6984	OWSDEN HALL	4-6-0	OC	Sdn		1948	
	SWORDFISH	0-6-0ST	OC	AB	2138	1941	
	RICHARD TREVITHICK	0-4-0ST	OC	AB	2354	1954	
3135	SPARTAN	0-6-0T	OC	Chrz	3135	1953	
70		0-6-0T	IC	HC	1464	1921	a
	"IVOR"	0-4-0ST	OC	P	1555	1920	
(D)2022	(03022)	0-6-0DM		Sdn		1958	
D2152	(03152)	0-6-0DM		Sdn		1960	
D3261	(13261)	0-6-0DE		Derby		1956	
D3668	(09004)	0-6-0DE		Dar		1959	
E6003	(73003) SIR HERBERT WALKER	Bo-BoRE/DE		Elh		1962	
PWM 651	(97651)	0-6-0DE		RH	431758	1959	
	WOODBINE	0-4-0DM		JF	21442	1936	
–		0-4-0DM		JF	4210137	1958	b
	BLUNSDON	0-4-0DH		JF	4220031	1964	
51074	595	2-2w-2w-2DMR		GRC&W		1959	
51104	119 021	2-2w-2w-2DMR		GRC&W		1959	
60127	1302 (207203)	4-4wDER		Afd/Elh		1962	
DR 98504		4wDHR		Plasser	52792	1985	
68009	(PWM 4305)	2w-2PMR		Wkm	7508	1956	
–		2w-2PMR		Wkm	8089	1958	

a based here, but visits other locations
b incorporates parts of JF 4210082 / 1953

SWINDON COLLEGE, DEPARTMENT OF ENGINEERING, NORTH STAR SITE, NORTH STAR AVENUE, SWINDON **SN2 1DY**

Gauge 1ft 2in www.swindon.ac.uk **SU 148855**

	NORTH STAR	2-2-2	IC	SdnCol		1978

SWINDON DESIGNER OUTLET CENTRE, KEMBLE DRIVE, SWINDON **SN2 2DY**

(part of the McArthurglen Group)

Gauge 4ft 8½in www.mcarthurglen.com **SU 138847**

7821	DITCHEAT MANOR	4-6-0	OC	Sdn		1950

P.S. WEAVER, NEW FARM, LACOCK, near CORSHAM

Gauge 1ft 9in **ST 899691**

–		0-4-0VBT	VCG	WeaverP		1978

WORCESTERSHIRE

INDUSTRIAL SITES

UK LOCO LTD,
UNITS 2 & 3, HEATH PARK, MAIN ROAD, CROPTHORNE, PERSHORE **WR10 3NE**
www.uk-loco.com **SO 987446**
New UK LOCO vehicles under construction or repair occasionally present

PRESERVATION SITES

IAN ASHBY, VALLEY NURSERIES, PERSHORE ROAD, EVESHAM **WR11 2PX**
Visitors welcome - but first telephone I. Ashby on 07845 527880
Gauge 2ft 0in

12	1 ARCHER	4wDM		MR	4709	1936	
–		4wDM		MR	7128	1936	Dsm

Gauge 1ft 6in

	CLIFF	4wDH		Ashbyl		2015	
	a rebuild of	4wPM		TEE		2010	
ML-2-17		4wBH		Tunn		1980	
			reb	Tunn		1996	a
–		4wBH		Tunn		1980	
			reb	Tunn		1996	a

a two of Tunn TQ121 to TQ126

DINMORE MANOR LOCOMOTIVE LTD, HONEYBOURNE
Gauge 4ft 8½in www.dinmoremanor.co.uk

3845		2-8-0	OC	Sdn		1942

EVESHAM VALE LIGHT RAILWAY,
THE VALLEY, (EVESHAM COUNTRY PARK), TWYFORD, EVESHAM **WR11 4TP**
Gauge 1ft 3in www.evlr.co.uk **SP 044465**

	MONTY	0-4-2T	OC	ESR	300	1996
	ST EGWIN	0-4-0TT	OC	ESR	312	2003
3	DOUGAL	0-6-2T	OC	SL		1970
	"THE BISCUIT"	4wPMR		Eddy/Knowell		2004
	CROMWELL	4wDM		RH	452280	1960
	rebuilt as	4wDH		AK	13R	1984

LONGLANDS RAILWAY, PRIVATE LOCATION
Gauge 2ft 0in

–		4wDM		RH	200512	1940

ROSS-ON-WYE STEAM ENGINE SOCIETY – GREAT WELLAND RAILWAY, WELLAND STEAM RALLY, WOODSIDE FARM, WELLAND, MALVERN **WR13 6LN**

Gauge 4ft 8½in www.wellandsteamrally.co.uk **SO 803410**

Visiting locomotives may be present on rally days

SEVERN VALLEY RAILWAY CO LTD,
KIDDERMINSTER and BEWDLEY STATIONS

Gauge 4ft 8½in **SO 836757, 793753**

For details of locomotives see under Shropshire entry.

D. TURNER, "FAIRHAVEN", WYCHBOLD

Gauge 2ft 0in **SO 922660**

–	4wDM		MR	8600	1940

WEST MIDLAND SAFARI PARK, SPRING GROVE, BEWDLEY **DY12 1LF**

Gauge 400mm - Simba Kiddies Train www.wmsp.co.uk **SO 801755**

–	4-6wRE	s/o	_(Dotto		1991
			(SL	365.3.91	1991

EAST YORKSHIRE

INDUSTRIAL SITES

SOUTH CAVE TRACTORS LTD,
COMMON LANE, NEWPORT, BROUGH, near KINGSTON UPON HULL **HU15 2RD**

UK agents for Zagro. www.southcavetractors.com **SE 872311**

Unimog vehicles for resale/conversion occasionally present.

OMYA UK LTD, HUMBER PLANT, HUMBER INDUSTRIAL ESTATE, GIBSON LANE, MELTON, NORTH FERRIBY, near KINGSTON-UPON-HULL **HU14 3LJ**

Gauge 4ft 8½in www.omya.com **SE 965258**

–	0-6-0DH	JF	4240017	1966	OOU

WANSFORD TROUT FARM, LTD,
WHIN HILL, WANSFORD, near DRIFFIELD **(YO25 5NW)**

Gauge 2ft 0in **TA 051568**

–	4wPH	(? Denmark)
–	4wPH	(? Denmark)
–	4wDH	(? Driffield)

PRESERVATION SITES

EAST WRESSLE AND BRIND RAILWAY, WRESSLE BRICKYARD FARM, WRESSLE, near HOWDEN

Gauge 4ft 8½in — **SE 124313**

	JANE	4wDM		RH	371971	1954
PWM 4313 (TR 23 B52)		2w-2PMR		Wkm	7516	1956

HULL CITY MUSEUM & ART GALLERIES, 'STREETLIFE' – MUSEUM OF TRANSPORT, 40 HIGH STREET, KINGSTON UPON HULL — HU1 1PS

Gauge 4ft 8½in www.humbermuseums.com — **TA 103287**

	FRANK GALBRAITH	4wVBT	VCG	S	9629	1957

Gauge 3ft 0in

1		0-4-0Tram	OC	K	2551 T56	1882

YORKSHIRE WOLDS RAILWAY RESTORATION PROJECT, FIMBER HALT, BEVERLEY ROAD, near WETWANG

Gauge 4ft 8½in www.yorkshirewoldsrailway.org.uk — **SE 911607**

5576	SIR TATTON SYKES	0-4-0DH		GECT	5576	1979

NORTH YORKSHIRE

LOCOMOTIVES WHICH ARE LOCATED WITHIN THE COUNTY BOROUGHS OF MIDDLESBROUGH AND REDCAR & CLEVELAND ARE LISTED UNDER A SEPARATE HEADING OF TEESSIDE.

INDUSTRIAL SITES

AQUARIUS RAILROAD TECHNOLOGIES LTD, OLD SLENINGFORD FARM, MICKLEY, RIPON — HG4 3JB

Road/Rail vehicles under conversion/repair or for hire/resale usually present. **SE 263769**
www.railrover.com Private site. Permission is required prior to visiting.

ATKINSON VOS LTD, WENNING AVENUE, HIGH BENTHAM — LA2 7LW

Unimog vehicles for resale/conversion occasionally present. **SD 664688**
www.unimogs.co.uk

G.C.S. JOHNSON LTD, BARTON PARK, BARTON, RICHMOND — DL10 6NF

Locomotives in transit may be stored in this yard for short periods of time **NZ 217079**
www.gcsjohnson.co.uk

NETWORK RAIL, HOLGATE DEPOT, YORK OLD WORKS, HOLGATE ROAD, YORK — YO24 4EH

Gauge 4ft 8½in www.networkrail.co.uk **SE 585518**

Network Rail owned vehicles based here. See Section 7 for details.

PLASMOR LTD, CONCRETE BLOCK MANUFACTURERS, BLOCK WORKS RAIL TERMINAL, HECK WORKS, GREEN LANE, GREAT HECK — DN14 0BZ

Gauge 4ft 8½in www.plasmor.co.uk **SE 597213**

–		0-4-0DH	JF	4220038	1966

A.C. PRICE (ENGINEERING) LTD, INGLETON INDUSTRIAL ESTATE, NEW ROAD, INGLETON — LA6 3NU

Unimog vehicles for resale, repair or conversion occasionally present. **SD 691724**

www.acprice.co.uk

TARMAC plc – A CRH COMPANY, SWINDEN QUARRY, LINTON, near SKIPTON — BD23 6BE

Gauge 4ft 8½in www.tarmac.com **SD 983614**

	CRACOE	6w-6wDH	RFSD	067/GA/57000/001	1994

PRESERVATION SITES

TIM ACKERLEY, PRIVATE LOCATION

Gauge 4ft 8½in

D3255		0-6-0DE	Derby		1956

DERWENT VALLEY LIGHT RAILWAY SOCIETY, YORKSHIRE MUSEUM OF FARMING, MURTON PARK, MURTON LANE, YORK — YO19 5UF

Gauge 4ft 8½in www.dvlr.org.uk **SE 650524**

(D2079)	03079	0-6-0DM	Don		1960
D2245	(11215)	0-6-0DM	_(RSHN	7864	1956
			(DC	2577	1956
	"CHURCHILL"	0-4-0DM	JF	4100005	1947
–		0-4-0DM	JF	4200022	1948
No.165	BRITISH SUGAR YORK	0-4-0DM	RH	327964	1953
(No.1)		4wDM	RH	417892	1959
(No.2)		4wDM	RH	421419	1958
No.3	KEN COOKE	4wDM	RH	441934	1960
088		4wDM	RH	466630	1962
–	(DB 965050)	2w-2PMR	Wkm	7565	1957

EMBSAY & BOLTON ABBEY STEAM RAILWAY, EMBSAY STATION, EMBSAY, near SKIPTON

Locomotives are kept at :–

Embsay Station	BD23 6QE	SE 007533
Bolton Abbey	BD23 6AF	SE 060533

Gauge 4ft 8½in www.embsayboltonabbeyrailway.org.uk

5643		0-6-2T	IC	Sdn		1925	
No.22		0-4-0ST	OC	AB	2320	1952	
	ILLINGWORTH	0-6-0ST	OC	HC	1208	1916	
–		0-6-0ST	OC	HC	1450	1922	
	rebuilt as	0-6-0T	OC	Byworth		2000	
SLOUGH ESTATES LTD No.5		0-6-0ST	OC	HC	1709	1939	
No.140		0-6-0T	OC	HC	1821	1948	
	(AIREDALE No.3)	0-6-0ST	IC	HE	1440	1923	
No.7	BEATRICE	0-6-0ST	IC	HE	2705	1945	
S.134	WHELDALE	0-6-0ST	IC	HE	3168	1944	
S.121	PRIMROSE No.2	0-6-0ST	IC	HE	3715	1952	
–		0-6-0ST	IC	HE	3785	1953	
	(MONCKTON No.1)	0-6-0ST	IC	HE	3788	1953	
–		0-4-0ST	OC	RSHN	7661	1950	
7164	ANN	4wVBT	VCG	S	7232	1927	
No.3		0-4-0ST	OC	YE	2474	1949	
D1524	(47004)	Co-CoDE		BT	419	1963	
(D2203)	11103	0-6-0DM		_(VF	D145	1952	
				¯(DC	2400	1952	
(D3067)	08054	0-6-0DE		Dar		1953	
D3941	(08773)	0-6-0DE		Derby		1960	
(D5537)	31119	A1A-A1ADE		BT	136	1959	
D5600	(31435)	A1A-A1ADE		BT	200	1960	
(D6994	37294)	Co-CoDE		_(EE	3554	1965	
				¯(EEV	D983	1965	
(D9513)	N.C.B. 38	0-6-0DH		Sdn		1964	
(MEAFORD POWER STATION LOCO No.1)		0-4-0DH		AB	440	1958	
–		4wDM		Bg/DC	2136	1938	Dsm
–		0-6-0DM		HC	D1037	1958	
–	(H.W.ROBINSON)	0-4-0DM		JF	4100003	1946	
–		0-4-0DM		JF	4200003	1946	
887		4wDM		RH	394009	1955	
(GR 5093)		4wDM		Robel	21.12 RN5	1973	
No.3170		4-4wDMR		EBAR/GCR		2016	
	a rebuild of	4-4wPMR		NER		1903	
	DB 965095	2w-2PMR		Wkm	7610	1957	

Embsay Narrow Gauge Project, Embsay

BD23 6QE
SE 007533

Gauge 2ft 6in

–		4wDM		SMH	60SD755	1980	

Gauge 2ft 0in

–		4wPM		L	10225	1938	OOU
P 1215		4wDM		MR	5213	1930	
–		4wDM		MR	8979	1946	Dsm
–		4wDM		RH	175418	1936	Dsm
4	VULCAN A.M.W. No.197	4wDM		RH	198287	1940	

Y.W.A. L2	4wDM		RH			OOU
754	4wDM		SMH	60SD754	1980	
RTT/767172	2w-2PM		Wkm	3164	c1943	OOU

FLAMINGO LAND LTD, FLAMINGO LAND RESORT, KIRBY MISPERTON, near PICKERING — YO17 6UX

Gauge 2ft 0in – Daktari Express www.flamingoland.co.uk **SE 778800**

97	C.P.HUNTINGTON	4w-2-4wPH	s/o	Chance	73 5097-24	1973
1863		4w-2-4wPH	s/o	Chance	76 50141 24	1976

Gauge monorail – Zoo monorail **SE 780796**

–	2w-2PH	BCM	1984

LIGHTWATER VALLEY LEISURE LTD, LIGHTWATER VALLEY FARM, RIPON — HG4 3HT

Gauge 1ft 3in www.lightwatervalley.co.uk **SE 285756**

278	7	2-8-0DH	s/o	SL	1/84	1984

J.LLOYD, CASTLETON LIGHT RAILWAY, CASTLETON

Gauge 2ft 0in

–		0-4-0DM		Dtz	47414	1951
–		4wDM		Eclipse		c1956
No.10		4wDM		Moës		
	BECKY	4wDM		MR	7215	1938
2		4wDM		MR	7333	1938
7494	ALNE	4wDM		MR	7494	1940
			reb	York(BRE)		1991
	IRTHING	4wDM		MR	8655	1941

NATIONAL RAILWAY MUSEUM, LEEMAN ROAD, YORK — YO26 4XJ

Some of the collection is exhibited at 'Locomotion – at Shildon' and is rotated quite frequently.
Some locomotives may be used on 'Mainline Runs' and also exhibited at other sites.

Gauge 4ft 8½in www.railwaymuseum.org.uk **SE 593519**

	THE AGENORIA	0-4-0	VC	FosterRastrick		1829	
	ROCKET	0-2-2	OC	RS	4089	1934	a
No.1		4-2-2	OC	Don	50	1870	
No.3		0-4-0	IC	BuryC&K		1846	
No.66	AEROLITE	2-2-4T	IC	Ghd		1869	
82	BOXHILL	0-6-0T	IC	Bton		1880	
214	GLADSTONE	0-4-2	IC	Bton		1882	
673		4-2-2	IC	Derby		1897	
990	HENRY OAKLEY	4-4-2	OC	Don	769	1898	
No.1275		0-6-0	IC	D	708	1874	
4003	LODE STAR	4-6-0	4C	Sdn	2231	1907	

(30245)	No.245	0-4-4T	IC	9E		1897	
(31737)	No. 737	4-4-0	IC	Afd		1901	
(33001)	C1	0-6-0	IC	Bton		1942	
35029	ELLERMAN LINES	4-6-2	3C	Elh		1949	a
(42500)	2500	2-6-4T	3C	Derby		1934	
(42700	2700) 13000	2-6-0	OC	Hor		1926	
(46229)	6229 DUCHESS OF HAMILTON	4-6-2	4C	Crewe		1938	
(50621)	No.1008	2-4-2T	IC	Hor	1	1889	
60007	SIR NIGEL GRESLEY	4-6-2	3C	Don	1863	1937	
(60022)	4468 MALLARD	4-6-2	3C	Don	1870	1938	
60103	(4472) FLYING SCOTSMAN	4-6-2	3C	Don	1564	1923	
(68846)	1247	0-6-0ST	IC	SS	4492	1899	
92220	EVENING STAR	2-10-0	OC	Sdn		1960	
	BAUXITE No.2	0-4-0ST	OC	BH	305	1874	
	ROCKET	0-2-2	OC	FlourMill		2010	
		rebuild of		LocoEnt	No.2	1979	
KF7	(607)	4-8-4	OC	VF	4674	1935	
D200	(40122)	1Co-Co1DE		_(EE	2367	1957	
				(VF	D395	1957	
D1023	WESTERN FUSILIER	C-CDH		Sdn		1963	
(D1656	47072, 47609,47834) 47798 PRINCE WILLIAM						
		Co-CoDE		Crewe		1965	
D2860		0-4-0DH		YE	2843	1961	
(D3079	08064) 13079	0-6-0DE		Dar		1954	
(D4105)	09017	0-6-0DE		Hor		1961	
(D5500)	31018	A1A-A1ADE		BT	71	1957	
D6700	(37350)	Co-CoDE		_(EE	2863	1960	
				(VF	D579	1960	
D8000	(20050)	Bo-BoDE		_(EE	2347	1957	
				(VF	D375	1957	
D9002	(55002)	Co-CoDE		_(EE	2907	1960	
	THE KINGS OWN YORKSHIRE LIGHT INFANTRY			(VF	D559	1960	
26020	(76020)	Bo-BoWE		Gorton	1027	1951	
87001	STEPHENSON / ROYAL SCOT	Bo-BoWE		Crewe		1973	
3308		Bo-BoWE		GEC-Alsthom		1995	
22-141		4w-4wWER				1976	
7050		0-4-0DM		_(EEDK	874	1934	
				(DC	2047	1934	
(BEL 2)	No.1	4wBE		Stoke		1917	
(W4W)	No.4	4w-2w+2DMR		_(AEC	(852004?)	1934	
				(PRoyal	B3550	1934	
M51562		2-2w-2w-2DMR		DerbyC&W		1959	
(51922)		2-2w-2w-2DMR		DerbyC&W		1960	
8143	1293	4w-4RER		_(MV		1925	
				(MetC&W		1925	
28249		4w-4wRER		Oerlikon		1915	

a locomotive is sectioned

Gauge 3ft 0in

	"HANDYMAN"	0-4-0ST	OC	HC	573	1900

Gauge 900mm

RA 36		4wBE/WE		HE	9423	1990

Gauge 2ft 0in

809	(215 216)	2w-2-2-2wRE		EEDK	809	1931	a

a built 1931 but originally carried plates dated 1930

Gauge 1ft 11½in

No.3	LIVINGSTON THOMPSON	0-4-4-0T	4C	FRCo		1885
			reb	FRCo		1905

Gauge 1ft 6in

	WREN	0-4-0STT	OC	BP	2825	1887
	PET	0-4-0ST	IC	Crewe		1865

Gauge 1ft 4½in

–		2-4-0	OC	Young&Co		1856

Gauge 1ft 3in

	LITTLE GIANT	4-4-2	OC	BL	10	1905

Gauge monorail

W.D. No.1		0-2-2-0BER		Brennan		1907

NORTH BAY RAILWAY CO LTD, NORTH BAY RAILWAY, NORTHSTEAD MANOR GARDENS, NORTH BAY, SCARBOROUGH — YO12 6PF

Gauge 1ft 8in www.nbr.org.uk — TA 035898

	STEAMPLEX		4wVBT	VCG	AK	93R	2013
		rebuild of	4wDM		MR	5877	1935
	GEORGINA		0-4-0ST	OC	NBRES		2016
1931	NEPTUNE		4-6-2DH	s/o	HC	D565	1931
No.570	ROBIN HOOD		4-6-4DH	s/o	HC	D570	1932
				reb	AK		1982
1932	TRITON		4-6-2DH	s/o	HC	D573	1932
1933	POSEIDON		4-6-2DM	s/o	HC	D582	1933
		rebuilt as	4-6-2DH	s/o	#		1991
–			4wDM		RH	375360	1955

\# rebuilt by Lenwade Hydraulic Services, locomotive was probably DH from new and not as shown.

NORTH YORKSHIRE MOORS HISTORICAL RAILWAY TRUST LTD, NORTH YORKSHIRE MOORS RAILWAY

Locomotives are kept at :-

Grosmont YO22 5QE NZ 828049, 828053
Levisham SE 818909
New Bridge YO18 8JL SE 803854
Pickering YO18 7AJ SE 797842

Gauge 4ft 8½in www.nymr.co.uk

3814		2-8-0	OC	Sdn		1940	
(30825)	825	4-6-0	OC	Elh		1927	
(30830)		4-6-0	OC	Elh		1930	
30841		4-6-0	OC	Elh		1936	Dsm
30926	REPTON	4-4-0	3C	Elh		1934	
34101	HARTLAND	4-6-2	3C	Elh		1950	
44806		4-6-0	OC	Derby		1944	
(45428)	5428 ERIC TREACY	4-6-0	OC	AW	1483	1937	

61264	(1264)	4-6-0	OC	NBQ	26165	1947	
63395	(No.2238)	0-8-0	OC	Dar		1918	
(65894)	2392	0-6-0	IC	Dar		1923	
75029	THE GREEN KNIGHT	4-6-0	OC	Sdn		1954	
76079		2-6-0	OC	Hor		1957	a
80135		2-6-4T	OC	Bton		1956	
80136		2-6-4T	OC	Bton		1956	
92134		2-10-0	OC	Crewe		1957	
No.3672	DAME VERA LYNN	2-10-0	OC	NBH	25458	1944	
	"LUCIE"	0-4-0VBT	OC	Cockerill	1625	1890	
			reb	Dorothea		1988	
No.29		0-6-2T	IC	K	4263	1904	
No.5		0-6-2T	IC	RS	3377	1909	
D2207		0-6-0DM		_(VF	D208	1953	
				(DC	2482	1953	
(D3723)	08556	0-6-0DE		Dar		1959	
(D3610)	08495	0-6-0DE		Hor		1958	
(D4018)	08850	0-6-0DE		Hor		1961	
D5032	(24032) HELEN TURNER	Bo-BoDE		Crewe		1959	
D5061	(24061) (97201) IAN JOHNSON	Bo-BoDE		Crewe		1960	
(D5338)	26038 TOM CLIFT 1954 - 2012	Bo-BoDE		BRCW	DEL83	1959	
(D6964)	37264	Co-CoDE		_(EE	3524	1965	
				(EEV	D953	1965	
D7628	(25278) SYBILLA	Bo-BoDE		BP	8038	1965	
	DEPARTMENTAL LOCOMOTIVE No.16	0-4-0DM		_(DC	2164	1941	
				_(EEDK	1195	1941	
				(VF		1941	
12139	REDCAR	0-6-0DE		EEDK	1553	1948	
(50160)	53160	2-2w-2w-2DMR		MetCam		1956	
M50164	(53164) DAISY 1956-2003	2-2w-2w-2DMR		MetCam		1956	
E50204	(53204)	2-2w-2w-2DMR		MetCam		1957	
E51511		2-2w-2w-2DMR		MetCam		1959	
–		2w-2PMR		Wkm	417	1931	DsmT
–		2w-2PMR		Wkm	1305	1933	DsmT
–		2w-2PMR		Wkm			DsmT

a based here but often visits other locations

POPPLETON COMMUNITY RAILWAY NURSERY LTD, STATION ROAD, POPPLETON, YORK YO26 6QA

Gauge 2ft 0in www.poppletonrailwaynursery.co.uk **SE 558536**

No.1	LOWECO	4wDM		L	20449	1942
			reb	ALR	No.1	1978
	TERRY STANHOPE	4wDM		StanhopeT		

PRIVATE OWNER, SLEIGHTS, near WHITBY

Gauge 2ft 0in

P1200	4wPM	TeasdaleS		20xx

RIPON & DISTRICT LIGHT RAILWAY, CANALSIDE, DALLAMIRES LANE, RIPON

Gauge 1ft 11½in www.riponlightrailway.co.uk **SE 321705**

7	4wBE	GB	2848	1957	
3	4wDM	L	7954	1936	
1	4wDM	L	50191	1957	

Gauge 1ft 3in

–	4wDM	StanhopeT		c1977

WENSLEYDALE RAILWAY ASSOCIATION, LEEMING BAR STATION

Locomotives are kept at :- Leeming Bar DL7 9AR SE 013889
Bedale DL8 1AN SE 268884

Gauge 4ft 8½in www.wensleydalerail.com

92219	2-10-0	OC	Sdn		1960	
(D1909 47232 47665 47820) 47785	Co-CoDE		BT	671	1965	
(D1945 47502) 47715 HAYMARKET	Co-CoDE		BT	707	1966	a
(D2144) 03144	0-6-0DM		Sdn		1961	
(D5684 31256) 31459	A1A-A1ADE		BT	285	1961	a
(D6553) 33035	Bo-BoDE		BRCW	DEL145	1961	
(D6869 37169) 37674	Co-CoDE		(EE (EEV	3347 D833	1963 1963	
(D6950) 37250	Co-CoDE		(EE (EEV	3507 D938	1964 1964	
(D7663 25213 25313)	Bo-BoDE		Derby		1966	
(D8169 20169)	Bo-BoDE		(EE (EEV	3640 D1039	1966 1966	
D9523	0-6-0DH		Sdn		1964	
01526 (265)	4wDH		TH	307V	1983	a
01529 (268)	4wDH		TH	310V	1984	a
01544 (252)	4wDH		TH	270V	1977	a
01545 (253)	4wDH		TH	271V	1977	a
E50256 (53256)	2-2w-2w-2DMR		MetCam		1957	
(50746 53746)	2-2w-2w-2DMR		MetCam		1957	
51210	2-2w-2w-2DMR		MetCam		1958	
51353 117301	2-2w-2w-2DMR		PSteel		1960	
W51400 117420	2-2w-2w-2DMR		PSteel		1960	
M51572 ROBIN	2-2w-2w-2DMR		DerbyC&W		1960	
55032 121032	2-2w-2w-2DMR		PSteel		1961	
LEV 1 RDB 975874	4wDMR		Leyland		1978	
RTU 6898 99709 901038-8	2w-2PMR		Geismar	ST/01/33	2001	
WR 3012 MPP 0010	2w-2PMR		Wkm	10731	1974	

a property of Harry Needle Railroad Co Ltd, Derbyshire

Aysgarth Station, Aysgarth DL8 3TH

Gauge 4ft 8½in **Private site** **SE 013890**

–	0-6-0DH	EEV	3870	1969
WENSLEY	4wDM	RH	476141	1963

YORKSHIRE DALES NATIONAL PARK AUTHORITY, DALES COUNTRYSIDE MUSEUM, HAWES STATION, HAWES — DL8 3NT

Gauge 4ft 8½in www.yorkshiredales.org.uk **SD 875899**

67345		0-6-0T	OC	RSHN	7845	1955	

SOUTH YORKSHIRE

INDUSTRIAL SITES

C.F. BOOTH LTD, SCRAP MERCHANTS, CLARENCE METAL WORKS, ARMER STREET, ROTHERHAM — S60 1AF

Locomotives for scrap are also usually present at this location.

Gauge 4ft 8½in www.cfbooth.com **SK 421924**

506/1		0-4-0DH		AB	506/1	1969	OOU a
			reb	AB		1989	
506/2		0-4-0DH		AB	506/2	1969	OOU a
			reb	AB		1989	
	LAURA	0-6-0DH		AB	646	1979	
			reb	HAB	6767	1990	
			reb	HE		2004	
–	"LITTLE BLUE"	0-6-0DH		EEV	D1194	1967	
	MADDIE	0-6-0DH		HE	6662	1966	
01565		0-6-0DH		S	10144	1963	
426		4wDH		TH	170V	1966	OOU

a rebuild of 0-8-0DH AB 506/1965

DAVY MARKHAM, PRINCE OF WALES ROAD, DARNALL, SHEFFIELD — S9 4EX

(In administration)

Gauge 4ft 8½in R.T.C. www.davymarkham.com **SK 395875**

(7600)	(RUNAR)	4wDH		TH	189C	1967	
	a rebuild of	4wVBT	VCG	S	9536	1952	

ELECTRO-MOTIVE DIESEL LTD, ROBERTS ROAD LOCOMOTIVE MAINTENANCE DEPOT, ROBERTS ROAD, DONCASTER — DN4 0JT

(Subsidiary of Progress Rail; A Caterpillar Company)

Gauge 4ft 8½in www.progressrail.com **SE 566023**

–		4wBE	Niteq	B284	2009	

EUROPEAN METAL RECYCLING LTD, EMR SHEFFIELD, EAST COAST ROAD, ATTERCLIFFE, SHEFFIELD — S9 3YD

Locomotives for scrap or resale are occasionally present.

Gauge 4ft 8½in www.emrgroup.com **SK 373888**

(D3966) 08798	0-6-0DE		Derby		1960	
(D4040) 08872	0-6-0DE		Dar		1960	OOU
(D4111) 09023	0-6-0DE		Hor		1961	
1137	0-6-0DH		EEV	D1137	1966	
–	0-6-0DE		YE	2641	1957	
–	0-6-0DE		YE	2714	1958	

HITACHI RAIL EUROPE LTD, DONCASTER CARR DEPOT, TEN POUND WALK, DONCASTER — DN4 5xx

Gauge 4ft 8½in www.hitachirail-eu.com **SE 576016**

–	4wBE	R/R	Zephir	2589	2016
–	4wBE		Zephir	2590	2016

RON HULL Jnr LTD, RON HULL GROUP, MANGHAM WORKS, BARBOT HALL INDUSTRIAL ESTATE, MANGHAM ROAD, PARKGATE, ROTHERHAM — S62 6EF

Locomotives for scrap or resale occasionally present. www.ronhull.co.uk **SK 433950**

LIBERTY STEELS, LIBERTY SPECIALITY STEELS, ALDWARKE WORKS, ROTHERHAM — (S65 3SR)

www.libertyhousegroup.com

Gauge 4ft 8½in **SK 447951, 449953, 449962, 451954, 452963, 456957, 459958**

31	0-6-0DE	YE	2904	1964	
32	0-6-0DE	YE	2935	1964	
34	0-6-0DE	YE	2594	1956	OOU
35	0-6-0DE	YE	2635	1957	
37	0-6-0DE	YE	2736	1959	Dsm
93	0-6-0DE	YE	2889	1962	
94	0-6-0DE	YE	2890	1962	
96	0-6-0DE	YE	2905	1963	

LIBERTY STEELS, LIBERTY SPECIALITY STEELS, STOCKSBRIDGE WORKS, STOCKSBRIDGE, SHEFFIELD — S36 2JA

Gauge 4ft 8½in www.libertyhousegroup.com **SK 260990, 267992**

No.1 714/37	4wDM	Robel	21.12RK3	1969
30	0-6-0DE	YE	2750	1959
33	0-6-0DE	YE	2740	1959
38	0-6-0DE	YE	2798	1961

MECHAN LTD, DAVY INDUSTRIAL PARK, PRINCE OF WALES ROAD, SHEFFIELD — S9 4EX

UK suppliers of Zwiehoff rail shunters and depot equipment. www.mechan.co.uk **SK 395875**

THE SCOTTS COMPANY (UK) LTD, HATFIELD PEAT WORKS, STAINFORTH MOOR ROAD, THORNE, DONCASTER DN5 8TE

Gauge 3ft 0in R.T.C. www.scotts.com **SE 713084**

S20	THE THOMAS BUCK	4wDH		Schöma	5130	1990	
			reb	AK		1998	Pvd

STAGECOACH SUPERTRAM MAINTENANCE LTD, NUNNERY SUPERTRAM DEPOT, WOODBOURN ROAD, SHEFFIELD S9 3LS

(Subsidiary of Stagecoach Holdings)

Gauge 4ft 8½in www.supertram.com **SK 374878**

08	M992 NNB	4wDM	R/R	_(Multicar	000339	1995		
				(Perm		1995	Dsm	a
09	YM02 DJY	4wDM	R/R	_(Hako	000870	2002		
				(Harsco		2002		
	YS63 FMG	4wDM	R/R	_(Hako		2014		
				(APEL		2014		

a rail wheels removed

WABTEC RAIL LTD, DONCASTER WORKS, HEXTHORPE ROAD, DONCASTER DN4 0BF

This is understood to be a complete FLEET LIST of the locomotives owned by (or in the care of) this company, which operates from the above address. Some locomotives are for hire or resale, and may be in use or stored at a number of locations. Locomotives under repair usually present.

Gauge 4ft 8½in www.wabtec.com **SE 569031**

(D1690	47514)	47703	Co-CoDE	BT	622	1967	a
(D3587)	08472		0-6-0DE	Crewe		1958	b
(D3738)	08571		0-6-0DE	Crewe		1959	c
(D3763)	08596		0-6-0DE	Derby		1959	b
(D3782)	08615		0-6-0DE	Derby		1959	d
(D3836)	08669	BOB MACHIN	0-6-0DE	Crewe		1960	
(D3892)	08724		0-6-0DE	Crewe		1960	
(D4021)	08853		0-6-0DE	Hor		1961	
	PAMMY		4wDH	TH	166V	1966	

a property of Harry Needle Railroad Co Ltd, Derbyshire
b currently at London North Eastern Railway Ltd, Craigentinny Depot, Edinburgh
c currently at Prologis R.F.I., D.I.R.F.T, Daventry, Northamptonshire
d currently at LH Group Services Ltd, Barton-under-Needwood, Staffordshire

PRESERVATION SITES

COUNTY BOROUGH OF DONCASTER, MUSEUM & ART GALLERY, CHEQUER ROAD, DONCASTER DN1 2AE

Gauge 2ft 6in www.doncaster.gov.uk **SE 579030**

–	4wDM	LB	53977	1964	a

a on loan from C.& D.Lawson, Hertfordshire

ELSECAR STEAM RAILWAY, ELSECAR HERITAGE CENTRE, WATH ROAD, ELSECAR, BARNSLEY — S74 8HJ

Gauge 4ft 8½in www.elsecarrailway.co.uk **SE 386998**

	EARL FITZWILLIAM	0-6-0ST	OC	AE	1917	1923	
No.15	HASTINGS	0-6-0ST	IC	HE	469	1888	
2150	MARDY MONSTER	0-6-0ST	OC	P	2150	1954	
	BIRKENHEAD	0-4-0ST	OC	RSHN	7386	1948	
	GERVASE	0-4-0VBT	VCG	S	6807	1928	
	a rebuild of	0-4-0ST	OC	MW	1472	1900	a
–		4wVBT	VCG	S	9374	1947	b
–		4wVBT	VCG	S	9376	1947	
			reb	TH		1960	
LMS 7165	WILLIAM	4wVBT	VCG	S	9599	1956	
	LOUISE	0-6-0DH		HE	6950	1967	
	ELIZABETH	4wDH		TH	138C	1964	
	a rebuild of	4wVBT	VCG	S	9584	1955	
2895	EARL OF STRAFFORD	0-6-0DH		YE	2895	1964	
DB965949	DX 68005	2w-2PMR		Wkm	10645	1972	
–		2w-2DM					c

a carries plate S 6710
b locomotive under restoration elsewhere
c dumper truck with fixed rail wheels

Gauge 2ft 2in

–		4wDM		RH	382808	1955

MRW RAILWAYS LTD, SHEFFIELD

Miniature locomotives for resale and under construction usually present

Gauge 1ft 3in

	WENDY	4-4wDM		ColebySim		1972	
–		2-8-0DH	s/o	SL	73.35	1973	
	rebuilt as	2-6-0DH	s/o			1994	
	rebuilt as	2-8-0DH	s/o	MRWRS		c2009	a

a currently at Thorpe Light Railway, Whorlton, Co. Durham

NATIONAL COLLEGE for HIGH SPEED RAIL, CAROLINA WAY, DONCASTER — DN4 5PN

Gauge 4ft 8½in www.nchsr.ac.uk **SE 593012**

3101	#DONNYSTAR	Bo-BoWE		GEC-Alsthom		1992
T244 CNN	99709 918002-5	4wDM	R/R	(Volvo	219691	1999
				(SRS		1999

SHEFFIELD INDUSTRIAL MUSEUMS TRUST, KELHAM ISLAND INDUSTRIAL MUSEUM, ALMA STREET, SHEFFIELD — S3 8RY

This locomotive is not on public display.

Gauge 4ft 8½in www.simt.co.uk **SK 352882**

BSC 1	0-4-0DE	YE	2481	1950

SOUTH YORKSHIRE TRANSPORT MUSEUM,
UNIT 9, WADDINGTON WAY, ALDWARKE, ROTHERHAM **S65 3SH**
Gauge 4ft 8½in www.sytm.co.uk **SK 440944**

(BROWN BAYLEY No.7)	0-4-0ST	OC	HC	1689	1937	

WEST YORKSHIRE

INDUSTRIAL SITES

ARRIVA TRAINS UK LTD - ARRIVA TRAINS NORTH LTD,
and EAST MIDLANDS TRAINS LTD (part of the Stagecoach Group),
NEVILLE HILL TRAINCARE DEPOT, OSMONDTHORPE LANE, LEEDS **LS9 9BJ**
Gauge 4ft 8½in www.northernrailway.co.uk / www.eastmidlandstrains.co.uk

See Section 7 for details **SE 329331, 327329, 330330**

CROSSLEY EVANS LTD, METAL PROCESSORS,
STATION ROAD, SHIPLEY **BD18 2LY**
Gauge 4ft 8½in **SE 148372**

42 M	PRINCE OF WALES	0-4-0DH	HE	7159	1969	
0040		4wDM	RH	284838	1950	OOU
9	BETH	4wDM	RH	425483	1958	OOU
01507	425 VENOM	0-6-0DH	RH	459519	1961	
	KATIE	4wDH	S	10023	1960	OOU

W.H. DAVIS LTD, WAGON REPAIRERS, FERRYBRIDGE DEPOT,
c/o GMOS FERRYBRIDGE MAINTENANCE & ENGINEERS,
RWE GENERATION UK, OLD GREAT NORTH ROAD, KNOTTINGLEY **WF11 8NG**
Gauge 4ft 8½in www.whdavis.co.uk **SE 481252**

01520	(274)	4wDH	TH	V322	1987	a
01543	(263)	4wDH	TH	303V	1982	a

a on hire from Harry Needle Railroad Co Ltd, Barrow Hill, Derbyshire

PETER DUFFY LTD, CIVIL ENGINEERING & UTILITY SERVICES,
PARK VIEW, LOFTHOUSE, WAKEFIELD **WF3 3HA**
Gauge 2ft 0in www.peterduffyltd.com **SE 334248**

–	4wBE	CE	5956	1972	

FREIGHTLINER MAINTENANCE LTD, MIDLAND ROAD, HUNSLET, LEEDS LS10 2RJ

(part of the America Genesee & Wyoming Railway Co)

Gauge 4ft 8½in www.freightliner.co.uk **SE 312311**

See Section 7 for details

TERBERG DTS (UK) LTD, LOWFIELDS WAY, LOWFIELDS BUSINESS PARK, ELLAND, HALIFAX (Part of the Terberg Group) www.terbergdts.co.uk HX5 9DA

Terberg/Zagro Road-Rail shunting locomotives may occasionally be present. **SE 116218**

PRESERVATION SITES

ALLAN BAMFORD, MODEL FARM, TOFTSHAW FOLD, off TOFTSHAW LANE, EAST BIERLEY, BRADFORD BD4 6QR

Gauge 4ft 8½in www.modelfarmshopbradford.co.uk **SE 189296**

–		0-4-0F	OC	WB	2473	1932

CITY OF BRADFORD METROPOLITAN COUNCIL ART GALLERIES & MUSEUMS, BRADFORD INDUSTRIAL & HORSES AT WORK MUSEUM, MOORSIDE MILLS, MOORSIDE ROAD, BRADFORD BD2 3HP

Gauge 4ft 8½in www.bradfordmuseums.org **SE 184353**

	NELLIE	0-4-0ST	OC	HC	1435	1922

THE CHILDRENS MUSEUM LTD, EUREKA !, THE NATIONAL CHILDRENS MUSEUM, DISCOVERY ROAD, HALIFAX HX1 2NE

Gauge 4ft 8½in www.eureka.org.uk **SE 097247**

02641		0-4-0DM		HE	2641	1941

KEIGHLEY & WORTH VALLEY LIGHT RAILWAY LTD

Locomotives are kept at :-

Haworth BD22 8NJ SE 034371
Ingrow BD21 5AX SE 058399
Oxenhope BD22 9LB SE 032355

Gauge 4ft 8½in www.kwvr.co.uk

5775		0-6-0PT	IC	Sdn		1929	
41241		2-6-2T	OC	Crewe		1949	
43924		0-6-0	IC	Derby		1920	
45212		4-6-0	OC	AW	1253	1935	a
45596	BAHAMAS	4-6-0	3C	NBQ	24154	1935	
			reb	HE	5596	1968	b
47279		0-6-0T	IC	VF	3736	1924	
48431		2-8-0	OC	Sdn		1944	
51218	(68)	0-4-0ST	OC	Hor	811	1901	
(52044)	957	0-6-0	IC	BP	2840	1887	

(58926	7799) 1054	0-6-2T	IC	Crewe	2979	1888	
75078		4-6-0	OC	Sdn		1956	
78022		2-6-0	OC	Dar		1954	
80002		2-6-4T	OC	Derby		1952	
No.2258	TINY	0-4-0ST	OC	AB	2258	1949	
	LORD MAYOR	0-4-0ST	OC	HC	402	1893	
(31)	(HAMBURG)	0-6-0T	IC	HC	679	1903	
No.1704	NUNLOW	0-6-0T	OC	HC	1704	1938	
118	BRUSSELS	0-6-0ST	IC	HC	1782	1945	
5820		2-8-0	OC	Lima	8758	1945	
No.85		0-6-2T	IC	NR	5408	1899	
–		0-4-0CT	OC	RSHN	7069	1942	
90733		2-8-0	OC	VF	5200	1945	
D2511		0-6-0DM		HC	D1202	1961	
(D3336	13336) 08266	0-6-0DE		Dar		1957	
(D3759	08592) 08993 ASHBURNHAM	0-6-0DE		Crewe		1959	
			reb	Landore		1985	
(D5209)	25059	Bo-BoDE		Derby		1963	
(D6775)	37075	Co-CoDE		_(EE	3067	1962	
				(RSHD	8321	1962	
(D8031)	20031	Bo-BoDE		_(EE	2753	1959	
				(RSHD	8063	1959	
D0226	VULCAN	0-6-0DE		_(EE	2345	1956	
				(VF	D226	1956	
23	MERLIN	0-6-0DM		HC	D761	1951	
32	HUSKISSON	0-6-0DM		HE	2699	1944	
	JAMES	0-4-0DE		RH	431763	1959	
M50928		2-2w-2w-2DMR		DerbyC&W		1959	
M51189		2-2w-2w-2DMR		MetCam		1958	
M51565		2-2w-2w-2DMR		DerbyC&W		1959	
Sc51803		2-2w-2w-2DMR		MetCam		1959	
E79962		2w-2DMR		WMD	1267	1958	
M79964		2w-2DMR		WMD	1269	1958	

a currently at Riley & Sons (E) Ltd, Heywood, Greater Manchester
b currently at Tyseley Locomotive Works Ltd, Birmingham, West Midlands

KIRKLEES LIGHT RAILWAY CO LTD,
PARK MILL WAY, CLAYTON WEST, near HUDDERSFIELD **HD8 9XJ**
Gauge 1ft 3in www.kirkleeslightrailway.com **SE 258112**

	KATIE	2-4-2	OC	Guest		1956	a
	FOX	2-6-2T	OC	TaylorB	No.9	1987	
	BADGER	0-6-4ST	OC	TaylorB	No.10	1991	
	HAWK	0-4-4-0T	4C	TaylorB		2007	b
	OWL	4w-4wT	VC	TaylorB	No.12	2000	
7		2-2wPH	s/o	TaylorB		1991	
	JAY	4wDH		TaylorB		1992	

a carries plate Guest 14/1954
b rebuilt using parts from TaylorB No.11 1998

LEEDS CITY COUNCIL, DEPARTMENT OF LEISURE SERVICES, LEEDS MUSEUMS & GALLERIES, LEEDS INDUSTRIAL MUSEUM, ARMLEY MILLS, CANAL ROAD, LEEDS

LS12 2QF

Gauge 4ft 8½in www.leeds.gov.uk **SE 275342**

(GWR 252)	0-6-0	IC	EBW		1855	Dsm
ELIZABETH	0-4-0ST	OC	HC	1888	1958	
R.A.F.No.111 ALDWYTH	0-6-0ST	IC	MW	865	1882	
–	4wBE		GB	1210	1930	
		reb	HE	9146	1987	
S 1986.0028	0-4-0WE		GB	2543	1955	
SOUTHAM 2	0-4-0DM		HC	D625	1942	
B16 ND 3066	0-4-0DM		HE	2390	1941	
–	0-4-0DM		JF	22060	1937	
S 1990.0013 No.10 MP 351	0-4-0DM		JF	22893	1940	

Gauge 3ft 6in

PIONEER	0-6-0DMF		HC	DM634	1946	
S 1985.0008	0-6-0DMF		HC	DM733	1950	

Gauge 3ft 0in

S 1985.0014 4057	0-6-0DMF		HE	4057	1953	
S 1985.0020	2-4-0DM		JF	20685	1935	

Gauge 2ft 11in

–	4wDM		HC	D571	1932	

Gauge 2ft 8in

–	0-4-0DMF		HE	3200	1945	

Gauge 2ft 6in

S 1990.0015 "JUNIN"	2-6-2DM		HC	D557	1930	
S 1992.0009	4wBE		HT	9728	1985	

Gauge 2ft 1½in

No.5	0-4-0DMF		HE	4019	1949	

Gauge 2ft 0in

S 1992.0020 FAITH	0-4-0DMF		HC	DM664	1952	
–	0-4-0DMF		HC	DM749	1949	
1368	4wDMF		HC	DM1368	1965	
S.2000.0002	0-4-0DMF		HE	2008	1939	
8	4wDMF		HE	4756	1954	
–	0-4-0DMF		HE	5340	1957	
21294 LAYER	4wDM		JF	21294	1936	Dsm

Gauge 1ft 6in

–	4wBE		GB	1325	1933	Dsm
4 5A	4wBE		GB	1326	1933	

MIDDLETON RAILWAY TRUST, MOOR ROAD STATION, TUNSTALL ROAD, HUNSLET, LEEDS — LS10 2JQ

Gauge 4ft 8½in www.middletonrailway.org.uk **SE 305310**

No.1310		0-4-0T	IC	Ghd	(38?)	1891
68153	59	4wVBT	VCG	S	8837	1933
Nr.385		0-4-0WT	OC	Hart	2110	1895
	HENRY DE LACY II	0-4-0ST	OC	HC	1309	1917
67		0-6-0T	IC	HC	1369	1919
	SLOUGH ESTATES No.3	0-6-0ST	OC	HC	1544	1924
	MIRVALE	0-4-0ST	OC	HC	1882	1955
–		0-4-0ST	OC	HE	1493	1925
	"PICTON"	2-6-2T	OC	HE	1540	1927
–		0-4-0T	OC	HE	1684	1931
	"BROOKES No.1"	0-6-0ST	IC	HE	2387	1941
	rebuilt as	0-6-0T	IC	Middleton		1999
	rebuilt as	0-6-0ST	IC	Middleton		2007
	SWANSCOMBE No.6	0-4-0ST	OC	HL	3860	1935
	SIR BERKELEY	0-6-0ST	IC	MW	1210	1891
	MATTHEW MURRAY	0-6-0ST	IC	MW	1601	1903
–		0-4-0ST	OC	P	2103	1950
D2999		0-4-0DE		_(BT	91	1958
				(BP	7856	1958
	MARY	0-4-0DM		HC	D577	1932
	CARROLL	0-4-0DM		HC	D631	1946
MDHB 45		0-6-0DH		HC	D1373	1965
7051	(WD 27 70027)	0-6-0DM		HE	1697	1932
	"JOHN ALCOCK"		reb	HE		1949
	"COURAGE" "SWEET PEA"	4wDM		HE	1786	1935
	CONOCO	0-4-0DH		HE	6981	1968
–		0-4-0DM		JF	3900002	1945
	HARRY	0-4-0DH		JF	4220033	1965
	AUSTINS No.1	0-4-0DM		P	5003	1961
	(03-03-PO300)	4wWE		GB	420452	1979
DB 998901	"OLIVE"	2w-2DMR		Bg/DC	2268	1950

Gauge 3ft 0in

BEM 402		4wDHF		HE	8505	1981

Gauge 2ft 2in

	"FLYING SCOTSMAN"	4wDM		HE	7274	1973

OUTGANG MICRO BREWERY, THE KINSLEY HOTEL, WAKEFIELD ROAD, KINSLEY, near PONTEFRACT — WF9 5EH

Gauge 2ft 0in www.kinsleyhotel.com **SE 419144**

713009		4wBE		CE	B0182A	1974

Mr TAYLOR, CALDERDALE

Gauge 2ft 0in

	4470	4wPM		OK	4470	1931

A.J. WILSON, 35 HOLT PARK ROAD, LEEDS

Gauge 3ft 0in

06/22/6/2		4wDM		RH	224337	1944	

Gauge 2ft 0in

	THE WASP	2-2wPM		WilsonAJ		1969	
			reb	WilsonAJ		1979	a

a currently at Leander Architectural, Dove Holes, Derbyshire

YORKSHIRE MINING MUSEUM TRUST, NATIONAL COAL MINING MUSEUM FOR ENGLAND, CAPHOUSE COLLIERY, NEW ROAD, OVERTON, WAKEFIELD WF4 4RH

Gauge 4ft 8½in www.ncm.org.uk **SE 248161, 249162, 253164**

	PROGRESS	0-6-0ST	IC	RSHN	7298	1946
NCB 44		0-6-0DH		HE	6684	1968
–		0-6-0DH		HE	7307	1973
(No.47)		0-6-0DH		TH	249V	1974

Gauge 2ft 6in

No. 1	4471 COMPO	4w-4wBEF		CE	B3538	1989
No.7	SICK NOTE	4w-4wBEF		CE		
	KIRSTIN	4w-4wDHF	RACK	GMT	0592	1981
	(ANNA) YKSMM 2001.831	4w-4wDHF	RACK	GMT		1984
20		4wDHF	RACK	HE	9271	1987
(E682)		0-4-0DMF		HC	DM746	1951
No.2	"DEBORAH"	0-4-0DMF		HC	DM1356	1965
	"LARK"	0-6-0DMF		_(HC	DM1433	1978
				(HE	8581	1978
2	T198	4wDMF		RH	480679	1961

Gauge 2ft 4in

–		4wBEF		Atlas	2463	1944
	YKSMM 1997.857	4wDM		RH	375347	1954

Gauge 2ft 3in

	CAPHOUSE FLIER	4wDHF		HE	8832	1978

Gauge 2ft 2in

	YKSMM 1992.156	4wDH		HE	7530	1977

Gauge 2ft 1½in

T199		4wDM		RH	379659	1955

Gauge 2ft 0in

	YKSMM 1986.54	0-4-0DMF		HC	DM655	1949
713007		4wBE		CE	B0182B	1974

SECTION 2 — SCOTLAND

Consequent upon the fragmentation of the County Geography of Scotland, due to the creation of a number of autonomous Unitary Authorities and similar bodies and, in view of the relatively few remaining locations hosting locomotives, this Section of this Handbook is now presented in two parts - Industrial sites and Preservation sites.

Defining Areas in accordance with the 1974-1996 "large counties" era, now provides a direct relationship between this volume and the "Historic Handbook" series which is also published by the Industrial Railway Society. The historic books contain the recorded details of all past and recent locomotives, at all known past and present locations, in the areas they cover, together with extensive texts describing the locations and the businesses they served.

INDUSTRIAL SITES

1ST BATTALION, THE ROYAL HIGHLAND FUSILIERS, FORT GEORGE RANGE, INVERNESS (IV2 7TD)

Gauge 600mm **Highland NH 783571**

BEN	2w-2PM	Wkm	11682	1990
BRUCE	2w-2PM	Wkm	11683	1990

ABELLIO SCOTRAIL LTD, t/a SCOTRAIL (subsidiary of Nederlandse Spoorwegen)

Inverness Rail Depot, Longman Road, Inverness IV1 1RY

Gauge 4ft 8½in www.scotrail.co.uk **Highland NH 668457**

See Section 7 for details

Shields Electric Traction Depot, 35 St. Andrews Drive, Pollockshields, Glasgow G41 5SG

Gauge 4ft 8½in www.scotrail.co.uk **City of Glasgow NS 371638, 367639**

See Section 7 for details

ALSTOM TRANSPORT, GLASGOW TRAINCARE DEPOT, 109 POLMADIE ROAD, POLMADIE, GLASGOW (part of Alstom Holdings S.A.) G5 0BA

Gauge 4ft 8½in www.alstom.com **City of Glasgow NS 598625**

(D3932) 08764	0-6-0DE	Hor		1961
(D4117) 08887	0-6-0DE	Hor		1962
(D4184) 08954	0-6-0DE	Dar		1963

BABCOCK INTERNATIONAL GROUP plc, ROSYTH BUSINESS PARK, ROSYTH (KY11 2YD)

Gauge 4ft 8½in www.babcockinternational.com **Fife NT 108821**

236	0-4-0DM		AB	372	1945
		reb	YEC	L123	1994
BRIL 001 "JINTY"	0-6-0DM		AB	385	1952

BRODIE ENGINEERING LTD, BONNYTON RAIL DEPOT,
1 BONNYTON INDUSTRIAL ESTATE, MUNRO PLACE, KILMARNOCK **KA1 2NP**
Gauge 4ft 8½in www.brodie-engineering.co.uk **East Ayrshire NS 421383**

18 TINY II	0-4-0DH		YE	2676	1959	

R. & N. CESSFORD, WHANLAND FARM, FARNELL, BRECHIN **DD9 6UF**
Gauge 4ft 8½in **NO 621543**

-	4wDM	R/R	S&H	7505	1967	OOU

stored off site. Visitors by appointment only.

CHEMRING ENERGETICS UK LTD, LUNDHOLM ROAD, ARDEER,
STEVENSTON (part of the Chemring Group plc) **KA20 3NF**
Gauge 2ft 6in www.chemringenergetics.co.uk **North Ayrshire NS 290401, 290405**

05/582	4wDH		AK	21	1987	
05/583	4wDH		AK	22	1987	

DSM NUTRITIONAL PRODUCTS (UK) LTD, DRAKEMYRE, DALRY **KA24 5JJ**
Gauge 4ft 8½in www.dsm.com **North Ayrshire NS 295503**

50 GC 8	0-6-0DH		RR	10267	1967	
		reb	RFSK		1990	

EDINBURGH TRAMS LTD,
GOGAR DEPOT, off GLASGOW ROAD, GOGAR, EDINBURGH **EH12 9DH**
Gauge 4ft 8½in www.edinburghtrams.com **City of Edinburgh NT 172726**

–	4wBE	R/R	Niteq	B300	2009	
SN12 DWG	4wDM	R/R	Unimog	224283	2010	

GEMINI RAIL SERVICES UK LTD, SPRINGBURN DEPOT,
79 CHARLES STREET, off SPRINGBURN ROAD, GLASGOW **G21 2PS**
(part of the Gemini Rail Group)
Gauge 4ft 8½in www.geminirailgroup.co.uk **City of Glasgow NS 605665**

(D3898) 08730 THE CALEY	0-6-0DE		Crewe		1960	
2777	4wDM	R/R	Unilok	2091	1982	OOU

INEOS, GRANGEMOUTH REFINERY, BO'NESS ROAD, GRANGEMOUTH
Gauge 4ft 8½in www.ineos.com **Falkirk NS 942817, 944814, 952822**

10	0-6-0DH		AB	600	1976	a
11	0-6-0DH		AB	649	1980	
		reb	HE	750215	2006	a
3	0-6-0DH		EEV	D1232	1968	
		reb	AB		1980	a

a property of LH Group Services Ltd, Staffordshire, England

MINISTRY OF DEFENCE, DEFENCE MUNITIONS, GLEN DOUGLAS
See Section 6 for details

LESMAC (FASTENERS) LTD,
73 DYKEHEAD STREET, QUEENSLIE INDUSTRIAL ESTATE, GLASGOW G33 4AQ
www.lesmac.co.uk **City of Glasgow NS 658654**

New Lesmac railcars under construction or repair occasionally present

LITHGOWS LTD, NETHERTON, LANGBANK PA14 6YG
Gauge 2ft 0in **R.T.C.** **Renfrewshire NS 393722**

–	4wPM		MR	2097	1922	OOU
–	4wPM		MR	2171	1922	OOU
No.2	4wDM		MR	8700	1941	OOU

LONDON NORTH EASTERN RAILWAY LTD, t/a LNER,
Craigentinny Maintenance Depot, 167 Mountcastle Crescent, Edinburgh EH8 7TE
Gauge 4ft 8½in www.lner.co.uk **City of Edinburgh NT 298738**

See Section 7 for details

Craigentinny Wheel Lathe Depot, Stanley Street, Edinburgh EH15 1JJ
Gauge 4ft 8½in **City of Edinburgh NT 306732**

See Section 7 for details

W & D McCULLOCH, CRAIGIE MAINS, MAIN STREET, BALLANTRAE KA26 0NB
Gauge 4ft 8½in www.mccullochrail.com **Ayrshire NX 083829**

L482 LNU		4wDM	R/R	_(Multicar (Perm	1993 1993	OOU a
FH06 LBG	99709 979073-4 UNIMOG 04	4wDM	R/R	_(Unimog 209478 (LH Access	2006 2006	
	THE BIG GIRL	4w-4wDHR		_(McCulloch (McDowall	2016 2016	

a located off site

Other road/rail vehicles usually present at this location

THOS MUIR HAULAGE & METALS LTD, DEN ROAD, KIRKCALDY KY1 2ER
(part of the Thomas Muir Group)

Gauge 4ft 8½in www.thomasmuirltd.co.uk **Fife NT 282926**

No.3		0-4-0ST	OC	AB	946	1902	OOU
No.22		0-4-0ST	OC	AB	1069	1906	OOU
–		0-4-0ST	OC	AB	1807	1923	OOU
No.7		0-4-0ST	OC	AB	2262	1949	OOU
No.12	H 662	0-4-0DH		NBQ	27732	1957	OOU

NUCLEAR DECOMMISSIONING AGENCY, CHAPELCROSS WORKS, ANNAN (operated by Magnox Ltd) DG12 6RF

Gauge 5ft 4in (Closed) www.magnoxsites.com **Dumfries & Galloway NY 216695**

No.1	PETER	4wDM	RH	411320	1958	
No.2	JIM	4wDM	RH	411321	1958	

QTS RAIL LTD, RENCH FARM, DRUMCLOG, near STRATHAVEN ML10 6QJ

(part of the QTS Group) Plant depot with various road-rail and rail plant present between contracts. Prior permission is required before visiting this location.

Gauge 4ft 8½in www.qtsgroup.com **Lanarkshire NS 630386**

RRU 09	AD02 FKU 99709 979060-9	4wDM	R/R	Unimog	197737	2002	

JOHN G. RUSSELL (TRANSPORT) LTD, t/a RUSSELL LOGISTICS, DEANSIDE ROAD, HILLINGTON, GLASGOW G52 4XB

(a member of The Russell Group)

Gauge 4ft 8½in www.johngrussell.co.uk **City of Glasgow NS 522659**

(D3562)	08447	0-6-0DE	Derby		1958	

SCOTTISH WATER, STORNOWAY WATERWORKS, ISLE OF LEWIS

Gauge 2ft 0in www.scottishwater.co.uk **Western Isles NB 410375**

–		4wPM	(MR ?)			Dsm

STADLER BUSSNANG AG / ANSALSO STS CONSORTIUM, GLASGOW METRO CONTRACT

Gauge 4ft 0in www.stadlerrail.com / www.ansaldo-sts.com

–		4wBE	CE	B4624A	2017
–		4wBE	CE	B4624B	2017

E.G. STEELE & CO LTD, WINTON WAGON WORKS, 25 DALZELL STREET, HAMILTON ML3 9AU

The following is a FLEET LIST of the locomotives owned by this contractor.

Gauge 4ft 8½in www.egsteele.com **North Lanarkshire NS 708561**

–		4wDH	R/R	NNM	75511	1979	
–		4wDH	R/R	NNM	81504	1983	
T4	05/273	4wDH	R/R	NNM	83503	1984	
–		4wDH	R/R	NNM			a b

a one of NNM 82503/1983, NNM 83501/1983 or NNM 83504/1984
b on hire to DB Cargo (UK) Ltd, Motherwell C&W, Motherwell

STRATHCLYDE PARTNERSHIP FOR TRANSPORT, BROOMLOAN DEPOT, ROBERT STREET, GOVAN — G51 3HB

Glasgow underground railway maintenance locomotives.

Gauge 4ft 0in www.spt.co.uk **City of Glasgow** **NS 555655**

No.	Name	Type		Builder	Works No.	Year	Notes
L2	LOBEY DOSSER	4wBE		CE	B0965B	1977	
L3	RANK BAJIN	4wBE		CE	B0965A	1977	
W5		4wBE		CE	B0186	1974	Dsm a
L6		4wBE		CE	B4477A	2010	
L7		4wBE		CE	B4477B	2010	
–		4wBE	R/R	NNM	78101E	1979	
–		4wBE	R/R	Zephir	2766	2018	
–		2w-2BER		Bance	ERV2 276	2015	
–		2w-2BER		Bance	ERV2 278	2015	

a rebuilt into non-powered permanent way vehicle

TARMAC plc – A CRH Company, OXWELLMAINS CEMENT WORKS, DUNBAR — EH42 1SL

Gauge 4ft 8½in www.tarmac.com **East Lothian** **NT 708768**

Name	Type		Builder	Works No.	Year
DOON HILL	0-6-0DH		HE	7304	1972
		reb	HE	9374	2009
BLACK AGNES	0-6-0DH		HE	8979	1979
		reb	HE	9373	2009

TRAC ENGINEERING LTD, DOVECOTE ROAD, EUROCENTRAL, HOLYTOWN, Near MOTHERWELL — ML1 4GP

Gauge 4ft 8½in www.tracengineering.com **Lanarkshire** **NS 761616**

No.			Type		Builder	Works No.	Year
SV 090	AE06 PFN	99709 977014-8	4wDM	R/R	Unimog	208529	2006
SV 210			4wDM	R/R	Unimog	083216	1982

Other road/rail vehicles usually present at this location

UPM-KYMMENE (UK) LTD, CALEDONIAN PAPER MILL, MEADOWHEAD ROAD, SHEWALTON, IRVINE — KA11 5AT

Gauge 4ft 8½in www.upmpaper.com **North Ayrshire** **NS 335354**

Name	Type		Builder	Works No.	Year
–	0-6-0DH		HE	9092	1988
CHRISTIAN	0-6-0DH		RR	10217	1965
		reb	HE	9371	2006

WABTEC RAIL SCOTLAND LTD, CALEDONIA WORKS, WEST LANGLANDS STREET, KILMARNOCK — KA1 2QD

Vehicles under repair or overhaul are usually present.

Gauge 4ft 8½in www.wabtec.com **East Ayrshire** **NS 425382**

Name	Type	Builder	Works No.	Year
–	0-4-0DH	AB	482	1963

PRESERVATION SITES

ABERDEEN CITY COUNCIL, SEATON PARK, DON STREET, ABERDEEN AB24 1XQ

Gauge 4ft 8½in www.aberdeencity.gov.uk **Aberdeenshire NJ 943092**

	MR THERM	0-4-0ST	OC	AB	2239	1947	

ALFORD VALLEY RAILWAY CO LTD, ALFORD STATION, MURRAY PARK, ALFORD AB33 8DG

Gauge 2ft 0in (Closed) Aberdeenshire NJ 579159

	JAMES GORDON	0-4-0DH	s/o	AK	63	2001	
AVR No.1	HAMEWITH	4wDM		L	3198	c1930	Pvd
	THE BRA'LASS	4wDM	s/o	MR	9381	1948	
87022	THE WEE GORDON HIGHLANDER / BYDAND	4wDM		MR	22221	1964	

ALMOND VALLEY HERITAGE TRUST, ALMOND VALLEY HERITAGE CENTRE, MILLFIELD, LIVINGSTON EH54 7AR

Gauge 2ft 6in www.almondvalley.co.uk **West Lothian NT 034667**

–		4wDH	AB	557	1970
–		4wDH	BD	3752	1980
	"OAKBANK No.2"	4wWE	BLW	20587	1902
20		4wBE	BV	612	1972
38		4wBE	BV	698	1974
42		4wBE	BV	700	1974
7A	TAM	4wBE	BV	1143	1976
–		4wBEF	GB	1698	1940
	YARD No.B10	0-4-0DM	HE	2270	1940
–		4wDM	HE	7330	1973
–		4wDM	SMH	40SPF522	1981

AYRSHIRE RAILWAY PRESERVATION GROUP, SCOTTISH INDUSTRIAL RAILWAY CENTRE, DUNASKIN HERITAGE CENTRE, DALMELLINGTON ROAD, WATERSIDE, PATNA KA6 7JF

Gauge 4ft 8½in www.scottishindustrialrailwaycentre.org.uk **East Ayrshire NS 443083, (NS 475073)**

No.16	0-4-0ST	OC	AB	1116	1910	
(19)	0-4-0ST	OC	AB	1614	1918	
–	0-4-0F	OC	AB	1952	1928	
No.10	0-4-0ST	OC	AB	2244	1947	
NCB No.23	0-4-0ST	OC	AB	2260	1949	
–	0-6-0ST	OC	AB	2358	1954	
–	0-4-0ST	OC	AB	2368	1955	
No.1	0-4-0DM		AB	347	1941	
M3571	0-4-0DM		AB	366	1943	Dsm

–		0-4-0DM		AB	399	1956	
–		4wDMR		Donelli	163	1979	
–	Yd. No.107	0-4-0DM		HE	3132	1944	
–		0-4-0DM		JF	22888	1939	
–		0-4-0DH		NBQ	27644	1959	
DY322	YARD No. BE 1116 YARD No.AK1	4wDM		RH	224352	1943	
M/C 324		4wDM		RH	284839	1950	
–		4wDM		RH	417890	1959	
–		0-4-0DM		RH	421697	1959	
–		4wDH		S	10012	1959	

Gauge 3ft 0in

–		4wDH		HE	8816	1981	

Gauge 2ft 6in

05/579		4wDH		AB	561	1971	
3		4wBE		BV	307	1968	
43		4wBE		BV	701	1974	
7329		4wDM		HE	7329	1973	
2		4wDM		RH	183749	1937	Dsm
No.3		4wDM		RH	210959	1941	
No.1		4wDM		RH	211681	1942	
1		4wDM		RH	422569	1959	

BARCLAY HOUSE, CALEDONIA WORKS OFFICE BLOCK SITE, WEST LANGLANDS STREET, KILMARNOCK KA1 2PR

Gauge 4ft 8½in **East Ayrshire NS 424381**

	DRAKE	0-4-0ST	OC	AB	2086	1940	

CALEDONIAN RAILWAY (BRECHIN) LTD, BRECHIN, near MONTROSE

Locomotives are kept at :
Brechin DD9 7AF Angus NO 603603
Bridge of Dun DD10 9LH Angus NO 663587

Gauge 4ft 8½in www.caledonianrailway.co.uk

46464		2-6-0	OC	Crewe		1950	
–		0-4-0ST	OC	AB	1863	1926	
1		0-6-0T	OC	AB	2107	1941	
–		0-6-0ST	IC	HE	2879	1943	
6		0-4-0ST	OC	P	1376	1915	
	MENELAUS	0-6-0ST	OC	P	1889	1934	
–		0-6-0ST	IC	WB	2749	1944	
–		0-6-0ST	IC	WB	2759	1944	
(D)3059	(08046 13059) BRECHIN CITY	0-6-0DE		Derby		1954	
(D5222)	25072	Bo-BoDE		Derby		1963	
(D5233)	25083	Bo-BoDE		Derby		1963	
D5301	(26001)	Bo-BoDE		BRCW	DEL46	1958	
D5314	(26014)	Bo-BoDE		BRCW	DEL59	1959	
(D5335)	26035	Bo-BoDE		BRCW	DEL80	1959	

(D5353 27007) 27015	UNIVERSITY OF WIBBLEFROTH						
		Bo-BoDE		BRCW	DEL196	1961	
D5370	(27024)	Bo-BoDE		BRCW	DEL213	1962	
(D6797)	37097 OLD FETTERCAIRN	Co-CoDE		(EE	3226	1962	
				(VF	D751	1962	
12052		0-6-0DE		Derby		1949	
12093		0-6-0DE		Derby		1951	
–		4wDM		FH	3747	1955	
–		4wDM		RH	458957	1961	
211	ROLLS	0-4-0DE		YE	2628	1956	
(212)		0-4-0DE		YE	2684	1958	

CARNEGIE DUNFERMLINE TRUST / FIFE COUNCIL, PITTENCREIFF PARK, DUNFERMLINE — KY12 8QH

Gauge 4ft 8½in www.fifedirect.org.uk/pittencrieff **Fife NT 086872**

No.29		0-4-0ST	OC	AB	1996	1934	

CLYDE VALLEY FAMILY PARK, CROSSFORD, CARLUKE, near LANARK — ML8 5NJ

Gauge 600mm **NS 831461**

No.49	SANDY	4wDH		AK	49	1994	
1863		4w-4wDH	s/o	Chance			
					64-5031-24	1964	
–		4-4wBE	s/o	CityTrain		2018	
	rebuild of	4wDH		Schöma			

COUNTESS OF SUTHERLAND, DUNROBIN STATION, BRORA, near GOLSPIE — KW10 6SF

Gauge 2ft 0in www.dunrobincastle.co.uk **Highland NC 849013**

	BRORA	0-4-0PM	s/o	Bg	1797	1930	
			reb	Bg	2083	1934	

JOHN DEWAR & SONS LTD, DEWAR'S WORLD OF WHISKY, ABERFELDY DISTILLERY, ABERFELDY — PH15 2EB

Gauge 4ft 8½in www.dewars.com **Perth & Kinross NN 865496**

–		0-4-0ST	OC	AB	2073	1939	

DUNDEE MUSEUM OF TRANSPORT, MARKET MEWS, MARKET STREET, DUNDEE — DD1 3LA

Gauge 4ft 8½in www.dmoft.co.uk **City of Dundee NO 417309**

–		4wWE		BTH		1908	Dsm a

a currently located at Maryfield Tram Depot, Forfar Street, Dundee

EAST LINKS FAMILY PARK,
JOHN MUIR PARK, BELTONFORD, near DUNBAR — EH42 1XF

Gauge 2ft 0in www.eastlinks.co.uk **East Lothian NT 648786**

	THE EXPRESS	0-6-0DM	s/o	Bg	3014	1938	
	rebuilt as	0-6-0DH	s/o	BES		1993	

EAST LOTHIAN COUNCIL, MUSEUM SERVICES,
PRESTONGRANGE MUSEUM, COAST ROAD, PRESTONGRANGE — EH32 9RX

Gauge 4ft 8½in www.prestongrange.org **East Lothian NT 374737**

17		0-4-0ST	OC	AB	2219	1946	
No.7	PRESTONGRANGE	0-4-2ST	OC	GR	536	1914	
–		4wDH		EEV	D908	1964	
–		4wDM		MR	9925	1963	
No.33		4wDM		RH	221647	1943	
No.2	IVOR GEORGE EDWARDS	4wDM		RH	398613	1956	a
–		4wDM		RH	458960	1962	

a carries plate 398163 in error

Locomotives are stored in a secure shed on site with no public access

EASTRIGGS & GRETNA HERITAGE GROUP, DEVIL'S PORRIDGE MUSEUM,
STANFIELD, ANNAN ROAD, EASTRIGGS, ANNAN — DG12 6TF

Gauge 4ft 8½ www.devilsporridge.org.uk **Dumfries & Galloway NY 252662**

	"SIR JAMES"	0-6-0F	OC	AB	1550	1917

FALKIRK DISTRICT COUNCIL, DEPARTMENT OF LIBRARIES & MUSEUMS,
MUSEUM WORKSHOP, ABBOTSINCH COURT, 7–11 ABBOTSINCH ROAD,
ABBOTSINCH INDUSTRIAL ESTATE, GRANGEMOUTH — FK3 9UX

Gauge 4ft 8½in www.falkirk.gov.uk **Falkirk NS 936814**

No.1		0-4-0DM		JF	22902	1943

FERRYHILL RAILWAY HERITAGE TRUST,
FERRYHILL RAILWAY HERITAGE CENTRE,
off POLMUIR AVENUE, ABERDEEN — AB11 7TH

Gauge 4ft 8½in www.frht.org.uk **Aberdeenshire NJ 941045**

	NORTH DOWNS No.3	0-6-0T	OC	RSHN	7846	1955	
No.6	144-6	0-4-0DM		RH	421700	1959	
(900338	LNER 338)	2w-2PMR		Wkm	626	1934	a
A37W		2w-2PMR		Wkm	8502	1960	
DX 68002	DB 965330	2w-2PMR		Wkm	10180	1968	

a rebuilt using parts from Wkm 1583

Gauge 600mm

RTT/767162	2w-2PM		Wkm	3235	1943

FIFE REGIONAL COUNCIL,
LOCHORE MEADOWS COUNTRY PARK, LOCHORE **KY5 8AL**
Gauge 4ft 8½in www.fifedirect.org.uk **Fife NT 172963**

–	0-4-0ST	OC	AB	2259	1949

FRASERBURGH HERITAGE SOCIETY LTD,
FRASERBURGH HERITAGE CENTRE, QUARRY ROAD, FRASERBURGH AB43 9DT
Gauge 2ft 0in www.fraserburghheritage.com **Aberdeenshire NJ 997675**

677 KESSOCK KNIGHT	4wDM	s/o	LB	53541	1963	Pvd

FRIENDS OF CRAIGTOUN COUNTRY PARK RAILWAY, CRAIGTOUN PARK RAILWAY,
CRAIGTOUN COUNTRY PARK, near ST.ANDREWS **KY16 8NX**
Gauge 1ft 3in www.friendsofcraigtoun.org.uk **Fife NO 482150**

278	2-8-0DH	s/o	SL	R8	1976

THE GARDEN OF COSMIC SPECULATION,
PORTRACK SCOTTI GARDEN OF RAILS 2004,
PORTRACK HOUSE, HOLYWOOD, near DUMFRIES **DG2 0RW**
Gauge 4ft 8½in www.charlesjencks.com **Dumfries & Galloway NX 939830**

–	0-4-0DM		RH	418790	1958

GLASGOW CITY COUNCIL, CULTURAL AND SPORT GLASGOW
Glasgow Museums Resource Centre,
200 Woodhead Road, South Nitshill Industrial Estate, Nitshill, Glasgow G53 7NN
Gauge 4ft 8½in www.glasgowlife.org.uk **City of Glasgow NS 518601**

–	0-4-0VBT	VCG	Chaplin	2368	1885
No.1	0-6-0F	OC	AB	1571	1917
–	2w-2PM		Albion		c1916

Gauge 4ft 0in

–	4wBE		(JF	16559	1925
			(WR	583	1927
SUBWAY CAR No.1	4w-4wRER		OldburyC&W		1896

Riverside Museum, 100 Pointhouse Road, Glasgow **G3 8RS**
Gauge 4ft 8½in www.glasgowlife.org.uk **City of Glasgow NS 557659**

103	4-6-0	OC	SS	4022	1894
123	4-2-2	IC	N	3553	1886

(62469)	No.256 GLEN DOUGLAS	4-4-0	IC	Cowlairs		1913
9		0-6-0T	OC	NBH	21521	1917

Gauge 3ft 6in

3007		4-8-2	OC	NBQ	25546	1945

GRAMPIAN TRANSPORT MUSEUM, MONTGARRIE ROAD, ALFORD AB33 8AE

Gauge 4ft 8½in www.gtm.org.uk **Aberdeenshire NJ 577161**

No.3		0-4-0ST	OC	AB	1889	1926

D. HERBERT, GLASGOW AREA

Locomotives stored at a private location.

Gauge 2ft 6in **City of Glasgow**

2209		4wDM		HE	2209	1941
No.4	ND 10394	0-4-0DM		HE	2243	1941
R9	ND 6473	4wDM		RH	235727 .	1944

THE HIGHLAND LIGHT RAILWAY, THE RAILWAY FARM, MILL OF LOGIERAIT FARM, BALLINLUIG, near PITLOCHRY PH9 0LH

Gauge 2ft 0in www.railwayfarm.co.uk **Perth & Kinross NN 975516**

–		4wDM		AK	No.5	1979
–		4wDM		RH	283513	1949
	rebuilt as	4wDH		AK	20R	1986

I. HUGHES, PRIVATE LOCATION

Gauge 2ft 0in **Perth & Kinross**

T7	DOLLY	4wDH		AB	560	1971	
1	5	4wBE		CE	B2905	1981	
			reb	CE	B3550A	1988	a
–		4wDM		MR	9846	1952	
–		4wDM		MR	40S383	1971	

a plate reads CE B2903

THE INVERGARRY & FORT AUGUSTUS RAILWAY MUSEUM LTD, INVERGARRY STATION, SOUTH LAGGAN, near SPEAN BRIDGE PH34 4EA

Gauge 4ft 8½in www.invergarrystation.org.uk **Highland NN 303983**

–		4wDM		RH	236364	1946
	99709 901002-4	2w-2PMR		Lesmac	LMS012	2006

KEITH & DUFFTOWN RAILWAY ASSOCIATION, DUFFTOWN — AB55 4BA

Gauge 4ft 8½in www.keith-dufftown-railway.co.uk **Moray NJ 323414**

–		0-4-0DH		AB		1979
	a rebuild of	0-4-0DM		AB	415	1957
KDR 40	THE WEE MAC	4wDH		CE	B1844	1979
(KDR 41)		0-6-0DH		EEV	D1193	1967
51568	(KDR 6) SPIRIT OF BANFFSHIRE	2-2w-2w-2DMR		DerbyC&W		1959
52053	(KDR 7) SPIRIT OF BANFFSHIRE	2-2w-2w-2DMR		DerbyC&W		1960
(50628)	53628 (KDR 4) SPIRIT OF SPEYSIDE	2-2w-2w-2DMR		DerbyC&W		1958
(55500)	140.001	4wDMR		(DerbyC&W (Leyland	 (R2.001)	1981 1981
(55501)	140.001	4wDMR		(DerbyC&W (Leyland	 (R2.002)	1981 1981
CAR 1	C N RAIL 144-5	2-2wPMR		Fairmont	244095	
(CAR 2	KDR 43) 144-55	2-2wPMR		Fairmont		
CAR 3		2-2wPMR		Fairmont	252180	
PQ 364		4wDH	R/R	NNM	80505	1980

KINGDOM OF FIFE PRESERVATION SOCIETY, KIRKLAND YARD, BURNMILLS INDUSTRIAL ESTATE, LEVEN — KY8 4RB

Gauge 4ft 8½in **Fife NO 373007**

No.10	"LOCHGELLY"	0-4-0ST	OC	AB	1890	1926
No.17		0-4-0ST	OC	AB	2292	1951
(400)	RIVER EDEN	0-4-0DH		NBQ	27421	1955
	N.C.B.No.10	0-6-0DH		NBQ	27591	1957
1	YARD No.DP35	0-4-0DM		RH	313390	1952
No.4	"NORTH BRITISH"	4wDM		RH	421415	1958
2	THE GARVIE FLYER	0-4-0DE		RH	431764	1960
1	"LARGO LAW"	0-4-0DE		RH	449753	1961

LOMOND HILLS NARROW GAUGE RAILWAY, PRIVATE SITE, FREUCHIE

Gauge 2ft 0in **Fife**

	N.C.B. No.10	4wDM	HE	4440	1952
L3	SYLVIA	4wDM	MR	22128	1961

LOWTHERS RAILWAY SOCIETY LTD, LEADHILLS & WANLOCKHEAD RAILWAY, THE STATION, STATION ROAD, LEADHILLS — ML12 6XS

Gauge 2ft 6in www.leadhillsrailway.co.uk **South Lanarkshire NS 888145**

–		4wBE		WR	(1614	1940)?	Dsm

Gauge 2ft 0in

–		0-4-0WTT	OC	Decauville	917	1917
(9)	"CHARLOTTE"	0-4-0WT	OC	OK	6335	1913
(12)		4wDH		CE	B1819D	1978

(8)	NITH	0-4-0DMF		HC	DM1002	1956	
(6)	CLYDE	4wDH		HE	6347	1975	
(10)	(MENNOCK)	4wDH		HE	9348	1994	
250	8 8564	4wDM		MR	8564	1940	Dsm
251		4wDM		MR	8863	1944	a Dsm
(253)	53	4wDM		MR	8884	1944	b Dsm
(2)	ELVAN	4wDM		MR	9792	1955	
(5)	LITTLE CLYDE	4wDM		RH	7002/0467/2	1966	
(4)	LUCE	4wDM		RH	7002/0467/6	1966	

a converted into a coach.
b converted into a brake van

G. MANN, SAUGHTREE STATION, NEWCASTLETON TD9 0SP

Gauge 4ft 8½in www.saughtreestationbb.co.uk **Borders NT 565981**

275882	MEG OF SAUGHTREE	4wDM		RH	275882	1949	
	99709 909135-4	2w-2PMR		Geismar	ST/04/03	2004	

NATIONAL MINING MUSEUM SCOTLAND, RAIL & MINING HERITAGE CENTRE, LADY VICTORIA COLLIERY, NEWTONGRANGE EH22 4QN

Gauge 4ft 8½in www.nationalminingmuseum.com **Mid Lothian NT 332638**

No.21		0-4-0ST	OC	AB	2284	1949	

Gauge 3ft 6in

	TRAINING LOCO No.2	0-6-0DMF		HE	4074	1955	a

Gauge 2ft 6in

–		4wBE		CE	5871A	1971	a
–		4wBEF		CE	B3325A	1986	a

a These locomotives are in storage and can only be viewed from public tours

NATIONAL MUSEUM OF SCOTLAND, CHAMBERS STREET, EDINBURGH EH1 1JF

Gauge 5ft 0in www.nms.ac.uk **City of Edinburgh NT 258734**

	"WYLAM DILLY"	4wG	VC	Hedley		1827-1832	a

a incorporates parts of locomotive of same name built c1814 to c1815

Gauge 4ft 8½in

	"ELLESMERE"	0-4-0WT	OC	H(L)	244	1861	

Gauge 1ft 7in

	WYLAM DILLY	4wG	VC	RSM		1885	

ORKNEY ISLANDS COUNCIL, ORKNEY ARTS, MUSEUMS and HERITAGE, SCAPA FLOW VISITOR CENTRE and MUSEUM, LYNESS, HOY KW16 3NU

Gauge 600mm **Orkney Islands ND 310947**

–		2w-2PM		Wkm	3030	1941

D. RITCHIE & FRIENDS

Gauge 3ft 0in **Private Site** **City of Edinburgh**

–		4wDM		RH	466591	1961

Gauge 2ft 6in

–		4wDM		RH	189992	1938
–		4wDM		RH	242916	1946
–		4wDM		RH	273843	1949
10553		0-4-0DM		RH	338429	1955
P 9303	YARD No.1018	4wBE		VE	7667	

Gauge 2ft 0in

–		4wDM		HE	2654	1942
	TERRAS	4wDM		MR	7189	1937
CCC 51		4wDM		MR	7330	1938
–		4wDM		MR	9982	1954
–		4wDM		RH	179005	1936
–		4wDM		RH	249530	1947

F. ROACHE, SLEEPERZZZ, ROGART STATION, PITTENTRAIL, near GOLSPIE IV28 3XA

Gauge 4ft 8½in www.sleeperzzz.com **Highland NC 725020**

8016	ERIC	4wDM		RH	294263	1950

THE ROYAL DEESIDE RAILWAY, MILTON OF CRATHES, BANCHORY AB31 5QH

Gauge 4ft 8½in www.deeside-railway.co.uk **Aberdeenshire NO 740962, 724964**

	BON ACCORD	0-4-0ST	OC	AB	807	1897
	SALMON	0-6-0ST	OC	AB	2139	1942
RRM 10	8 "WELBECK No.6"	0-4-0ST	OC	P	2110	1950
(D2037	03037)	0-6-0DM		Sdn		1959
D2094	(03094)	0-6-0DM		Don		1960
D2134	(03134)	0-6-0DM		Sdn		1960
	rebuilt as	0-6-0DH				
(Sc 79998	RDB 975003)	4w-4wBER		DerbyC&W		1956
			reb	Cowlairs		1958
–		2w-2BER		TS&S	N/P1023	1985
	(PWM 2830)	2w-2PMR		Wkm	5008	1949
A34W		2w-2PMR		Wkm	8501	1960

SCOTTISH RAILWAY PRESERVATION SOCIETY, BO'NESS & KINNEIL RAILWAY, BO'NESS STATION, UNION STREET, BO'NESS EH51 9AQ

Gauge 4ft 8½in www.bkrailway.co.uk **West Lothian** **NT 003817**

(55189)	419	0-4-4T	IC	StRollox		1907	
61994	THE GREAT MARQUESS	2-6-0	3C	Dar	(1761?)	1938	
(62277)	No.49 GORDON HIGHLANDER	4-4-0	IC	NBH	22563	1920	
65243	MAUDE	0-6-0	IC	N	4392	1891	
68095	No.42	0-4-0ST	OC	Cowlairs		1887	
80105		2-6-4T	OC	Bton		1955	a
(45170	WD 554)	2-8-0	OC	NBH	24755	1942	
3	LADY VICTORIA	0-6-0ST	OC	AB	1458	1916	
No.3		0-4-0ST	OC	AB	1937	1928	
	(LORD ASHFIELD)	0-6-0F	OC	AB	1989	1930	
No.6		0-4-0ST	OC	AB	2043	1937	
	(B.A.CO.LTD No.3)	0-4-0ST	OC	AB	2046	1937	
THE WEMYSS COAL CO LTD No.20		0-6-0T	IC	AB	2068	1939	
GLENGARNOCK WORKS No.6		0-4-0CT	OC	AB	2127	1942	
No.24		0-6-0T	OC	AB	2335	1953	
	CITY OF ABERDEEN	0-4-0ST	OC	BH	912	1887	
(17)		0-6-0ST	IC	HE	2880	1943	
No.19		0-6-0ST	IC	HE	3818	1954	
No.5		0-6-0ST	IC	HE	3837	1955	
6		0-4-0ST	OC	HL	3640	1926	
No.13	N.C.B.13	0-4-0ST	OC	N	2203	1876	
(No.1	LORD ROBERTS)	0-6-0T	IC	NR	5710	1902	
	(JOHN)	4wVBT	VCG	S	9561	1953	a
	(RANALD)	4wVBT	VCG	S	9627	1957	
	(DENIS)	4wVBT	VCG	S	9631	1958	a
68007	(WD 75254)	0-6-0ST	IC	WB	2777	1945	
(D1970)	47643	Co-CoDE		Crewe		1965	
D2767		0-4-0DH		NBQ	28020	1960	
			reb	AB		1968	
D3558	(08443)	0-6-0DE		Derby		1958	
(D5324)	26024	Bo-BoDE		BRCW	DEL69	1959	
(D5338)	26038 TOM CLIFT 1954 - 2012	Bo-BoDE		BRCW	DEL83	1959	b
(D5347)	27001	Bo-BoDE		BRCW	DEL190	1961	
(D5351)	27005	Bo-BoDE		BRCW	DEL194	1961	
(D6607)	(37307) 37403	Co-CoDE		(EE	3567	1965	
				(EEV	D996	1965	c
(D6725)	37025 INVERNESS TMD	Co-CoDE		(EE	2888	1961	
				(VF	D604	1961	d
(D6767)	37067 (37703)	Co-CoDE		(EE	3059	1962	
				(VF	D721	1962	e
(D6914)	37214	Co-CoDE		(EE	3392	1963	
				(EEV	D858	1963	
(D6961)	37261	Co-CoDE		(EE	3521	1965	
				(EEV	D950	1965	
(D7585)	25235	Bo-BoDE		Dar		1964	

(D8020)	20020	Bo-BoDE		_(EE	2742	1959	
				(RSHD	8052	1959	
(E3036)	84001	Bo-BoWE		NBH	27793	1960	
No.1		0-6-0DM		AB	343	1941	
	F.G.F.	0-4-0DH		AB	552	1968	
F.82		4wWE/BE		EEDK	1131	1940	
	(TIGER)	0-4-0DH		NBQ	27415	1954	
	(KILBAGIE)	4wDM		RH	262998	1949	
–		4wDM		RH	275883	1949	
P6687		0-4-0DE		RH	312984	1951	
–		4wDM		RH	321733	1952	
–		0-4-0DE		RH	421439	1958	
	(ST MIRREN)	0-4-0DE		RH	423658	1958	
–		0-4-0DH		RH	457299	1962	
Sc51017		2-2w-2w-2DMR		Sdn		1959	
Sc51043		2-2w-2w-2DMR		Sdn		1959	
61503	(303 032)	4w-4wRER		PSteel		1960	
	99709 909131-3	2w-2PMR		Geismar	ST/03/57	2003	
SV138	99709 909264-2	2w-2PMR		Geismar	ST/09/04	2009	
SV142	99709 909305-3	2w-2PMR		Geismar	ST/06/03C	2006	
	(HCT 022)	4wDH		Perm	022	1988	f
(970213)		2w-2PMR		Wkm	6049	1952	
–		2w-2PMR		Wkm	10482	1970	

a property of the Locomotive Owners Group (Scotland) Ltd
b currently at North Yorkshire Moors Railway, Grosmont, North Yorkshire
c currently on hire to Direct Rail Services Ltd, Carlisle, Cumbria
d currently on hire to Colas Rail
e property of Direct Rail Services Ltd, Carlisle, Cumbria
f property of Northumbria Rail Ltd, Bedlington, Northumberland

Gauge 4ft 0in

55		4w-4wRER		OldburyC&W		1901	
			reb	Govan		1935	

Gauge 3ft 0in

	(FAIR MAID OF FOYERS)	0-4-0T	OC	AB	840	1899	
–		4wDH		MR	110U082	1970	

SHED 47 RAILWAY RESTORATION GROUP, LATHALMOND RAILWAY MUSEUM, SCOTTISH VINTAGE BUS MUSEUM, M90 COMMERCE PARK, LATHALMOND **KY12 OSJ**

Gauge 4ft 8½in www.shed47.org **Fife** **NT 093922**

	N.C.B. No.29	0-4-0ST	OC	AB	1142	1908	
No.17		0-4-0ST	OC	AB	2296	1950	
D2650		0-4-0DH		HE	9045	1980	
			reb	YEC	L135	1994	
(251)		0-4-0DH		HE	9046	1980	
	TEXACO	0-4-0DM		JF	4210140	1958	
A315885	(A241059)	2w-2PMR		Geismar	ST/02/02	2002	a

a rebuilt using parts from Geismar ST/03/52

Gauge 2ft 0in – The West Fife Munitions Railway

(T9	T4)	4wDM		AK	28	1989
L1	SANDRA	4wDH		AK	47	1994
(T11)	"ELOUISE"	4wDM		MR	21505	1955

STRATHSPEY RAILWAY CO LTD

Locomotives are kept at :–

Aviemore PH22 1PY NH 898131
Boat of Garten PH24 3BQ NH 943189
Broomhill (PH26 3LU) NH 997226

Gauge 4ft 8½in www.strathspeyrailway.co.uk **Highland**

(45025)	5025	4-6-0	OC	VF	4570	1934	
46512	E.V.COOPER ENGINEER	2-6-0	OC	Sdn		1952	
(57566)	828	0-6-0	IC	StRollox		1899	
No.17	BRAERIACH	0-6-0T	IC	AB	2017	1935	
6		0-4-0ST	OC	AB	2020	1936	
No.9	CAIRNGORM	0-6-0ST	IC	RSHN	7097	1943	
D2774		0-4-0DH		NBQ	28027	1960	
			reb	AB		1968	
D3605	(08490)	0-6-0DE		Hor		1958	
(D5302	26002)	Bo-BoDE		BRCW	DEL47	1958	
(D5325	26025)	Bo-BoDE		BRCW	DEL70	1959	
D5394	(27050)	Bo-BoDE		BRCW	DEL237	1962	
D5862	31327	A1A-A1ADE		BT	398	1962	
–		0-4-0DH		AB	517	1966	
–		4wDM		MR	5763	1957	
14		0-4-0DH		NBQ	27549	1956	
–		0-4-0DM		RH	260756	1950	
	QUEEN ANNE	4wDM		RH	265618	1948	
–		4wDH		TH	277V	1977	
Sc51367		2-2w-2w-2DMR		PSteel		1960	
Sc51402		2-2w-2w-2DMR		PSteel		1960	
(51990	960 932) 977830	2-2w-2w-2DMR		DerbyC&W		1960	
Sc 52008		2-2w-2w-2DMR		DerbyC&W		1961	
(52030	960 932 977831)	2-2w-2w-2DMR		DerbyC&W		1961	
99709 909005-9	T111719 S45373	2w-2PMR		Geismar	ST/02/03	2002	
99709 909153-7	T114925 65058	2w-2PMR		Geismar	ST/02/21	2002	
99709 909157-8	T113892 65095	2w-2PMR		Geismar	ST/03/36	2003	
99709 909158-6	T111720 S45401	2w-2PMR		Geismar	ST/03/37	2002	
99709 909162-8	T111722	2w-2PMR		Geismar	ST/03/60	2003	
99709 909265-9	T117259	2w-2PMR		Geismar	ST/04/17	2004	
99709 901047-9		2w-2PMR		Lesmac	LMS010	2006	
99709 901003-2		2w-2PMR		Lesmac	LMS013	2006	
813	CALE	2w-2PMR		Wkm	1288	1933	a

a located at Mountain Rescue Centre, Boat of Garten

SUMMERLEE – THE MUSEUM OF SCOTTISH INDUSTRIAL LIFE, HERITAGE WAY, COATBRIDGE — ML5 1QD

Gauge 4ft 8½in www.culturenl.co.uk/summerlee **North Lanarkshire NS 728655**

No.11			0-4-0ST	OC	GH		1898
No.9			0-6-0T	IC	HC	895	1909
	ROBIN		4wVBT	VCG	S	9628	1957
(62174)	(936 103)	977845	4w-4wWER		Cravens		1967
W280			0-4-0DH		AB		1966
		a rebuild of	0-4-0DM		AB	472	1961

Gauge 3ft 6in

4112	(SPRINGBOK)	4-8-2+2-8-4T	4C	_(BP	7827	1957
				(NBH	27770	1957

TWEEDDALE RAILWAY SOCIETY, c/o NEWLANDS ACTIVITY CENTRE ROMANNO BRIDGE, near LEADBURN — EH46 7BZ

Gauge 2ft 6in www.newlandscentre.org.uk **Peeblesshire NT 158479**

CHRISTINE	4wDH	Byers	1998	a

a incorporates frame of MR 115U094 / 1970

Loco may be stored off site

WAVERLEY ROUTE HERITAGE ASSOCIATION, WHITROPE HERITAGE CENTRE, WHITROPE SUMMIT, near HAWICK — TD9 9TY

Gauge 4ft 8½in www.wrha.org.uk **Roxburghshire NT 525001**

(D5340)	26040	Bo-BoDE	BRCW	DEL85	1959	
–		0-6-0DH	JF	4240015	1962	
	ARMY 110 A2 EQ	4wDM	RH	411319	1958	
RB004		4wDMR	_(DerbyC&W		1984	
			(Leyland	RB004	1984	a
(DX 68811)	99709 018044-6 ETI 41	4wDHR	Perm	001	1987	

a property of Northumbria Rail Ltd, Bedlington, Northumberland

WEST LOTHIAN COUNCIL, POLKEMMET COUNTRY PARK, WHITBURN — EH47 0AD

Gauge 4ft 8½in www.visitwestlothian.co.uk **NS 924649**

No.8	DARDANELLES	0-6-0ST	OC	AB	1175	1909

SECTION 3 — WALES

Consequent upon the fragmentation of the County Geography of Wales, due to the creation of a large number of autonomous Unitary Authorities and similar bodies and, in view of the relatively few remaining locations hosting locomotives, this section of this handbook is now presented in two areas of coverage.

Defining Areas in accordance with the 1974-1996 "large counties" era, now provides a direct relationship between this volume and the "Historic Handbook" series which is also published by the Industrial Railway Society. The historic books contain the recorded details of all past and recent locomotives, at all known past and present locations, in the areas they cover, together with extensive texts describing the locations and the businesses they served.

The Areas can, in the main, be summarised as follows:

NORTH & MID WALES
Breconshire (apart from a narrow strip at its southern boundary which was transferred to Gwent and Mid-Glamorgan – note that Breconshire was also known as Brecknock, and Brecknockshire), Caernarfonshire, Cardiganshire, Ceredigion, Clwyd, Conwy, Denbighshire, Dyfed (north; i.e. Ceredigion), Flintshire, Flintshire Detatched, Gwynedd, Isle of Anglesey (Ynys Môn), Merionethshire, Montgomeryshire, Powys, Radnorshire and Wrexham.

SOUTH WALES
Blaenau Gwent, Bridgend, Caerphilly, Cardiff, Carmarthenshire, Dyfed (excluding Ceredigion), Glamorgan (prior to 1974), Gwent, Merthyr Tydfil, Mid Glamorgan, Monmouthshire, Neath Port Talbot, Newport, Pembrokeshire, Rhondda Cynon Taff, Swansea, Torfaen, South Glamorgan, West Glamorgan and Vale of Glamorgan.

The relevant Historic Series Handbooks are as follows:

NORTH & MID WALES

Handbook NW	–	Industrial Locomotives of North Wales	(1992)
Handbook DP	–	Industrial Locomotives of Dyfed & Powys	(1994)

SOUTH WALES

Handbook DP	–	Industrial Locomotives of Dyfed & Powys	(1994)
Handbook WG	–	Industrial Locomotives of West Glamorgan	(1996)
Handbook GT	–	Industrial Locomotives of Gwent	(1999)
Handbook GL	–	Industrial Locomotives of Glamorgan (Mid and South)	(2007)

Update Bulletins for the published Historic Handbooks are available. For information see the Society website at www.irsociety.co.uk or write to the address on page 2.

NORTH AND MID WALES

INDUSTRIAL LOCATIONS

ANGLESEY ALUMINIUM METAL LTD,
PENRHOS WORKS, LONDON ROAD, HOLYHEAD
(Rio Tinto and Kaiser Aluminium & Chemical Corporation Joint Venture)
Gauge 4ft 8½in R.T.C. www.riotinto.com **Isle of Anglesey SH 264807**

–	0-4-0DH	HE	7183	1970	a

a property of private owner

BNM ALLIANCE - BARHALE plc and NORTH MIDLAND CONSTRUCTION plc JOINT VENTURE, BIRMINGHAM RESILIENCE PROJECT, ELAN VALLEY AQUEDUCT REHABILITATION PROJECT, KNIGHTON

(for Severn-Trent Water Authority - Knighton Tunnel : scheduled completion 30/8/2019)

Gauge 2ft 0in www.barhale.com / www.nmcn.com

–	4wBE	CE	
–	4wBE	CE	

GREAVES WELSH SLATE CO LTD, BLAENAU FFESTINIOG

Gauge 2ft 0in **Gwynedd**

–	4wDM	RH	174536	1936	OOU

locomotive stored at private location

HANSON QUARRY PRODUCTS EUROPE LTD, PENMAENMAWR QUARRY, BANGOR ROAD, PENMAENMWAR (part of the Heidelberg Cement Group) LL34 6NA

Locomotive abandoned in a remote location on "Level 2", officially known as "Bottom Bank East Quarry".

Gauge 3ft 0in **R.T.C.** www.heidelbergcement.com **Conwy SH 701758**

1878	PENMAEN	0-4-0VBT	VC	DeW	1878	Dsm

TATA STEEL EUROPE, TATA STEEL COLORS, SHOTTON WORKS CH5 2NH

(part of the Tata Group)

Gauge 4ft 8½in www.tatasteeleurope.com **Flintshire SJ 302704, 305705**

(D3956)	08788	0-6-0DE	Derby		1960	a
(D3977)	08809 24	0-6-0DE	Derby		1960	a
H 014		0-6-0DH	RR	10262	1967	a OOU
H 003	ROSEDALE	4wDH	S	10070	1961	a

a on hire from British American Railway Services Ltd, Stanhope, Co. Durham

UNITED UTILITIES GROUP plc, MILWR TUNNEL, HALKYN MINE, RHYDYMWYN

Locomotives underground – no public access

Gauge 1ft 10½in www.unitedutilities.com **Flintshire SJ 296536**

–	4wDM	RH	182138	1936	OOU
–	4wDM	RH	226309	1943	Dsm a
–	4wDM	RH	354029	1953	OOU
774	4wBE	WR	744	1929	OOU
3	4wBE	WR	773	1930	OOU
–	2-2wDM	EdwardsE&J		2004	b

a dumped underground in workshops adjacent to Pen-y-Bryn shaft
b owned and operated by the Grosvenor Caving Club

PRESERVATION LOCATIONS

ANGLESEY CENTRAL RAILWAY, // LEIN AMLWCH

Llannerchymedd **LL71 8EU**

Gauge 4ft 8½in www.leinamlwch.co.uk **Anglesey SH 416840**

	ELISEG	0-4-0DM		JF	22753	1939	

c/o Holyhead Truck Services Ltd,
Unit 10-15, Mona Industrial Park, Gwalchmai **LL65 4RJ**

Gauge 4ft 8½in **Anglesey SH 417753**

–		0-4-0DH		HE	7460	1977	
–		4wDM		RH	321727	1952	

BALA LAKE RAILWAY LTD, THE STATION, STATION ROAD, LLANUWCHLLYN, near BALA **LL23 7DD**

Gauge 1ft 11½in www.bala-lake-railway.co.uk **Gwynedd SH 881300**

	WINIFRED	0-4-0ST	OC	HE	364	1885	
1	GEORGE B	0-4-0ST	OC	HE	680	1898	
No.3	HOLY WAR	0-4-0ST	OC	HE	779	1902	
	ALICE	0-4-0ST	OC	HE	780	1902	
	MAID MARIAN	0-4-0ST	OC	HE	822	1903	
	TRIASSIC	0-6-0ST	OC	P	1270	1911	
	BOB DAVIES	4wDH		BD	3776	1983	
			reb	YEC	L125	1994	
	CHILMARK	4wDM		RH	194771	1939	
	LADY MADCAP	4wDM		RH	283512	1949	
	MEIRIONNYDD	4w-4wDH		SL	22	1973	
–		2w-2PMR		Bala		2011	
–		2w-2PMR		Wkm	1548	1934	DsmT
–		2w-2PMR		Wkm	10943	1976	

BRYMBO MINERAL RAILWAY, BRYMBO HERITAGE PROJECT, THE OLD IRON & STEEL WORKS, off GWALIA ROAD, BRYMBO, WREXHAM **LL11 5BT**

Gauge 2ft 0in www.brymboheritage.co.uk **Wrexham SJ 294535**

	CHARLOTTE	4wDM		MR	8937	1944	a

a carries incorrect worksplate MR 8905

CAERNARFON AIR MUSEUM, CAERNARFON AIRPORT AIRWORLD, DINAS DINLLE, CAERNARFON **LL54 5TP**

Gauge 2ft 0in www.airworldmuseum.com **Gwynedd SH 435584**

	RTT 767150	2w-2PM		Wkm	(3152 1943)?		

CAERNARFONSHIRE SLATE RAILWAY, PORTHMADOG

Gauge 2ft 0in **Private Site** **Gwynedd**

4		4wDM		RH	177638	1936	a
22		4wDM		RH	226302	1944	
–		0-4-0BE		WR	(3867	1948?)	
–		4wBE		WR	M7556	1972	
			reb	WR	10114	1984	
–		4wBE		WR			
–		4wBE		WR			
–		4wBE		WR			

a carries plate RH 177642/1938

Gauge 1ft 11½in

E2	No.7179 GWYNFYNYDD	0-4-0BE		WR	G7179	1967

CONWY VALLEY RAILWAY MUSEUM, BETWS-Y-COED **LL24 0AL**

Gauge 1ft 3in www.conwyrailwaymuseum.co.uk **Conwy** **SH 796565**

70000	BRITANNIA	4-6-2	OC	TMA	8733	1987	a

a construction begun by Longfleet Motor & Engineering Works Ltd, 46 Fernside Road, Poole, Dorset, in 1968; completed by TMA in 1988

CORRIS RAILWAY COMPANY LTD, MAESPOETH, near CORRIS **SY20 9RD**

Gauge 2ft 3in www.corris.co.uk **Gwynedd** **SH 753069**

	(SIR NEVILLE LUBBOCK)	0-4-2ST	OC	KS	857	1904	a
No.7		0-4-2ST	OC	Winson	17	2005	
No.8		4wDH		HE	6273	1965	
No.5	ALAN MEADEN	4wDM		MR	22258	1965	
11		0-4-0DH		OK	25721	1957	
	CORRIS RAILWAY No.6	4wDH		RH	518493	1966	
No. 9	ABERLLEFENNI	4wBE		CE	B0457	1974	

a not on public display

DOLGARROG RAILWAY SOCIETY LTD,
ALUMINIUM WORKS SIDING, CLARK STREET, DOLGARROG **LL32 8QE**

Gauge 4ft 8½in www.dolgarrograilway.co.uk **SH 774674**

2	TAURUS	0-4-0DM	(VF	D139	1951
			(DC	2273	1951

ERWOOD STATION GALLERY,
LLANDEILO GRABAN, near BUILTH WELLS **LD2 3SJ**

Gauge 4ft 8½in www.erwoodstation.com **Powys** **SO 089439**

A.W.M. No.169	ALAN	0-4-0DM	JF	22878	1939

FAIRBOURNE RAILWAY PRESERVATION SOCIETY, FAIRBOURNE RAILWAY, BEACH ROAD, FAIRBOURNE LL38 2EX

Gauge 1ft 3in www.fairbournerailway.com **Gwynedd SH 615128**

(DINGO)	4w-4wPM	Fairbourne	1951	a Dsm

a frames utilised in sector turnout

A section of 1ft 3in gauge railway is laid at this location with visiting locomotives occasionally present.

The locomotives of this 12¼in gauge railway are outwith the remit of the EL book, however the following locomotives based here were rebuilt from 1ft 3in and 2ft 0in gauges respectively :
TONY 4w-4wDM Guest/1961 and GWRIL 4wDH HE 9354/1994.

FFESTINIOG & WELSH HIGHLAND RAILWAYS

The Ffestiniog Railway Company, Porthmadog

Locomotives may be transferred to/from the Welsh Highland Railway on occasion, as required.

Locomotives are kept at :- Boston Lodge Shed & Works, Gwynedd LL48 6HT SH 584379, 585378
Glan-y-Pwll Depot, Blaenau Ffestiniog, Gwynedd LL41 3PF SH 693461
Minffordd P.W. Depot, Gwynedd (LL48 6HP) SH 599386
Porthmadog Station, Gwynedd LL49 9NF SH 571384

Gauge 1ft 11½in www.festrail.co.uk

NOTE Most of the locomotives have been "rebuilt" by the FRCo many times during their working lives. In this list we detail only those rebuilds which made significant alteration to the loco's appearance.

	MOUNTAINEER		2-6-2T	OC	Alco(C)	57156	1916	
				reb	FRCo		c1968	
No.10	MERDDIN EMRYS		0-4-4-0T	4C	FRCo		1879	
				reb	FRCo		1988	
	LIVINGSTON THOMPSON		0-4-4-0T	4C	FRCo		1885	a
				reb	FRCo		1905	
	EARL OF MERIONETH / IARLL MEIRIONNYDD		0-4-4-0T	4C	FRCo		1979	
12	DAFYDD LLOYD GEORGE / DAVID LLOYD GEORGE		0-4-4-0T	4C	FRCo		1992	
	TALIESIN		0-4-4T	OC	FRCo		1999	
	LYD 190		2-6-2T	OC	FRCo	14	2010	
No.1	PRINCESS		0-4-0TT	OC	GE	(200?)	1863	
		rebuilt as	0-4-0STT		FRCo		1895	
No.2	PRINCE		0-4-0TT	OC	GE	(199?)	1863	
		rebuilt as	0-4-0STT		FRCo		1891	
				reb	FRCo		1955	
				reb	FRCo		1979	
No.4	PALMERSTON		0-4-0TT	OC	GE		1864	
		rebuilt as	0-4-0STT		FRCo		1888	
				reb	FRCo		1910	
				reb	FRCo		1933	
(No.5)	WELSH PONY		0-4-0STT	OC	GE	234	1867	
				reb	FRCo		1891	
4			0-8-0	OC	Harbin	221	1988	
	LILLA		0-4-0ST	OC	HE	554	1891	
	BLANCHE		0-4-0ST	OC	HE	589	1893	
		rebuilt as	2-4-0STT		FRCo		1972	
	LINDA		0-4-0ST	OC	HE	590	1893	
		rebuilt as	2-4-0STT		FRCo		1970	

1	BRITOMART	0-4-0ST	OC	HE	707	1899	
	HUGH NAPIER	0-4-0ST	OC	HE	855	1904	b
	CASTELL HARLECH / HARLECH CASTLE	0-6-0DH		BD	3767	1983	
7011	MOELWYN	0-4-0PM		BLW	49604	1918	
	rebuilt as	0-4-0DM		FRCo		1956	
	rebuilt as	2-4-0DM		FRCo		1957	
	CRICCIETH CASTLE / CASTELL CRICIETH	0-6-0DH		FRCo		1995	
	VALE OF FFESTINIOG	4w-4wDH		Funkey		1968	
			reb	FRCo			
	MONSTER / AFANC	4wDM		FRSociety		1974	
	MOEL HEBOG	0-4-0DM		HE	4113	1955	
P13350	MOEL Y GEST	4wDH		HE	6659	1965	
	HAROLD	4wDM		HE	7195	1974	
4415		6wDM		KS	4415	1928	
(MPU 8)		4wDH		Matisa		1956	
	MARY ANN	4wPM		MR		1917	c
	rebuilt	4wPM		KC		1923	
	rebuilt as	4wDM		FRCo		c1957	
–		4wDM		MR	435	1917	
			reb	MR	3663	1924	
	THE COLONEL	4wDM		MR	8788	1943	Dsm
	ANDY	4wDM		MR	21579	1957	
			reb	FRCo		2010	
	DOLGARROG / INNOGY	4wDM		MR	22154	1962	
	"BUSTA"	2w-2PMR		FRCo		2006	
1543		2w-2PMR		Wkm	1543	1934	

a currently on display at National Railway Museum, York
b property of The National Trust; based here, but visits other locations
c carries incorrect plate 507/1917

Welsh Highland Railway // Rheilffordd Eryri, Dinas, near Caernarfon LL55 2YD

Locomotives may be transferred to/from the Festiniog Railway on occasion, as required.
Public railway operating from Caernarfon to Porthmadog.
Locomotives are kept at :- Dinas shed and workshops, Caernarfon LL54 5UP SH 476585, 478589

Gauge 3ft 0in — Static display at Dinas **Gwynedd SH 477587**

	LLANFAIR	0-4-0VBT	VC	DeW		1895	Pvd

Gauge 1ft 11½in www.festrail.co.uk

133		2-8-2	OC	AFB	2683	1952	OOU
134		2-8-2	OC	AFB	2684	1952	
	No.1K / No.2K	0-4-0+0-4-0T	4C	BP	5292	1909	
130		2-6-2+2-6-2T	4C	BP	7431	1951	
138		2-6-2+2-6-2T	4C	BP	7863	1958	
(140)		2-6-2+2-6-2T	4C	BP	7865	1958	
143		2-6-2+2-6-2T	4C	BP	7868	1958	
87		2-6-2+2-6-2T	4C	Cockerill	3267	1937	
9	RS0009	0-6-0DM		Bg/DC	2395	1952	
	CONWAY CASTLE / CASTELL CONWY	4wDM		FH	3831	1958	
	UPNOR CASTLE	4wDM		FH	3687	1954	
	CASTELL CAERNARFON	4w-4wDH		Funkey		1968	

BILL	4wDHF	HE	9248	1985	
(2)	4wDHF	HE	9262	1985	

FLINTSHIRE MUSEUM SERVICE, SHOTTON STORE, UNIT 3, ROWLEY PARK, EVANS WAY, SHOTTON CH5 1QJ

Gauge 1ft 10½in www.flintshire.gov.uk **Flintshire SJ 310687**

–	4wDM	RH	331250	1952	
–	4wBE	WR	898	1935	

GREENFIELD VALLEY TRUST LTD, GREENFIELD VALLEY HERITAGE PARK, HOLYWELL CH8 7GH

Gauge 1ft 10½in www.greenfieldvalley.com **Flintshire SJ 193773**

–	0-4-0BE	WR	(1080	1937?)	

GWYNEDD COUNCIL, DIFFWYS CAR PARK, HIGH STREET, BLAENAU FFESTINIOG LL41 3ES

Gauge 2ft 0in www.gwynedd.llyw.cymru **Gwynedd SH 702459**

2207	4wDM	HE	2207	1941	Pvd

INTERNAL FIRE - MUSEUM OF POWER, CASTELL PRIDD, TAN-Y-GROES SA43 2JS

Gauge 1000mm www.internalfire.com **Ceredigionshire SN 295497**

ND 3646	4wDM	RH	210961	1941	

Gauge 2ft 6in

BEV YARD No.4B/9B ND 3305 B40	4wBE	WR	3805	1948	

Gauge 600mm

–	4wDM	RH	229633	1944	

Locomotives currently in storage

I.B. JOLLY, MOLD

Locomotives are stored at two private locations.

Gauge 4ft 8½in **Flintshire**

MRTC 1944	4wDM	MR	1944	1919	

Gauge 2ft 7in

–	4wDM	MR	5025	1929	Dsm

Gauge 600mm

–	0-2-2GasE	HopleyCP		c1982	
(LR 2718) MRTC 997	4wPM	MR	997	1918	Dsm
–	4wDM	MR	4803	1934	Dsm
–	4wDM	MR	5852	1933	Dsm

–		4wDM		MR	8723	1941 Dsm
No.9		4wDM		MR	9547	1950

Gauge 1ft 11½in

–		4wDM		L	30233	1946
–		4wPM		MR	6013	1931 Dsm

INIGO JONES & CO LTD, TUDOR SLATE WORKS, Y GROESLON, near PENYGROES (subsidiary of Wincilate Ltd) LL54 7UE

Gauge 2ft 3in www.inigojones.co.uk **Gwynedd SH 471551**

–		4wBE		LMM		

LLANGOLLEN RAILWAY plc

Locomotives are kept at :–

Llangollen Shed, Denbighshire LL20 8SW SJ 212422
Carrog, Denbighshire LL21 9BD SJ 118435
Glyndyfrdwy, Denbighshire LL21 9HF SJ 150428
Pentrefelin C & W Shed LL20 8EE SJ 207434
Pentrefelin Sidings, Denbighshire SJ 209432

Gauge 4ft 8½in www.llangollen-railway.co.uk

3802		2-8-0	OC	Sdn		1938
4160		2-6-2T	OC	Sdn		1948
4709		2-8-0	OC	Llangollen		2012
5199		2-6-2T	OC	Sdn		1934
5532		2-6-2T	OC	Sdn		1928
5952	COGAN HALL	4-6-0	OC	Sdn		1935
6430		0-6-0PT	IC	Sdn		1937
6880	BETTON GRANGE	4-6-0	OC	Llangollen		2011
7754		0-6-0PT	IC	NBQ	24042	1930
7822	FOXCOTE MANOR	4-6-0	OC	Sdn		1950
45337		4-6-0	OC	AW	1392	1937
45551	5551 THE UNKNOWN WARRIOR	4-6-0	3C	Llangollen		2013
61673	SPIRIT OF SANDRINGHAM	4-6-0	OC	Llangollen		2015
62712	(246) MORAYSHIRE	4-4-0	3C	Dar	(1391?)	1928
80072		2-6-4T	OC	Bton		1953
No.17		0-6-0T	OC	AB	1338	1913
–		0-4-0ST	OC	AE	1498	1906
No.20	JENNIFER	0-6-0T	OC	HC	1731	1942
	JESSIE	0-6-0ST	IC	HE	1873	1937
	rebuilt as	0-6-0T	IC	Llangollen		2007
	rebuilt as	0-6-0ST	IC	Llangollen		2010
	rebuilt as	0-6-0T	IC	Llangollen		2012
68030		0-6-0ST	IC	HE	3777	1952
	AUSTIN 1	0-6-0ST	IC	K	5459	1932
(D)1566	(47449) ORION	Co-CoDE		Crewe		1964
(D2162)	03162	0-6-0DM		Sdn		1960
(D3265	08195) 13265	0-6-0DE		Derby		1956

D5310	(26010)	Bo-BoDE		BRCW	DEL55	1960	
(D)6940	(37240)	Co-CoDE		(EE (EEV	3497 D928	1964 1964	
–		0-6-0DE		EEDK	1901	1951	
	PILKINGTON	0-4-0DE		YE	2782	1960	
–		0-4-0DE		YE	2854	1961	
E50416		2-2w-2w-2DMR		Wkm	7346	1957	
M50447	M53447 610	2-2w-2w-2DMR		BRCW		1957	
M50454		2-2w-2w-2DMR		BRCW		1957	
M50528		2-2w-2w-2DMR		BRCW		1958	
M51618		2-2w-2w-2DHR		DerbyC&W		1959	
M51907	LO262	2-2w-2w-2DMR		DerbyC&W		1960	a
DR 90005	"NEPTUNE"	4wDMR		Matisa	PV6 627	1967	
	Q317 GRN	4wDM	R/R	Unimog	008983	1992	
	99709 909266-7	2w-2PMR		Geismar	ST/00/24	2000	

a currently at the Midland Railway - Butterley, Swanwick, Derbyshire

DAVID MITCHELL, LLANNERCHYMEDD, ANGLESEY

Gauge 1ft 3in **Anglesey**

–		2w-2PMR		Wkm	4816	1948	Dsm

NARROW GAUGE RAILWAY MUSEUM TRUST LTD, WHARF STATION, TYWYN **LL36 9EY**

Gauge 3ft 2¼in www.ngrm.org.uk **Gwynedd** **SH 586004**

No.5	WILLIAM FINLAY	0-4-0T	OC	FJ	173L	1880

Gauge 2ft 0in

	(OAKLEY)	0-4-0PM		BgC	774	1919

Gauge 1ft 10¾in

	GEORGE HENRY	0-4-0VBT	VC	DeW		1877
	ROUGH PUP	0-4-0ST	OC	HE	541	1891

Gauge 1ft 10in

13		0-4-0T	IC	Spence	13L	1895

Gauge 1ft 6in

	DOT	0-4-0WT	OC	BP	2817	1887

NATIONAL MUSEUM OF WALES, WELSH SLATE MUSEUM, GILFACHDDU, LLANBERIS **LL55 4TY**

Gauge 1ft 11½in www.museum.wales/slate **Gwynedd** **SH 586603**

	UNA	0-4-0ST	OC	HE	873	1905	
–		4wBE		BE		1917	
–		4wPMR		WilliamsWJ			a
–		4wDM		RH	175414	1936	

a vintage motorcycle converted for rail use

NATIONAL TRUST, INDUSTRIAL RAILWAY MUSEUM, PENRHYN CASTLE, LLANDYGÁI, near BANGOR

LL57 4HN

Gauge 4ft 8½in www.nationaltrust.org.uk/penrhyn-castle **Gwynedd SH 603720**

	HAWARDEN	0-4-0ST	OC	HC	526	1899	
	VESTA	0-6-0T	IC	HC	1223	1916	
No.1		0-4-0WT	OC	N	1561	1870	
	HAYDOCK	0-6-0T	IC	RS	2309	1876	a

a works plate reads 1879

Gauge 4ft 0in

	FIRE QUEEN	0-4-0	OC	AH		1848	

Gauge 3ft 0in

KETTERING FURNACES No.3		0-4-0ST	OC	BH	859	1885	
	(WATKIN)	0-4-0VBT	VC	DeW		1893	

Gauge 2ft 0in

	HUGH NAPIER	0-4-0ST	OC	HE	855	1904	a

a normally based at the Ffestiniog Railway, but occasionally visits other locations

Gauge 1ft 11½in

	ACORN	4wDM		RH	327904	1951	

Gauge 1ft 10¾in

	CHARLES	0-4-0ST	OC	HE	283	1882	

NORTH WALES MINERS ASSOCIATION TRUST LTD, BERSHAM COLLIERY, RHOSTYLLEN, WREXHAM

LL14 4EG

Gauge 1ft 10½in www.northwalesminers.com **Clwyd SJ 314481**

–	"HAWYS"	4wDM		RH	183727	1937	a

a currently stored off site

EXECUTORS of A. PILBEAM, LLANRHOS

Gauge 1ft 3in

–		2-2wBER		MossAJ		2000	
–		4wBE		Riordan	T6664	1967	

PRIVATE OWNER, CRIGGION, near OSWESTRY

Gauge 4ft 8½in **Powys SJ 284150**

12	ALEXANDRA	0-4-0ST	OC	AB	929	1902	a

a carries plate AB 979/1903

QUARRY TOURS LTD, LLECHWEDD SLATE MINE, LLECHWEDD SLATE MINE, BLAENAU FFESTINIOG LL41 3NB

Gauge 2ft 0in www.llechwedd-slate-caverns.co.uk **Gwynedd SH 699468**

(No.4	THE ECLIPSE)	0-4-0WE		Greaves		1927		
	a rebuild of	0-4-0ST	OC	WB	1445	1895	a	
	THE COALITION	0-4-0WE		Greaves		1930		
	a rebuild of	0-4-0ST	OC	WB	1278	1890	a	
–		4wBE		BEV	308	1921	b	Pvd

a stored off site
b on static display

RHEILFFORDD LLYN PADARN CYFYNGEDIG, LLANBERIS LAKE RAILWAY, GILFACHDDU, LLANBERIS LL55 4TY

Gauge 1ft 11½in www.lake-railway.co.uk **Gwynedd SH 586603**

No.1	ELIDIR	0-4-0ST	OC	HE	493	1889	
No.2	WILD ASTER // THOMAS BACH	0-4-0ST	OC	HE	849	1904	
No.3	DOLBADARN	0-4-0ST	OC	HE	1430	1922	
–		4wDM		MR	7927	1941	Dsm
No.9	GARRET	4wDM		RH	198286	1940	
No.8	YD No.AD689 TWLL COED	4wDM		RH	268878	1952	
–		4wDM		RH	425796	1958	Dsm
No.7	COED GORAU	4wDM		RH	441427	1961	
No.19	LLANELLI	4wDM		RH	451901	1961	

RHIW VALLEY LIGHT RAILWAY, LOWER HOUSE, MANAFON, near WELSHPOOL SY21 8BJ

Private railway; has occasional "open days".

Gauge 1ft 3in www.rvlr.co.uk **Powys SJ 143028**

JACK	0-4-0	OC	_(Rhiw		2003	
			(TMA	25657	2003	a
POWYS	0-6-2T	OC	SL	20	1973	
MONTY	4wPM		_(Jaco		1989	
			(Brunning		1989	

a construction commenced as an 0-4-2T in 1985, but completed 2003 as listed

RHYL STEAM PRESERVATION TRUST, RHYL MINIATURE RAILWAY // RHEILFFORDD FACH Y RHYL, MARINE LAKE, WELLINGTON ROAD, RHYL LL18 1AQ

Gauge 1ft 3in www.rhylminiaturerailway.co.uk **Denbighshire SJ 003805**

	JOAN	4-4-2	OC	Barnes	101	1920
	RAILWAY QUEEN	4-4-2	OC	Barnes	102	1921
105	MICHAEL	4-4-2	OC	Barnes	105	c1928
106	BILLY	4-4-2	OC	Barnes	106	c1934
44	CAGNEY	4-4-0	OC	McGarigle		c1910

	PRINCE EDWARD OF WALES	4-4-2	OC	_(WalkerG ‾(MossAJ		1991 1991	
	incorporates parts of			BL	15	1909	
1		0-4-2DM	s/o	Guest		1961	
	rebuilt as	0-4-2DH	s/o	Rhyl	2012		
–		4wDM		L	10498	1938	
–		2w-2-4BER		Hayne		1983	

SNOWDON MOUNTAIN RAILWAY, LLANBERIS — LL55 4TY

(Operated by Heritage Attractions Ltd)

Gauge 800mm www.snowdonrailway.co.uk **Gwynedd SH 582597**

All locomotives drive through the rack gear, with the "driving wheels" free to rotate on their axles.

2	ENID	0-4-2T OC	RACK	SLM	924	1895	
3	WYDDFA	0-4-2T OC	RACK	SLM	925	1895	
4	SNOWDON	0-4-2T OC	RACK	SLM	988	1896	
5	MOEL SIABOD	0-4-2T OC	RACK	SLM	989	1896	OOU
6	PADARN	0-4-2T OC	RACK	SLM	2838	1922	
7	RALPH	0-4-2T OC	RACK	SLM	2869	1923	OOU
8	ERYRI	0-4-2T OC	RACK	SLM	2870	1923	OOU
9	NINIAN	0-4-0DH	RACK	HE	9249	1986	
10	YETI	0-4-0DH	RACK	HE	9250	1986	
11	PERIS	0-4-0DH	RACK	_(HAB ‾(HE	775 9305	1991 1991	
12	GEORGE	0-4-0DH	RACK	HE	9312	1992	

TALYLLYN RAILWAY CO, TYWYN

Locomotive sheds at:- Tywyn Pendre Station, Gwynedd LL36 9EN SH 590008
Tywyn Wharf, Gwynedd LL36 9EY SH 585005

Gauge 2ft 6in www.talyllyn.co.uk

7	T 0009 00 NZ 35	4wDH		BD	3781	1984	OOU

Gauge 2ft 3in

(1)	TALYLLYN	0-4-2ST	OC	FJ	42	1865		
(No.2)	DOLGOCH	0-4-0WT	OC	FJ	63	1866		
No.3	(SIR HAYDN)	0-4-2ST	OC	HLT	323	1878		
(No.4)	EDWARD THOMAS	0-4-2ST	OC	KS	4047	1921		
(No.6	D)OUGLAS	0-4-0WT	OC	AB	1431	1918		
(No.7)	TOM ROLT	0-4-2T	OC	Pendre		1991		
	built using parts of	0-4-0WT	OC	AB	2263	1949		
No.11	TRECWN	4wDH		BD	3764	1983		
12	ST. CADFAN	4wDH		BD	3779	1983		
–		0-4-0DM		HE	4135	1950	a	Dsm
(No.9)	ALF	0-4-0DM		HE	4136	1950		
–		4wDH		HE	6292	1967	b	Dsm
5	MIDLANDER	4wDM		RH	200792	1940		
	“THE FLAIL MOWER”	2-2-0DH		RH	476109	1964	c	Dsm
(19)		2w-2PM		CurwenD		1952	d	Dsm

	TOBY		2w-2PMR		BateJ		1954	
		rebuilt as	2w-2DMR		Pendre		2001	d

a stored at Brynglas Station (SH 628031)
b frames utilised in a hydraulic press
c frames utilised in self-propelled flail mower
d currently off site for restoration

TEIFI VALLEY RAILWAY, VALE OF TEIFI NARROW GAUGE RAILWAY, STATION YARD, HENLLAN, near LLANDYSUL SA44 5TD

Gauge 2ft 0in www.teifivalleyrailway.wales **Ceredigion SN 357406**

No.	Name		Type	Cyl	Builder	Works No.	Year
	ALAN GEORGE		0-4-0ST	OC	HE	606	1894
	SGT MURPHY		0-6-0T	OC	KS	3117	1918
		rebuilt as	0-6-2T	OC	Winson	9	1991
(1835)			4wDM		HE	1835	1937
–			4wDM		HE	2433	1941
004	SAMMY		4wDM		MR	11111	1959
	JOHN HENRY		4wDM		RH	(433390	1959?)

JOHN TENNENT & P. SMITH, BRECON

These locomotives are in store at a private location.

Gauge 1ft 3in **Powys**

No.	Type	Builder	Works No.	Year
–	4wDM	AK	6	1981
–	4wDM	LB	52579	1961

UNKNOWN OWNER, BETTISFIELD, near WHITCHURCH SY13 2LB

Gauge 4ft 8½in **Flintshire SJ 460358**

No.	Type	Builder	Works No.	Year
–	4wDM	Bg/DC	2107	1937

VALE OF RHEIDOL LIGHT RAILWAY, PARK AVENUE, ABERYSTWYTH SY23 1PG

www.rheidolrailway.co.uk **Ceredigion SN 587812**

Gauge 1000mm

No.	Type	Builder	Works No.	Year
–	4wPM	RP	51168	1916

Gauge 2ft 0in [nominal] (Static display)

No.	Name	Type	Cyl	Builder	Works No.	Year	Note
–		0-4-0T	OC	Decauville	1027	1926	a
	MARGARET	0-4-0ST	OC	HE	605	1894	
SPSM 6		0-6-0TT	OC	JF	10249	1905	a
21		0-4-2T	OC	JF	11938	1909	a
23		0-6-2T	OC	JF	15515	1920	a
–		0-4-0ST	OC	KS	3114	1918	
No.31		0-8-0T	OC	Maffei	4766	1917	a
	JUBILEE 1897	0-4-0ST	OC	MW	1382	1897	
18BG		0-4-4T	OC	WB	2228	1924	a
D564		4wDM		HC	D564	1930	a

Gauge 1ft 11¾in (Operational)

7	OWAIN GLYNDWR	2-6-2T	OC	Sdn		1923	
8	(LLYWELYN)	2-6-2T	OC	Sdn		1923	
(9)	1213 (PRINCE OF WALES)	2-6-2T	OC	DM	2	1902	b
60	DRAKENSBURG	2-6-2+2-6-2T	4C	Hano	10551	1927	
10		0-6-0DH		BMR	002	1987	c
–		4wDH		HE	7495	1977	d
(68804)	THUNDERBIRD 4 No.5	4wDHR		Perm	005	1985	

Gauge 1ft 11½in

	TIGER	4-6-0T	OC	BLW	44699	1917	

Gauge 1ft 10¾in (Static display)

106	1877 KATHLEEN	0-4-0VBT	VC	DeW		1877	a

a stored at Capel Bangor Station
b possibly a new locomotive built by Sdn in 1924
c built with parts supplied by BD
d for use as a flail-mower

R.WATSON-JONES,
WOODLANDS, THE DINGLE, CONSTITUTION HILL, PENMAENMAWR

Gauge 3ft 0in **Conwy SH 720765**

–		4wDM		RH	202987	1941	

WELSH HIGHLAND RAILWAY // RHEILFFORDD ERYRI,
DINAS, near CAERNARFON LL54 5UP

(Operated by the Ffestiniog Railway Company) **Gwynedd SH 477587**
Public railway operating from Caernarfon to Porthmadog.

For details of locomotives, see entry under Ffestiniog & Welsh Highland Railways

WELSH HIGHLAND HERITAGE RAILWAY, WELSH HIGHLAND RAILWAY LTD,
TREMADOG ROAD, PORTHMADOG LL49 9DY

Public railway operating northwards from terminus in Porthmadog.
Locomotive shed and workshops at Gelert's Farm, Porthmadog.

Gauge 2ft 0in www.whr.co.uk **Gwynedd SH 571393**

"No.10"		0-4-0PM		BgC	736	1917
11		0-4-0PM		BgC	760	1918
No.5		4wDM		HE	6285	1968
3	ODIN	4wDM		MR	5859	1934
			reb	ALR	No.3	1989
–		4wDM		MR	22237	1965

Gauge 1ft 11½in

	GERTRUDE	0-6-0T	OC	AB	1578	1918
	RUSSELL	2-6-2T	OC	HE	901	1906
	KAREN	0-4-2T	OC	P	2024	1942

	GELERT	0-4-2T	OC	WB	3050	1953	
			reb	Winson		1991	
	(LADY MADCAP)	0-4-0ST	OC	WHR(GF)		2006	a
"2"		4wDH		AB	554	1970	
			reb	AB	6613	1987	
1	GOWRIE	4wDH		AB	555	1970	
No.58		0-6-0DH		U23A	23387	1979	
No.60		0-6-0DH		U23A	23389	1977	
No.08	LYD 2	0-6-0DH		U23A	24051	1980	
–		4wDM		HE	2024	1940	
59105	VILLAGE IDIOT	0-4-0DMF		HE	3510	1947	
	WEIGHITIN	4wDH		HE	7535	1977	
	EMMA	4wDH		HE	9346	1994	
2		4wDH		HE	9350	1994	
264		4wPM		MR	264	c1916	
36	CNICHT	4wDM		MR	8703	1941	
6		4wDM		MR	11102	1959	
4		4wDM		MR	60S333	1966	
9	KATHERINE	4wDM		MR	60S363	1968	
–		4wDM		RH	203031	1941	
1	GLASLYN	4wDM		RH	297030	1952	
			reb	WHR(GF)		1981	
	KINNERLEY	4wDM		RH	354068	1953	
L5	HOVERINGHAM	4wDM		RH	370555	1953	
No.2	BERWYN	4wDMF		RH	481552	1962	

a under construction. Incorporates some components from HE 652

WELSHPOOL & LLANFAIR LIGHT RAILWAY PRESERVATION CO LTD

Locomotives are kept at :- Llanfair Caereinion, Powys SY21 0SF SJ 106068
Welshpool (Raven Square), Powys SY21 7LT SJ 215074

Gauge 2ft 6in www.wllr.org.uk

822	THE EARL	0-6-0T	OC	BP	3496	1902	
823	COUNTESS	0-6-0T	OC	BP	3497	1902	
G.C.G.D. PROVAN WORKS No.1							
8	DOUGAL	0-4-0T	OC	AB	2207	1946	a
10	699.01 SIR DREFALDWYN	0-8-0T	OC	AFB	2855	1944	
12	JOAN	0-6-2T	OC	KS	4404	1927	
No.1	CHEVALLIER	0-6-2T	OC	MW	1877	1915	b
No.6	MONARCH	0-4-4-0T	4C	WB	3024	1953	Pvd
No.7	CHATTENDEN YD No.AD 690	0-6-0DM		Bg/DC	2263	1949	
17	175	6wDM		Diema	4270	1979	
11	FERRET YARD No.86	0-4-0DM		HE	2251	1940	OOU
"No.16"	ND 3082 SCOOBY / SCWBI	0-4-0DM		HE	2400	1941	
			reb	W&LLR		1992	
9150		4wDMR		BD	3746	1976	
–		2w-2PMR		Wkm	2904	1940	

a currently at Taisugar, Taiwan
b on loan from The Flour Mill, Bream, Gloucestershire

SOUTH WALES

INDUSTRIAL LOCATIONS

BEAVER POWER LTD, GOAT MILL ROAD, DOWLAIS, MERTHYR TYDFIL CF48 3TF

Gauge 5ft 3in **Merthyr Tydfil SO 062073**

1	0-6-0DH	EEV	(D1266	1969
			(3954	1969

CAF ROLLING STOCK UK LTD, 1 MONKS DITCH DRIVE, CELTIC BUSINESS PARK, NEWPORT NP19 4RH

Gauge 4ft 8½in www.caf.net

–	4wBE	R/R	Zephir	2828	2018

CELSA MANUFACTURING (UK) LTD, CARDIFF

Locomotives are kept at :-

Castle Works, Cardiff CF24 5NN ST 195755
Tremorfa Works, Cardiff CF24 2YE ST 209758

Gauge 4ft 8½in www.celsauk.com

(D3504) 08389	0-6-0DE	Derby	1958	a
(D3797 08630) CELSA 3 CELSA ENDEAVOR	0-6-0DE	Derby	1959	a
(D4134) 08904	0-6-0DE	Hor	1962	a
(D4154 08924) CELSA 2	0-6-0DE	Hor	1962	a

a property of Harry Needle Railroad Co Ltd, Derbyshire

Gauge 2ft 0in

–	2w-2CE	AS&W	1990	OOU

COAL PRODUCTS LTD, CWM COKING PLANT, LLANTWIT FARDRE, BEDDAU CF38 2PY

Gauge 4ft 8½in (Closed) www.coalproducts.co.uk **Rhondda Cynon Taff ST 066862**

2	0-4-0WE	GB	2180	1948	OOU
3	0-4-0WE	GB	2690	1957	OOU
1	0-4-0WE	GB	2691	1957	OOU

DOW CORNING LTD, CARDIFF ROAD, BARRY CF63 2YL

Gauge 4ft 8½in www.gb.dow.com **Vale of Glamorgan ST 142685**

DC 27584	4wDM	R/R	Trackmobile LGN963990492	1992	

FORD MOTOR CO LTD, BRIDGEND FACTORY, COWBRIDGE ROAD, WATERTON INDUSTRIAL ESTATE, BRIDGEND CF31 3PJ

Gauge 4ft 8½in www.ford.co.uk **Bridgend SS 931781**

–	0-6-0DH		GECT	5391	1973	
		reb	‡	LWO 2108	2002	
H 057	4wDH		S	10177	1964	a

a on hire from British American Railway Services Ltd, Stanhope, Co. Durham

‡ rebuilt by R.M.S.Locotec Ltd, Dewsbury, West Yorkshire

HANSON QUARRY PRODUCTS EUROPE LTD, HANSON AGGREGATES, MACHEN QUARRY, near NEWPORT CF83 8YP

(part of the Heidelberg Cement Group)

Gauge 4ft 8½in www.heidelbergcement.com **Newport ST 221886**

(D3955 08787) 08296	0-6-0DE	Derby		1960

INTERTISSUE LTD, (SOFIDEL GROUP), BRUNEL WAY, BAGLAN ENERGY PARK, BAGLAN BAY, PORT TALBOT SA11 2FP

Gauge 4ft 8½in www.sofidel.it **Neath Port Talbot SS 736928**

ZZ 44	0-6-0DH	EEV-AEI	D3989	1970	OOU

ORB STEEL TERMINAL, ORB, NEWPORT NP19 0RB

Gauge 4ft 8½in R.T.C. **Newport ST 326863**

JAMO	4wDH	RR	10198	1965	OOU

PULLMAN RAIL LTD, (CANTON TRACTION & ROLLING STOCK DEPOT), TRAIN MAINTENANCE DEPOT, LECKWITH ROAD, CARDIFF CF11 8HP

(part of Colas Rail Group, which is part of the Bouygues Group of companies)

Gauge 4ft 8½in www.pullmanrail.co.uk **Cardiff ST 171758**

(D3654) 08499 REDLIGHT	0-6-0DE	Don		1958

PUMA ENERGY (UK) LTD, HERBRANDSTON, MILFORD HAVEN

Gauge 4ft 8½in www.pumaenergy.com **Pembrokeshire SM 888085**

077 53 MR 77	0-6-0DH		EEV	D1198	1967
		reb	YEC	L122	1993
078 53 MR 78	0-6-0DH		HE	6663	1969
		reb	HAB	6586	1999
EDWARD	4wDH		TH	267V	1976

RWE NPOWER plc, ABERTHAW POWER STATION, THE LEYS, ABERTHAW CF62 4ZW

Gauge 4ft 8½in www.rwe.com **Vale of Glamorgan ST 026664**

–	0-6-0DM	P	5014	1959	Pvd
–	2w-2PMR	Bance	169/06	2006	

SIMEC ATLANTIS ENERGY LTD, (part of the SIMEC Group), USKMOUTH POWER STATION, WEST NASH ROAD, NEWPORT — NP18 2BZ

Gauge 4ft 8½in www.simecatlantis.com **Newport** **ST 327837**

–	0-6-0DH		RH	468046	1963	
		reb	YEC	L106	1992	OOU

TALYWAIN SALVAGE, STAR TRADING ESTATE, PONTHIR — NP18 1PQ

Gauge 1000mm **Torfaen** **ST 333922**

–	4wDH		MR	121UA117	1974	OOU

TARMAC plc – A CRH Company, ABERTHAW CEMENT WORKS — CF62 3ZR

Gauge 4ft 8½in R.T.C. www.tarmac.com **Vale of Glamorgan** **ST 033674**

125	4wDH		TH	213V	1969	OOU

TATA STEEL EUROPE, TATA STEEL PACKAGING, TROSTRE WORKS, LLANELLI (part of the Tata Group) — SA14 9SD

Gauge 4ft 8½in www.tatasteeleurope.com **Carmarthenshire** **SS 531994**

SAM	0-6-0DH		AB	659	1982	a
		reb	HAB	6768	1990	
		reb	HE		2004	
DH50-2	0-6-0DHF		TH	246V	1973	
		reb	HE	9377	2011	a
DH50-1	0-6-0DH		TH	278V	1978	
		reb	HE	9376	2011	a

a property of LH Group Services Ltd, Barton-under-Needwood, Staffordshire

TATA STEEL EUROPE, TATA STEEL STRIP PRODUCTS UK, LLANWERN WORKS, NEWPORT (part of the Tata Group) — NP19 4QZ

Gauge 4ft 8½in www.tatasteeleurope.com **Newport** **ST 385863**

302	0-6-0DH		GECT	5379	1972	
304	0-6-0DH		EEV	_(D1248	1968	
				(3946	1968	
		reb	LlanwernBSC		1996	OOU
DE1	6wDE		GECT	5409	1976	
DE2	6wDE		GECT	5410	1976	
DE4	6wDE		GECT	5412	1976	
DE5	6wDE		GECT	5413	1976	
EM 1	6wDE		GECT		1977	a

a property of Ed Murray & Sons Ltd, Hartlepool, Teesside

TATA STEEL EUROPE, TATA STEEL STRIP PRODUCTS UK, PORT TALBOT WORKS, PORT TALBOT (part of the Tata Group) SA13 2NG

Gauge 4ft 8½in www.tatasteeleurope.com **Neath Port Talbot SS 771859, 775861, 775876, 775878**

No	Name	Type		Builder	Works No	Year	Note
1		0-4-0WE		(BD	3748	1979	
				(GECT	5476	1979	
2		0-4-0WE		(BD	3749	1979	
				(GECT	5477	1979	
501		0-4-0DE		BBT	3066	1954	
504		0-4-0DE		BBT	3069	1954	
(505)	BT-2	0-4-0DE		BBT	3070	1954	a
506		0-4-0DE		BBT	3071	1954	
509		0-4-0DE		BBT	3099	1956	
(512)	BT-4	0-4-0DE		BBT	3102	1956	a
(513)	BT-3	0-4-0DE		BBT	3103	1957	
			reb	PortTalbotBSC		c1972	a
(514)	BT-1	0-4-0DE		BBT	3120	1957	
			reb	PortTalbotBSC		c1972	a
901		Bo-BoDE		BBT	3063	1955	
902		Bo-BoDE		BBT	3064	1955	
903		Bo-BoDE		BBT	3065	1955	
904		Bo-BoDE		(BT	92	1957	
				(WB	3137	1957	
905		Bo-BoDE		(BT	93	1957	
				(WB	3138	1957	
906		Bo-BoDE		(BT	94	1957	
				(WB	3139	1957	
07		Bo-BoDE		(BT	95	1957	
				(WB	3140	1957	
			reb	HAB		1993	
08		Bo-BoDE		(BT	96	1957	
				(WB	3141	1957	
			reb	HAB		1993	
09		Bo-BoDE		(BT	97	1957	
				(WB	3142	1957	
			reb	HAB	6063	1994	
910		Bo-BoDE		(BT	98	1957	
				(WB	3143	1957	OOU
920	BRANWEN	4w-4wDH		CNES	0001	2009	
		refurbished		HE	9381	2012	
921	RHIANNON	4w-4wDH		CNES	0002	2010	
		refurbished		HE	9380	2012	b
922	GUINEVERE	4w-4wDH		CNES	0003	2010	
		refurbished		HE	9379	2012	b
923	AURORA	4w-4wDH		CNES	0004	2011	
		refurbished		HE	9378	2012	b
953		Bo-BoDE		BBT	3113	1957	
305		0-6-0DH		GECT	5382	1973	
D.E.3		6wDE		GECT	5411	1976	

a converted to a brake tender runner

b worksplate dated 2009

WELSH WATER plc, SOLVA SEWAGE WORKS, SOLVA HARBOUR (SA62 6UT)
Gauge 600mm www.dwrcymru.com **Pembrokeshire SM 807244**

29-32	2-2wBE		Bradshaw			
		reb	Ash Dale Engineering		2005	
–	2-2wBE		Bradshaw	200374	2015	

PRESERVATION LOCATIONS

BRECON MOUNTAIN RAILWAY CO LTD, PANT STATION, MERTHYR TYDFIL CF48 2DD

Pontsticill Station, and the majority of the route of this railway, lies in Powys. The terminus, and principal locomotive shed and workshops are, however, at Pant, Merthyr Tydfil, and thus within our "South Wales Area".

Gauge 1ft 11¾in www.bmr.wales **Merthyr Tydfil SO 060098**

No.1	2-6-2	OC	BMR		2014	
a rebuild of	2-6-0	OC	BLW	15511	1897	
2	4-6-2	OC	BLW	61269	1930	
GRAF SCHWERIN-LÖWITZ	0-6-2WT	OC	Jung	1261	1908	
–	0-6-0DH		BMR	001	1987	a
–	4wDM		HE	6298	1964	
Tu46 001	4w-4wDH		Kambarka		1981	
Tu7 1698	4w-4wDH		Kambarka		1981	b

a built from parts supplied by BD
b originally built to 750mm gauge

Brecon Mountain Railway, Steam Museum, Pontsticill
Gauge 1ft 11¾in **Powys SO 063120**

"PENDYFFRYN"	0-4-0VBT	VC	DeW		1894
SYBIL	0-4-0ST	OC	HE	827	1903
"REDSTONE"	0-4-0VBT	VC	Redstone		1905

BRIDGEND VALLEYS RAILWAY RAILWAY COMPANY LTD, t/a GARW VALLEY RAILWAY // RHEILFFORD CWM GARW, PONTYCYMER LOCOMOTIVE WORKS, OLD STATION YARD, PONTYCYMER, BRIDGEND CF32 8AZ

Gauge 4ft 8½in www.garwvalleyrailway.co.uk **Bridgend SS 904914**

68070	(PAMELA)	0-6-0ST	IC	HE	3840	1956
–		0-4-0ST	OC	RSHN	7705	1952
–		4wDH		FH	3890	1959
–		4wDH		FH	4006	1963
(W51919)		2-2w-2w-2DMR		DerbyC&W		1960
W52048		2-2w-2w-2DMR		DerbyC&W		1960

CAMBRIAN TRANSPORT (CAMBRIANCO) LTD, BARRY TOURIST RAILWAY & RAIL CENTRE

Locomotives are kept at : Barry Island Station, Barry Island, Vale of Glamorgan CF62 5TH ST 117667
Barry Wagon Works, Barry, Vale of Glamorgan CF62 5QR ST 108672

Gauge 4ft 8½in www.barrytouristrailway.co.uk

5539		2-6-2T	OC	Sdn		1928	
6686		0-6-2T	IC	AW	974	1928	
92245		2-10-0	OC	Crewe		1958	
	SUSAN	4wVBT	VCG	S	9537	1952	
(D3658)	08503	0-6-0DE		Don		1958	a
(D8128	20228 2004)	Bo-BoDE		_(EE (EEV	3599 D998	1965 1965	
(E6024)	73118	Bo-BoDE/RE		_(EE (EEV	3586 E356	1966 1966	
(E6040)	73133	Bo-BoDE/RE		_(EE (EEV	3712 E372	1966 1966	 b
–		4wDM	R/R	Unilok	2183	1982	
(50222	53222 RDB977693) 901002 IRIS II	2-2w-2w-2DMR		MetCam		1957	
(50338	53338 RDB977694) 901002 IRIS II	2-2w-2w-2DMR		MetCam		1958	
DR 98308		4w-4wDHR		Geismar	826	1998	
F621 PWP	14201	4wDM	R/R	_(Renault (Bruff		c1988 c1988	
F591 RWP	68906	4wDM	R/R	_(Mercedes (Bruff		1988 1988	

a property of Railway Support Services Ltd, Wishaw, Warwickshire
b currently at South West Trains, Bournemouth Depot, Dorset

CARMARTHENSHIRE COUNTY COUNCIL, MUSEUM SERVICES, KIDWELLY INDUSTRIAL MUSEUM TRUST, KIDWELLY INDUSTRIAL MUSEUM, BROADFORD, KIDWELLY SA17 4LW

Gauge 4ft 8½in www.kidwellyindustrialmuseum.co.uk **Carmarthenshire SN 422078**

–	0-4-0ST	OC	AB	1081	1909
–	0-6-0ST	OC	P	2114	1951
No.2	0-4-0DM		AB	393	1954

Gauge 2ft 0in

–	4wDM	 reb	RH ESCA	398063	1956 1986

DARE VALLEY COUNTRY PARK, ABERDARE CF44 7PS

Gauge 3ft 0in www.darevalleycountrypark.co.uk **Rhondda Cynon Taf SN 983026**

–	0-4-0DMF		HC	DM1314	1963

GWILI RAILWAY CO LTD, BRONWYDD ARMS

Locomotives are kept at :- Abergwili Junction, Carmarthenshire (SA31 2AQ) SN 430213
Bronwydd Arms, Carmarthenshire SA33 6HT SN 417239
Llwyfan Cerrig, Carmarthenshire SN 405258

Gauge 4ft 8½in www.gwili-railway.co.uk

(No.28)		0-6-2T	IC	Cdf	306	1897	
–		0-6-0ST	IC	HE	3829	1955	
–		0-4-0ST	OC	P	1345	1914	
	OLWEN	0-4-0ST	OC	RSHN	7058	1942	
(71516)	WELSH GUARDSMAN	0-6-0ST	IC	RSHN	7170	1944	
47	MOORBARROW	0-6-0ST	OC	RSHN	7849	1955	
	HAULWEN	0-6-0ST	IC	VF	5272	1945	
			reb	HE	3879	1961	
D2178		0-6-0DM		Sdn		1962	
	ABIGAIL	4wDM		RH	312433	1951	
W 51347		2-2w-2w-2DMR		PSteel		1960	
W 51401		2-2w-2w-2DMR		PSteel		1960	
RTU 6910	99709 901037-0	2w-2PMR		Geismar	ST/01/23	2001	

LLANELLI & MYNYDD MAWR RAILWAY CO LTD, CYNHEIDRE SA15 5YG

Gauge 4ft 8½in www.llanellirailway.co.uk **Carmarthenshire SN 495071**

(625	690) H 043 16	0-6-0DE		_(EE	2122	1956	
				(VF	D312	1956	a
10222		4wDH		RR	10222	1965	
–		4wDM		RH	394014	1956	
(M55019)	975042 960015	2-2w-2w-2DMR		GRC&W		1958	

a property of British American Railway Services Ltd, Stanhope, Co. Durham

DAVID MATTHEW, PRIVATE SITE, CARDIFF

Gauge 2ft 0in

6	DRUID	4wDM		MR	8644	1941	
			reb	ALR	6	1999	
26	ND 6455	4wDM		RH	221625	1943	
–		4wDM		RH	487963	1963	

NATIONAL MUSEUM of WALES,
Big Pit National Coal Museum, Big Pit Colliery Site, Blaenavon NP4 9RL

Gauge 4ft 8½in www.museum.wales/bigpit **Torfaen SO 236088**

NORA No.5	THE BLAENAVON CO. LTD	0-4-0ST	OC	AB	1680	1920	
	P.D. 10	0-6-0ST	IC	HC	544	1900	

Gauge 2ft 0in

–		0-4-0DMF		HE	6049	1961	
15/10		4wBEF		_(EE	3147	1961	
				(RSHD	8289	1961	Dsm

Collection Centre, Heol Crochendy, Parc Nantgarw, Nantgarw **CF15 7QT**
Storage depot, visitors by prior appointment only.
Gauge 4ft 8½in www.museum.wales/collections-centre **Rhondda Cynon Taf** **ST 114861**

52/001	0-4-0F	OC	AB	1966	1929	
–	0-4-0F	OC	AB	2238	1948	

Gauge 3ft 0in

"OGGY"	4wDM		RH	187100	1937	

Gauge 2ft 0in

4	4wBE		GB			

Waterfront Museum, Oystermouth Road, Maritime Quarter, Swansea **SA1 3RD**
Gauge 4ft 4in www.museum.wales/swansea **Swansea** **SS 659927**

–	4wG	IC	NMW		1981	

NATIONAL TRUST, DOLAUCOTHI GOLD MINES, PUMSAINT, LLANWRDA **SA19 8US**

Gauge 1ft 10½in www.nationaltrust.org.uk **Carmarthenshire** **SN 666404**

–	4wBE		CE	B2944I	1982	
–	4wBE		WR	899	1935	OOU
–	0-4-0BE		WR	5311	1955	

NEATH PORT TALBOT COUNTY BOROUGH COUNCIL, Cefn Coed Colliery Museum, Neath Road, Crynant, Neath **SA10 8SN**

Gauge 4ft 8½in www.npt.gov.uk/cefncoedmuseum **Neath Port Talbot** **SN 786034**

2758 CEFN COED	0-6-0ST	IC	WB	2758	1944	

Gauge 2ft 0in

–	4wDH		HE	8812	1978	

Margam Country Park // Parc Gwledig, Off Margam Road, Groes, Margam **SA13 2AG**
Gauge 2ft 0in www.margamcountrypark.co.uk **Neath Port Talbot** **SS 806863**

MARGAM CASTLE	0-4-0DH	s/o	AK	65	2001	

OAKWOOD THEME PARK, CANASTON BRIDGE, NARBERTH **SA67 8DE**

(part of Aspro Parks Group S.L.) www.oakwoodthemepark.co.uk
Gauge 600mm — Brer Rabbit's Burrow **Pembrokeshire** **SN 072124**

–	4w-4wRE	s/o	SL	275/1.5.90	1990	
–	4w-4wRE	s/o	SL	275/2.5.90	1990	

Gauge 1ft 3in

LORNA	4-4wDHR		Goold		1989	
LINDY-LOU	0-8-0DH	s/o	SL	7218	1972	
LENKA	4-4wDHR		SL	7322	1973	a
–	0-8-0PH	s/o	SL	R9	1976	

a 4-wheel power unit incorporates the main frames of 4wDM L 7280 / 1936

D. PARFITT, PORTHCAWL

Gauge 1ft 3in **Bridgend**

1935	SILVER JUBILEE	4-6-4PE	s/o	SmithP		1935

PEMBROKESHIRE COUNTY COUNCIL, SCOLTON MANOR MUSEUM & COUNTRY PARK, SCOLTON, near HAVERFORDWEST **SA62 5QL**

Gauge 4ft 8½in www.culture4pembrokeshire.co.uk **Pembrokeshire** **SM 991222**

(No.1378)	0-6-0ST	IC	FW	410	1878	
(A 123 W)	2w-2PMR		Wkm	3361	1942	a

a not on public display

PONTYPOOL & BLAENAVON RAILWAY COMPANY (1983) LTD, PONTYPOOL & BLAENAVON RAILWAY, BLAENAVON **NP4 9SF**

Gauge 4ft 8½in www.pontypool-and-blaenavon.co.uk **Torfaen** **SO 237093, 234093**

9629		0-6-0PT	IC	Sdn		1945	
	ROSYTH No.1	0-4-0ST	OC	AB	1385	1914	
	"HARRY"	0-4-0ST	OC	AB	1823	1924	
–	"TOM PARRY"	0-4-0ST	OC	AB	2015	1935	
	LLANTANAM ABBEY	0-6-0ST	OC	AB	2074	1939	
	VICTORY	0-4-0ST	OC	AB	2201	1945	
2	"PONTYBEREM"	0-6-0ST	OC	AE	1421	1900	a
	(SIR JOHN)	0-6-0ST	OC	AE	1680	1914	
71515		0-6-0ST	IC	RSHN	7169	1944	
	(SWANSEA VALE No.1)	4wVBT	VCG	S	9622	1958	b
	EMPRESS	0-6-0ST	OC	WB	3061	1954	
(D2141	03141)	0-6-0DM		Sdn		1960	
D5627	(31203) STEVE ORGAN G.M.	A1A-A1A DE		BT	227	1960	
D6916	(37216)	Co-CoDE		(EE (EEV	3394 D860	1963 1963	
	BLAENAVON No.14	0-4-0DH		HC	D1344	1965	
–		0-4-0DH		HC	D1387	1967	
	JOHN RODEN	0-6-0DH		HE	5511	1960	
170		0-8-0DH		HE	7063	1971	
W50632	(53632) C398	2-2w-2w-2DMR		DerbyC&W		1958	
51351	(117418)	2-2w-2w-2DMR		PSteel		1960	
51397	(117418)	2-2w-2w-2DMR		PSteel		1960	
W52044	S942	2-2w-2w-2DMR		DerbyC&W		1960	
RTU 9218		2w-2PMR		Geismar	97/17	1997	
(PWM 3962)		2w-2PMR		Wkm	6947	1955	

a currently off site for restoration
b owned by Y Clwb Rheil Cymru

G. REES, UNKNOWN LOCATION

Gauge 750mm

13		4wDM		Moës		

Gauge 2ft 0in

263 051		4wBE		CE	B0119B	1973
11	BARGEE	4wDM		MR	8540	1940
–		4wDM		MR	9932	1972
			reb	AK		1988

RHONDDA CYNON TAFF COUNTY BOROUGH COUNCIL, RHONDDA HERITAGE PARK, LEWIS MERTHYR COLLIERY, COED CAE ROAD, TREHAFOD **CF37 2NP**

Gauge 4ft 8½in www.rctcbc.gov.uk **Rhondda Cynon Taff** **ST 040912**

–		4wDM		RH	441936	1960

Gauge 3ft 0in

–		0-4-0DMF		HE	6696	1966

R. STAPLE, ABERDARE MOTOR REPAIRS, OLD TRAMWAY, ROBERTSTOWN INDUSTRIAL ESTATE, ABERDARE **CF44 8HD**

Gauge 2ft 0in **Rhondda Cynon Taff** **SO 005028**

	ROBERT	4wDM	s/o	RH	393327	1956

SOUTH WALES LOCO CAB PRESERVATION GROUP, "THE CAB YARD", off HORSEFAIR ROAD, BRIDGEND **CF31 3YN**

Gauge 4ft 8½in www.traincabs.co.uk **Bridgend** **SS 940787**

DX 50002		4wDHR		Matisa	PV5 570	1957

SWANSEA MUSEUM, COLLECTIONS CENTRE, CROSS VALLEY LINK ROAD, LANDORE STORE, SWANSEA **SA1 2JT**

Open Wednesdays 10am to 4pm

Gauge 4ft 8½in www.swanseamuseum.co.uk **Swansea** **SS 662953**

	SIR CHARLES	0-4-0F	OC	AB	1473	1916
1426		0-6-0ST	OC	P	1426	1916
	incorporates parts of			P	1187	
–		0-4-0DM		_(RSHD	7910	1963
				‾(WB		1963

SECTION 4 — IRELAND

NORTHERN IRELAND

REPUBLIC OF IRELAND

NORTHERN IRELAND

LISTINGS OF LOCATIONS IN NORTHERN IRELAND ARE PRESENTED ON A SINGLE "LAND AREA" BASIS, RATHER THAN A COUNTY BASIS.

INDUSTRIAL SITES

BULRUSH PEAT CO LTD (Subsidiary of Pindstrup Mosebrug A/S)

Locomotives are kept at :- B - Newferry Road, Bellaghy, Magherafelt Co.Derry BT45 8ND H 986985
G - Grove Bog, Carrickmore Co.Tyrone H 600707
R - Randalstown Co.Antrim J 090935

Gauge 750mm www.bulrush.co.uk

2	4wDM		MR	40S307	1967		
		reb	Bulrush(B)		2009	R	a
L5	4wDH		Schöma	4979	1988	B	
–	4wDH		Schöma	4980	1988	B	b
–	4wDH		Schöma	4992	1989	R	b
L4	4wDH		Schöma	5601	1999	R	
L6	4wDH		Schöma	5602	1999	R	
–	4wDH		Schöma	5603	1999	R	b

a rebuilt using parts from MR 22220 and MR 40S307
b slave unit for use with a master unit.

NORTHERN EXCAVATORS LTD, t/a NE RAIL, 103 CULCAVEY ROAD, HILLSBOROUGH — BT26 6HH

Gauge 5ft 3in www.nerail.co.uk **Co. Down**

RR14	4wPMR	R/R	Kawasaki KAF950B/ 00729C/0359	2011

NORTHERN IRELAND RAILWAYS / TRANSLINK LTD YORK ROAD DEPOT, BELFAST — BT15 3RP

Gauge 5ft 3in www.translink.co.uk **Co.Antrim**

8094		4w-4wDER		Derby C&W		1977	a
8097	(97) SANDITE	4w-4wDER		Derby C&W		1977	a b
7015	(K-355 APT)	4w-4wDMR		P&T	46	1985	
			reb	P&T		2002	
(7017)	VMT 850GR	4wDHR		(Donelli	558	1998	
				(Geismar		1998	
	AFZ 7054	4wDM	R/R	(Mitsubishi		2009	
				(LH Group		2009	
11	99709 428011-9	4w-4wDHR		Windhoff		2016	
–		4wBE		Niteq	B177	2001	

a permanent coupled sandite unit formed of power cars 8094 and 8097
b unpowered

JAMES STEVENSON (QUARRIES) LTD,
CLINTY QUARRY, 215 DOURY ROAD, BALLYMENA BT43 6SS

Gauge 2ft 0in **Co. Antrim D 102073**

–		4wDM	MR	8684	1941	Dsm

SUNSHINE PEAT CO, 33 DERRYHUBBERT ROAD, DUNGANNON BT71 6NW

Gauge 2ft 6in **Closed** **Co.Armagh**

–		0-4-0DM	HE	(2242	1941?)	Dsm
L3	HENRY ABRAHAM	0-4-0DM	HE	2250	1940	OOU
L4	VICTORIA	0-4-0DM	HE	2252	1940	OOU
L1	ND3053 GAVIN	0-4-0DM	HE	2264	1940	OOU
L2	ND3061 W.P.O'KANE	0-4-0DM	HE	2399	1941	OOU

SEYMOUR SWEENEY CO LTD, COLERAINE

Gauge 3ft 0in **Co. Antrim**

–		2w-2PMR	Wkm	7441	1956	

PRESERVATION SITES

DEPARTMENT OF THE ENVIRONMENT (N.I.),
COUNTRYSIDE AND WILDLIFE BRANCH, BIRCHES PEATLANDS PARK,
DERRYHUBBERT ROAD, DUNGANNON BT71 6NW

Gauge 3ft 0in **R.T.C.** **Co.Armagh**

–		4wDH	AK	44	1993	
1		4wDM	FH	3719	1954	OOU
2	HENRY HEAL	4wDM	Schöma	1727	1955	OOU

DESTINED GROUP plc, FOYLE VALLEY RAILWAY MUSEUM,
FOYLE VALLEY RAILWAY COMPANY, FOYLE ROAD, LONDONDERRY BT48 6SQ

Track laid southwards on route of former GNR(I) broad gauge line, from Londonderry Foyle Road Station.

Gauge 3ft 0in **Co.Derry**

6	COLUMBKILLE	2-6-4T	OC	NW	830	1907
–		4wDM		MR	11039	1956
12		0-4-0+4DMR		_(WkB (Dundalk	M44719	1934 1934

DOWNPATRICK & COUNTY DOWN RAILWAY, DOWNPATRICK RAILWAY MUSEUM, DOWNPATRICK STATION, MARKET STREET, DOWNPATRICK — BT30 6LZ

Gauge 5ft 3in www.downrail.co.uk **Co.Down**

90		0-6-0T	IC	Inchicore		1875	
1		0-4-0T	OC	OK	12475	1934	
3		0-4-0T	OC	OK	12662	1935	
146		Bo-BoDE		GM	27472	1962	
A39R		Co-CoDE		MV	925	1955	
C231		Bo-BoDE		MV	977	1956	
G611		4wDH		Dtz	57225	1962	
G613		4wDH		Dtz	57226	1962	
G617		4wDH		Dtz	57229	1962	
E421	W F GILLESPIE OBE	6wDH		Inchicore		1961	
E432		6wDH		Inchicore		1962	
2509		0-4-0+4DMR		_(WkB	1747D	1947	
				(ELC		1947	OOU
6111	(2624)	4w-4wDMR		_(AEC	8031 046	1953	
				(PRoyal	B35171	1953	
(RB 3)	(RDB 977020)	4wDMR		_(Leyland		1981	
				(BRE(D)		1981	
8069		4w-4wDER		DerbyC&W		1977	
8090		4w-4wDER		DerbyC&W		1977	
8458	(ANTRIM CASTLE)	4w-4wDER		DerbyC&W		1986	
(713)	ROSIE	2w-2DMR		Wkm	8916	1962	
416A	(712)	2w-2PMR		Wkm	8919	1962	
	rebuilt as	2w-2DMR		Inchicore		1986	OOU
HC1		2w-2DMR		P&T	323	1971	
(7020)	JXI 2860	4wDM	R/R	_(Bedford		1986	
				(Bruff	531	1986	

GIANT'S CAUSEWAY & BUSHMILLS RAILWAY, CAUSEWAY STATION, RUNKERRY ROAD, near BUSHMILLS — BT57 8SZ

Gauge 3ft 0in www.freewebs.com/giantscausewayrailway **Co.Antrim**

No.3	SHANE	0-4-0WT	OC	AB	2265	1949	
No.1	TYRONE	0-4-0T	OC	P	1026	1904	OOU
–		4w-4wDH		SL	8165	2010	
	RORY	4wDH		SMH	102T016	1976	

RAILWAY PRESERVATION SOCIETY OF IRELAND, WHITEHEAD DEPOT, CASTLEVIEW ROAD, WHITEHEAD — BT38 9NA

Gauge 5ft 3in www.steamtrainsireland.com **Co.Antrim J 474922**

27		0-6-4T	IC	BP	7242	1949	OOU
No.85	MERLIN	4-4-0	3C	BP	6733	1932	
131		4-4-0	IC	NR	5757	1901	

171	SLIEVE GULLION	4-4-0	IC	BP	5629	1913	
			reb	Dundalk	42	1938	
No.184		0-6-0	IC	Inchicore		1880	OOU
186		0-6-0	IC	SS	2838	1879	
No.3		0-6-0ST	OC	AE	2021	1928	OOU
	GUINNESS	0-4-0ST	OC	HC	1152	1919	a
B142		Bo-BoDE		GM	27468	1962	
23		4wDM		FH	3509	1951	
–		4w-4wPMR		NCC		1933	
	rebuilt as	4w-4wDHR		NCC		1947	OOU
	CSE CARLOW No.1	4wDM		RH	382827	1955	
–		4wPM	R/R	_(Unilok		1965	
				(Jung	A114	1965	

Gauge 3ft 0in

(No.4)	(MEENGLAS)	2-6-4T	OC	NW	828	1907	
5	"DRUMBOE"	2-6-4T	OC	NW	829	1907	b

a ex works 30/9/1915 but worksplate dated 1919
b property of County Donegal Railway Restoration Ltd – here for restoration and storage

PETER SCOTT, FINNAGHY PARK CENTRAL, BELFAST

Private site. No casual visitors

Gauge 1ft 7¼in. **Co.Antrim**

–		2-4-0	OC	ScottP		c1969	
	rebuilt as	2-4-0T	OC	ScottP		1970	

ULSTER FOLK & TRANSPORT MUSEUM, 153 BANGOR ROAD, HOLYWOOD **BT18 0EU**

Gauge 5ft 3in www.nmni.com/uftm **Co.Down J 419806**

No.30		4-4-2T	IC	BP	4231	1901	
74	DUNLUCE CASTLE	4-4-0	IC	NBQ	23096	1924	
93		2-4-2T	IC	Dundalk	16	1895	
800	MAEDHBH	4-6-0	3C	Inchicore		1939	
No.1		0-6-0ST	OC	RS	2738	1891	
(102)	FALCON	Bo-BoDE		_(Don		1970	
				(HE	7198	1970	
B113		Bo-BoDE		Inchicore		1950	
1	(8178)	2w-2DMR		ADC		1928	
			reb	Dundalk		1933	
–		4wPM	R/R	_(Zagro	80159	2010	
				(Zwiehoff		2010	

Gauge 3ft 0in

2		0-4-0Tram	OC	K	T84	1883	a
2	BLANCHE	2-6-4T	OC	NW	956	1912	
2		0-4-0T	OC	P	1097	1906	
–		4-4-0T	OC	RS	2613	1884	
(1)		2-2-0PMR		A&O		1905	

11	PHOENIX		4wVBT		AtW	114	1928	
		rebuilt as	4wDM		Dundalk		1932	
3			2-4w-2PMR		DC/?	1519	1926	b
10			0-4-0+4DMR		_(WkB	M42830	1932	
					¯(NC		1932	
No.2			4+0-4-0RE/WER		_(M&P		1885	
					¯(Ashbury		1885	

a worksplate reads 84
b now unmotorised

Gauge 2ft 0in

–		4wDM		FH	3449	1950
–		4wDM		HE	3127	1944
246		4wPM		MR	246	1916
–		4wDM		MR	9202	1946

Gauge 1ft 10in

20		0-4-0T	IC	Spence		1905

REPUBLIC OF IRELAND

LISTINGS OF LOCATIONS IN THE REPUBLIC OF IRELAND ARE PRESENTED ON A COUNTY BASIS, WITH COUNTIES LISTED ALPHABETICALLY.

BORD NA MÓNA LOCATIONS ARE DESCRIBED UNDER THEIR COUNTY HEADINGS

A combined fleet list of Bord na Móna locomotives follows the county-based listings.

CARLOW

PRESERVATION SITES

PRIVATE OWNER, CARLOW AREA

Gauge 3ft 0in

–	4wDM		RH		

CAVAN

PRESERVATION SITES

BELTURBET RAILWAY AND VINTAGE RESTORATION ASSOCIATION, BELTURBET STATION, RAILWAY ROAD, BELTURBET

Gauge 5ft 3in www.belturbetheritagerailway/wordpress.com **H 363167**

–	4wDM		RH	312425	1951

Gauge 3ft 0in

LM 137	4wDM		RH	382817	1955

CLARE

PRESERVATION SITES

THE WEST CLARE RAILWAY CO LTD, MOYASTA JUNCTION, KILRUSH

Gauge 3ft 0in www.westclarerailway.ie

No.5 SLIEVE CALLAN	0-6-2T	OC	D	2890	1892
		reb	AK		2009
–	0-6-0DMF		HC	DM719	1950

–	4wDH	RFSD	L101	1989	
LM 60	4wDM	RH	259738	1948	
LM 64	4wDM	RH	259745	1948	

WEST OF IRELAND TRANSPORT MUSEUM, MOYASTA

Gauge 5ft 3in

A3R	Co-CoDE	MV	889	1955
A15	Co-CoDE	MV	901	1955
124	Bo-BoDE	GM	26274	1960
152	Bo-BoDE	GM	27478	1962
190	Bo-BoDE	GM	31257	1966

J. WHELAN LTD,
WHELAN TRACTOR SALES & AGRICULTURAL ENGINEERS, ENNIS ROAD, KILRUSH

Gauge 3ft 0in www.whelansgarage.com

LM 133	4wDM	RH	379090	1955	Dsm
LM 168	4wDM	RH	402980	1956	Dsm

J.WHELAN & ASSOCIATES, TULLYGOWER QUARRY, near KILRUSH

Gauge 5ft 3in **R 039603**

(C227 106) C202	Bo-BoDE	MV	973	1957	a
(710)	2w-2DMR	Wkm	8918	1962	Dsm a
(714)	2w-2DMR	Wkm	8920	1962	OOU a

Gauge 3ft 0in

LM 25	4wDM	RH	244871	1946	
LM 59	4wDM	RH	259737	1948	
LM 98	4wDM	RH	375335	1954	
LM 146	4wDM	RH	394024	1956	
LM 147	4wDM	RH	394025	1956	
LM 314	0-4-0DM	HE	8548	1977	b
C 53	4wPMR	_(BnM ‾(SM	9	1958 1958	 b
C 78	4wPMR	BnM		1972	b

a property of Irish Traction Group, Wickham Awareness Group
b stored for West Clare Railway Co Ltd, Moyasta Junction, Kilrush

CORK

INDUSTRIAL SITES

IARNRÓD ÉIREANN, MAINTENANCE DEPOT CORK, WATER STREET, CORK

Gauge 5ft 3in www.irishrail.ie

–	4wBH	R/R	AVB	AVB5741/1	2010	OOU

THE RAILROAD FACTORY LTD, UNIT E3, SOUTHLINK PARK, FRANKFIELD, CORK

Gauge 4ft 8½in www.trrf.ie

–	4wDH	R/R	_(TRRF		2015	
			(Mecalac	08801	2015	a

a demonstrator locomotive

new TRRF locomotives may be present

CHRIS SHILLING, t/a OIM LTD, OIBREACHA IARNROD NA MUMHAM, MUNSTER RAILWAY WORKS, KEALKILL, near BANTRY

Gauge 1ft 3in

–	4wPE	OIM	010	2015

Locomotives under construction may be present

PRESERVATION SITES

J. & R. BRADLEY, BALLYNAKILLWEST, NEWTOWNSHANDRUM, near CHARLEVILLE

Gauge 2ft 6in **R 461244, 482223**

SO 35 S 35	4wDM	MR	9709	1952	

Gauge 600mm

–	4wDM	Struver	60327	c1955	Dsm

DENNIS COLLINS, NEWMARKET ROAD, KANTURK

Gauge 3ft 0in

LM 360	4wDH	DunEW			
–	2w-2PM	Ferguson		c1938	a

a former road vehicle, converted to rail use 1985

Gauge 2ft 0in

BR 1949	(WANDERING WILLIE)	0-4-0VBT	VCG			(1949?)	
	rebuilt as	0-4-0PM					
	rebuilt as	4-2wPM		CollinsD		c2003	OOU
–		2-2-0PM		CollinsD		2011	OOU

TIM CROWLEY LTD, T/A CROWLEY ENGINEERING, UPPER GLANMIRE BRIDGE, GLANMIRE

Gauge 3ft 0in www.crowley.ie

LM 371	4wDH		DunEW			

TIM CROWLEY, YOUGHAL CASTLE, YOUGHAL

Gauge 3ft 0in

LM 140	4wDM		RH	392139	1955	

HALFWAY VINTAGE CLUB, c/o THE RAMBLE INN, HALFWAY, BALLINHASIG

Gauge 5ft 3in

–	4wDM		RH	252843	1948	

IARNRÓD ÉIREANN, CORK (CEANNT / KENT) STATION

Gauge 5ft 3in www.irishrail.ie

36	2-2-2	IC	BuryC&K		1847	

THE JAMESON EXPERIENCE, OLD DISTILLERY WALK, MIDLETON

Gauge 2ft 6in www.jamesonwhiskey.com

–	2-2-2	VC	Midleton		2012	

Non-working replica locomotive

TIM NAGLE, GREY GABLES, FEATHERBED LANE, off EASTERN ROAD, KINSALE

Gauge 600mm

–	4wDM		Strüver	60385	c1955	Dsm

JOHN TOUIG, LISSAGROOM, UPTON

Gauge 2ft 0in **R 522612**

–	4wDM		RH	264244	1949	OOU

WEST CORK MODEL RAILWAY VILLAGE,
CLONAKILTY STATION, INCHYDONEY ROAD, CLONAKILTY

Gauge 5ft 3in www.modelvillage.ie

–	4wDM		RH	305322	1951	

DONEGAL

INDUSTRIAL SITES

FOREST ENTERPRISES, BELLANAMORE, near FINTOWN

Gauge 2ft 0in

–	4wDM		MR	7944	1943	Dsm

PRESERVATION SITES

COMHLACHT TRAENACH na GAELTACHTA LAIR (2000) TEO //
CENTRAL GAELTACHT TRAIN COMPANY,
FINTOWN RAILWAY, FINTOWN STATION, FINTOWN

Gauge 3ft 0in www.fintownrailway.com

–	4wDH		MR	102T007	1974	
		reb	AK	78R	2007	
LM 77 H	4wDM		RH	329680	1952	OOU
18	0-4-0+4DMR		_(WkB	1361D	1940	
			(Dundalk		1940	

Gauge 2ft 0in

–	4wDM		RH	243387	1946	OOU

DIFFLIN LAKE RAILWAY, OAKFIELD PARK, RAPHOE

Open June to September inclusive.

Gauge 1ft 3in www.oakfieldpark.com

No.1 THE DUCHESS OF DIFFLIN	0-4-2T	OC	ESR	317	2003	
No.2 THE EARL OF OAKFIELD	0-4-0DH	s/o	AK	66	2002	
–	4wDM		L	26366	1944	
rebuilt as	4wDH		_(DLR		2018	
			(MossAJ		2018	

DONEGAL RAILWAY RESTORATION CLG, THE DONEGAL RAILWAY HERITAGE CENTRE, THE OLD STATION HOUSE, TYRCONNEL STREET, DONEGAL TOWN

Gauge 3ft 0in www.donegalrailway.com

(C 18)	2w-2PMR	RPSI(W)		2006	a

a incorporates parts from Wkm 4808 1948; kept in Goods Shed; not on public display

MICHAEL GALLAGHER, ARDRASOOL, BALLINDRAIT, near LIFFORD

Gauge 2ft 0in

–	4wDM	RH	371544	1954

ST. CONNELL'S MUSEUM & HERITAGE CENTRE SOCIETY LTD, ST. CONNELL'S MUSEUM & GLENTIES HERITAGE CENTRE, MILL ROAD, GLENTIES

Gauge 2ft 0in www.glentiesheritagemuseum.com

LM 263	4wDM	RH	7002/0600-1	1968

DUBLIN / BAILE ÁTHA CLIATH

INDUSTRIAL SITES

BREFFNI HIRE PLANT LTD, BALLYHACK FARM, KILSALLAGHAN

Gauge 5ft 3in www.breffnigroup.ie

R22		4wDHR	R/R	Dieci			a	OOU
RR 02	07D 53644	4wDM	R/R	Mecalac		2007	a	

a Teleporter/bridge inspection unit

Gauge 4ft 8½in

E1235	4wBER	Bance	133/04	2004	DsmT

DUBLIN LUAS, DUBLIN TRAMWAY SYSTEM

(Operated by Transdev Dublin Light Rail Ltd)

Gauge 4ft 8½in www.luas.ie

97 D 67550	4wDM	R/R	Landrover	1997
97 D 66701	4wDM	R/R	Nissan	1997

Broombridge Depot, Dublin

Gauge 4ft 8½in

–	4wBE	R/R	Zephir	2734	2017

Red Cow Depot, Dublin
Gauge 4ft 8½in

–	4wBH		BBM	155	2002	
–	4wBH		BBM	268	2009	
ENG 5873 13 006	4wDM		Richard		1960	Pvd
–	4wDH		Unilok	4019	2009	
162 D 16509	4wDM	R/R	Unimog		2016	

Sandyford Depot, Dublin
Gauge 4ft 8½in

–	4wBH		BBM	156	2002	
–	4wBE		SET	1007/1	2003	OOU
–	4wDH		Unilok	4013	2004	

IARNRÓD ÉIREANN
www.irishrail.ie

Departmental Stock

Railcars are based at :–

C - Connelly Station, Dublin
G - Greystones Station, Co. Wicklow
H - Heuston Carriage Sidings, Dublin
I - Inchicore Works, Dublin
K - Kildare Plant Depot, Co.Kildare
L - Limerick Station S&T Yard, Co. Limerick
LJ - Limerick Junction, Co.Tipperary
NW - North Wall, Dublin
P - Portlaoise Permanent Way Depot, Portlaoise, Co. Laois

Gauge 5ft 3in

700			4w-4wDMR	Plasser	26	1974	I
721	JESS		2w-2DMR	HPET	1068	1992	LJ
722	99609 429722-2		4wDMR	MatisaSPA	0214	1994	L
723			4wDMR	MatisaSPA	0215	1994	K
728			2w-2DH	Plasser		1974	
		rebuilt as	2w-2DHR	Inchicore		1999	LJ
734			2w-2DM	Plasser	1772	1980	
		rebuilt as	2w-2DMR	Inchicore		1998	K
780			4w-4wDHR	P&T	164	1979	P a
790			4w-4wDHR	Geismar	999	2009	K

a jointly owned by NIR / Iarnród Ėireann – Ballast Cleaner/Weed Sprayer

DART Overhead Maintenance Depot, Connolly Station, Dublin
Gauge 5ft 3in

25	02 D 13723	99609 919002-6	2w-2DM	R/R	_(Volvo ‾(SRS	2002 2002
(FR 220)	02 D 3124	99609 919001-8	2w-2DM	R/R	_(Volvo ‾(SRS	2002 2002

DART System Fairview Depot, Fairview Park, Dublin
Gauge 5ft 3in

–	2w-2PM		_(Zagro	80138	2010	
			(Zweihoff		2010	

Inchicore Works, St. Patrick's Terrace, Inchicore, Dublin
Gauge 5ft 3in

134	Bo-BoDE		GM	26284	1960	a
–	2w-2PM	R/R	_(Zagro		2010	
			(Zwiehoff		2010	OOU
–	4wDH	R/R	Zephir	2793	2018	

a owned by RPSI

SANLINE SYSTEMS LTD, 12th LOCK, GRAND CANAL, NEWCASTLE ROAD, LUCAN, DUBLIN
Gauge 5ft 3in

–	4wDMR	R/R	Bedford		1981	
92 MH 4823	4wDMR	R/R	Landrover		1992	Dsm
00 LH 9282	4wDM	R/R	Landrover		2000	
02 D 76075	4wDM	R/R	Landrover		2002	
06 D 84465	4wDMR		_(Landrover		2006	
			(Geismar		2006	
–	2w-2PMR		Sanline		1999	OOU

Gauge 4ft 8½in

09 LS 853	4wDM	R/R	_(Mitsubishi		2009	
			(Sanline		2018	

PRESERVATION SITES

THE CAVAN & LEITRIM RAILWAY CO LTD & IRISH NARROW GAUGE TRUST, c/o PRIVATE INDUSTRIAL ESTATE, NAAS ROAD, RED COW
Private site. No access without prior permission.

Gauge 3ft 0in www.cavanandleitrim.com

5		2w-2PMR	DC	1495	1927	
	rebuilt as	2w-2DHR	AK		1995	
W6/11-4		2w-2PMR	Wkm	9673	1964	

Locomotives and railcars for restoration occasionally present

DEPARTMENT OF MECHANICAL ENGINEERING, PARSONS BUILDING, TRINITY COLLEGE, DUBLIN
Gauge 1ft 9in www.tcd.ie/mecheng

–	0-6-0	IC	Kennan		1855	a

a working model; used for instruction only

GUINNESS IRELAND GROUP LTD, GUINNESS STOREHOUSE MUSEUM, ST JAMES'S GATE BREWERY, CRANE STREET, DUBLIN

Gauge 1ft 10in www.guinness-storehouse.com

17	0-4-0T	IC	Spence		1902	Pvd
47	4wDM		FH	3444	1950	Pvd

RAILWAY PRESERVATION SOCIETY OF IRELAND, CONNOLLY DEPOT, DUBLIN

Gauge 5ft 3in www.steamtrainsireland.com **No Public Access**

4	2-6-4T	OC	Derby		1947
461	2-6-0	IC	BP	6112	1922
B141	Bo-BoDE		GM	27467	1962
175	Bo-BoDE		GM	27501	1962

Other visiting preserved locomotives may occasionally be present

GALWAY

INDUSTRIAL SITES

BORD NA MÓNA, Derryfadda **M 800432**

Some locomotives are out based at the following locations – Daleysgrove, Gowla, Boughill Bog, Castlegar, Area 16 and Area 17 Killaderry Tea Centres.

For locomotive details see the Bord na Móna fleet list at the end of this Section.

DECLAN MAHER PLANT LTD, t/a CARRA PLANT SERVICES, CARRA, BULLAUN, LOUGHREA

Gauge 5ft 3in

T216 UGE	4wDM	R/R	_(Landrover (Perm	1999 1999

UNILOKOMOTIVE LTD, IDA INDUSTRIAL ESTATE, DUNMORE ROAD, TUAM

New locomotives under construction are usually present.

Gauge 5ft 6in www.unilok.ie

–	4wDH		Unilok	2353	2012	

Gauge 5ft 3in

–	4wDH	R/R	Unilok		2013	
–	4wDH					Dsm

Gauge 4ft 8½in

–	4wDH	R/R	Unilok		2013	Dsm

PRESERVATION SITES

BALLYGLUNIN STATION ASSOCIATION, BALLYGLUNIN

Gauge 5ft 3in www.ballyglunin.com

-	2w-2PMR		Bryce		2017	

GALWAY MINIATURE RAILWAY (LEISURELAND EXPRESS), SALTHILL, GALWAY

Operated by Galway City Council Leisure Services Department. Operates June to September

Gauge 2ft 0in www.leisureland.ie **M 278237**

LEISURELAND EXPRESS	4-4-0+4-4wDH	s/o	SL	2155	2003	

M. MITCHELL, DUNSANDLE STATION, DUNSANDLE

Gauge 5ft 3in **M 597242**

E428	6wDH		Inchicore		1962	

GARRY SKELTON, GALWAY

Private site. No casual visitors.

Gauge 1ft 10in

21	0-4-0T	OC	Spence		1905	Dsm

KERRY

PRESERVATION SITES

KERRY COUNTY COUNCIL, TRALEE & DINGLE STEAM RAILWAY CO LTD, BLENNERVILLE, near TRALEE

Gauge 3ft 0in (Closed) www.tdlr.info (unofficial website)

No.5T	2-6-2T	OC	HE	555	1892	OOU
LM 92 L	4wDM		RH	371967	1954	OOU

LARTIGUE MONORAILWAY, JOHN B KEANE ROAD, LISTOWEL

Gauge monorail www.lartiguemonorail.com

L.B.R. 4	0-2-0+2wDH	s/o	AK	62	2002	a

a carries worksplate dated 2000

KILDARE

INDUSTRIAL SITES

BORD NA MÓNA

Almhain North **N 801211**

Ballydermot **N 659216, 656215**

Supplies locomotives to the Ballydermot service sites, located at Black River, Glashbaun and Ballydermot North Tea Centres, in addition to the main workshops. Locomotives can occasionally be found at Lullymore Lodge Bog and at Derrybrennan Fuelling Point. Service locomotives work over a wide area including Codd Bog North and South, Ballydermot South, Leonards Corner, Ticknevin and Lullymore East

Gilltown Landsale **N 796338**

Kilberry **S664998**

Lullymore **N 711292, 708286, 706286**

Prosperous **N 831292**

Timahoe R.T.C. **N 781284, 754335**

Includes Timahoe North Bog, Drumachon.

Ummeras **N 623147**

For all locomotive details see the Bord na Móna fleet list at the end of this Section.

DEPARTMENT OF DEFENCE, IRISH ARMY, ENGINEER MECHANICAL WORKSHOPS, No.1 MAINTENANCE YARD, CURRAGH CAMP, near KILDARE

Gauge 2ft 0in R.T.C. www.military.ie

–		4wDM	MR	8970	1945	OOU

LYONS & BURTON NEW HOLLAND AGRICULTURAL, THE MOUNT, CLANE ROAD, BALLYBRACK, near KILCOCK

Gauge 5ft 3in www.lyonsandburton.ie **N 864374**

99 KK 723	99609 945012-3	2w-2DM	R/R	Zetor 112450302	1992	

PRESERVATION SITES

LULLYMORE HERITAGE & DISCOVERY PARK, RATHANGAN

(Note : Rathangan is located in Co. Offaly, but the park is located in Co. Kildare)

Gauge 3ft 0in www.lullymoreheritagepark.com

LM 75		4wDM	RH	326047	1952	Dsm
LM 85		4wDM	RH	329693	1952	
LM 139		4wDM	RH	392137	1955	
LM 309	LM 3	0-4-0DM	HE	8545	1977	Dsm
C 72	LM 9	4wPMR	BnM		1972	OOU

KILKENNY

INDUSTRIAL SITES

DON BUTLER COMMERCIALS LTD, t/a DON BUTLER TRANSPORT & PLANT HIRE, SKEARD, BISHOPSHALL TOWNLAND, KILMACOW

Gauge 3ft 0in **S 583183**

LM 189		0-4-0DM	Dtz	57125	1960
(LM 257)		0-4-0DM	Dtz	57836	1965
LM 356	SEAN	4wDH	DunEW		
LM 125	O	4wDM	RH	379084	1954
LM 117	MÍCHAÉL	4wDM	RH	379910	1954

LAOIS

INDUSTRIAL SITES

BORD NA MÓNA, Coolnamóna S 456947

Supplies locomotives to the surrounding area bogs, with some locomotives outbased at the following locations – Togher Bog Tea Centre, Cashel Bog Tea Centre, Coolcarton Bog and Dogs Head Bog.

For all locomotive details see the Bord na Móna fleet list at the end of this Section.

IARNRÓD ÉIREANN, CHIEF MECHANICAL ENGINEERS DEPARTMENT, LAOIS TRAINCARE FACILITY, near PORTLAOISE

Gauge 5ft 3in www.irishrail.ie

621 RBL-020-400 DRIVER TOM LYNAM	4wBE	Scul		2008

IARNRÓD ÉIREANN, ENGINEERING SERVICES DIVISION, PORTLAOISE RAIL SUPPLY & INFRASTRUCTURE FACILITY, PORTLAOISE

Gauge 5ft 3in www.irishrail.ie

02 D 78046	4wDM	R/R	Unimog	201419	2002

LINMAG RAIL GmbH, c/o IARNRÓD ÉIREANN, ENGINEERING SERVICES DIVISION, PORTLAOISE RAIL SUPPLY & INFRASTRUCTURE FACILITY, PORTLAOISE

Gauge 5ft 3in / 4ft 8½in (Dual Gauge)

GM 987 GF 99809 900005-6 ERIN	4w+4w-6wDM	R/R	MAN		2016	a

a MAN tractor unit couples to Milling Machine (Dual Gauge for IR/LUAS)

PRESERVATION SITES

JIM DEEGAN, CROOKED VALLEY RAILWAY, LUGGACURREN, near STRADBALLY

Private Site. No casual visitors.

Gauge 3ft 0in

LM 194		0-4-0DM		Dtz	57136	1960	

WILLIAM DEEGAN, STRADBALLY

Gauge 3ft 0in

(C 42)		2w-2PMR		Wkm	7129	1955	DsmT

IRISH STEAM PRESERVATION SOCIETY, STRADBALLY WOODLAND EXPRESS, STRADBALLY HALL, STRADBALLY

Gauge 3ft 0in www.irishsteam.ie **S 568966**

No.2	LM 44	0-4-0WT	OC	AB	2264	1949	
	NIPPY	4wDM		FH	2014	1936	
(4)		4wDM		RH	300518	1950	
1		4wDM		RH	326051	1952	
3		4wDM		RH	326052	1952	
LM 55		4wDM		RH	259205	1948	Dsm
(LM 167)	Q	4wDM		RH	402978	1956	
(LM 191)		0-4-0DM		Dtz	57134	1960	Dsm
(LM 317)		4wDM		SMH	60SL742	1980	

STRADBALLY STEAM MUSEUM, MAIN STREET, STRADBALLY

Gauge 1ft 10in www.irishsteam.ie **S 575961**

15		0-4-0T	IC	Spence		1912	a

a carries incorrect plate dated 1895

LEITRIM

PRESERVATION SITES

THE CAVAN & LEITRIM RAILWAY CO LTD and IRISH NARROW GAUGE TRUST, DROMOD STATION, STATION ROAD, DROMOD

Gauge 5ft 3in www.cavanandleitrim.com

720		2w-2DMR		P&T	322	1971	
			reb	BRE		1984	a

	JZA 979	2-2wDM		_(Scammell (Cold Chon		1959 1963	b

a stored in IE Dromod yard and not on public display
b conversion of a Scammell road lorry by Cold Chon Ltd, Oranmore, Co. Galway

Gauge 3ft 0in

1	DROMAD	0-4-2ST	OC	KS	3024	1916	
	rebuilt as	0-4-2T	OC	AK		1993	
(LM 180)		0-4-0DM		Dtz	57122	1960	
(LM 186)		0-4-0DM		Dtz	57132	1960	
LM 253		0-4-0DM		Dtz	57834	1965	
LM 258		0-4-0DM		Dtz	57839	1965	
LM 260		0-4-0DM		Dtz	57841	1965	
LM 262		0-4-0DM		Dtz	57843	1965	
–		4wDM		HE	2280	1941	Dsm
–		4wDM		HE	6075	1961	Dsm
F511	DINMOR	4wDM		JF	3900011	1947	OOU
9		4wDH		MR	115U093	1970	
1		4wDM		RH	249526	1947	
LM 49		4wDM		RH	259189	1948	
(213)	L2 2 "MARIE"	4wDM		RH	283896	1949	
2		4wDM		RH	314223	1951	
(LM 87)		4wDM		RH	329696	1952	Dsm
LM 91		4wDM		RH	371962	1954	
LM 101		4wDM		RH	379059	1954	
LM 106		4wDM		RH	375345	1954	
LM 114	JOE ST LEGER	4wDM		RH	379070	1954	
LM 116	M 15	4wDM		RH	379077	1954	Dsm
(LM 131)		4wDM		RH	379086	1955	Dsm
(LM 138)		4wDM		RH	382819	1955	
212		4wDM		RH	394022	1956	
LM 161		4wDM		RH	402175	1956	
LM 174		4wDM		RH	402986	1957	OOU
LM 175		0-4-0DM		RH	420042	1958	
LM 11		0-4-0DM		Ruhrthaler	1348	1934	
LM 349		4wDM		SMH	60SL743	1980	
(LM 351)		4wDM		SMH	60SL745	1980	Dsm
LM 361		4wDH		DunEW	LM361	1984	
LM 362		4wDH		DunEW	G4464	1984	
LM 369		4wDH		DunEW	S4484	1984	
C 11		2w-2PHR		SolHütte		1934	
	rebuilt as	2w-2PMR		BnM(Cs)		c1963	
C 47		4wPMR		_(BnM (SM	3	1958 1958	
(C 51)		4wPMR		_(BnM (SM	7	1958 1958	
C 56		2w-2PMR		Wkm	7681	1957	DsmT
C 66		4wDMR		BnM		1972	
C 74		4wPMR		BnM		1972	

Gauge 2ft 0in

(No.1)	4wDM		HE	2239	1940	
		reb	HE	7340	1972	
(No.3)	4wDM		HE	2304	1940	
		reb	HE	7341	1972	
–	4wDM		HE	2659	1942	
(No.2)	4wDM		HE	2763	1943	
		reb	HE	7342	1972	
LM 16 B	4wDM		RH	200075	1940	
LM 42 C	4wDM		RH	252849	1947	Dsm
LM 198	4wDM		RH	398076	1956	OOU

Gauge 1ft 10in

–	0-4-0T	IC	Spence		1912	Dsm
–	4wDM		FH	3068	1947	
31	4wDM		FH	3446	1950	
–	4wDM		FH	3447	1950	Dsm

ANDREW MARSHALL, STRACARNE, MOHILL

Gauge 3ft 0in

LM 178	0-4-0DM		Dtz	57120	1960	
LM 187	0-4-0DM		Dtz	57133	1960	Dsm

LIMERICK

INDUSTRIAL SITES

IARNRÓD ÉIREANN,
LIMERICK WAGON WORKS, ROXBOROUGH ROAD, LIMERICK

Gauge 5ft 3in www.irishrail.ie

42-048-087	4wBH	R/R	AVB	AVB5735/1	2010	
42-048-088	2w-2wBH		Mechan	6063/3	2008	OOU

LONGFORD

INDUSTRIAL SITES

BORD NA MÓNA, Mountdillon — N 048688

Supplies locomotives for use at Lough Rea Power Station; some locomotives are outbased at that site with additional stabling points as follows : Derraghan, Mountdillon p-way and track repair depot, Old wagon repair facility (known as Mountdillon Old Works), Derrycolumb Bog, Corlea Bog, Derryaroge Bog, Begnagh Branch, Erenagh Tea Centre, Clontusket Tea Centre, Derryad Bog No 1, Derryad Bog No 2, Derryad Bog No 3, Killashee, Derryshangoe Bog Tea Centre, Coolumber Tea Centre, Edera, Loughbannow, Cloonbonny (various spellings), Knappoge, Cloneeny, Clondara, Moher, Curraghroe, Derycashel, Grannaghan, Cloonshannagh and Derrymoylin Bogs.

For all locomotive details see the Bord na Móna fleet list at the end of this Section.

McCORMICK BROS. RAIL & MACHINERY HIRE LTD, LAUREL LODGE, TERLICKEN, BALLYMAHON

Gauge 5ft 3in

98 LD 1484		4wDM	R/R	Case		1998
98 LD 1594	99609 945008-1	4wDM	R/R	Case	JAJ2000	1998

LOUTH

INDUSTRIAL SITES

S. DUFFY PLANT HIRE LTD, JENKINSTOWN, near DUNDALK

Gauge 5ft 3in www.sduffyplanthire.ie

RRJ 02	P580 JVC	99609 976006-7	4wDM	R/R	_(Landrover (Perm	999690	1997 1997	
RRJ 03	T586 RGE	99609 976007-5	4wDM	R/R	_(Landrover (Perm		2000 2000	
RRD 02	04 LH 3097	99609 943055-4	2w-2DH	R/R	Dieci		2004	a
	77 LH 521	(RO M 186)	4wDMR	R/R	IFA		1977	OOU
TPR 01	99609 919003-4		4wDMR	R/R	Manitou	739502	2010	
	NK54 SDY 99609 943084-4		2-2wDH	R/R	_(SchörlingB (Bergman Baufix	05-1949	2005 2005	
ACL 126	NK54 SDZ 99609 943083-6		2-2wDH	R/R	_(SchörlingB (Bergman Baufix	04-1761	2005 2005	
	V493 EBC		4wDM	R/R	_(Thwaites (Rexquote	3-95209 1229	1999	

a converted dumper

Gauge 4ft 8½in

	R913 MAU	4wDM	R/R	_(Landrover	1998	
				(Perm	1998	OOU
1228	V494 EBC 99709 943047-9	4wDM	R/R	_(Thwaites 7-95400	1998	
				(Rexquote 1228		OOU

PRESERVATION SITES

RIVERSTOWN MILL RAILWAY, OLD RIVERSTOWN CORN MILL, RIVERSTOWN, near DUNDALK

Closed. Casual visitors welcome.

Gauge 4ft 8½in **J 168063**

No.4	ROBERT NELSON No.4	0-6-0ST	IC	HE	1800	1936	OOU
	(BREL 75)	4wDMR		_(BRE		c1984	
				(Leyland RB002/001		c1984	OOU

MAYO

PRESERVATION SITES

MAYO NORTH OLD ENGINE AND TRACTOR CLUB, ENNISCOE HERITAGE CENTRE, ENNISCOE HOUSE AND GARDENS, near CROSSMOLINA

Gauge 3ft 0in www.mayovintage.org

LM 129	4wDM	RH	383264	1955

WESTPORT HOUSE & COUNTRY ESTATE, QUAY ROAD, WESTPORT

Gauge 1ft 3in www.westporthouse.ie

WESTPORT HOUSE EXPRESS	2-6-0DH	s/o	SL	80.10.89	1989

MEATH

INDUSTRIAL SITES

BORD NA MÓNA, Kinnegad **N 600441**

This site is known locally by BNM employees as the Rosen Bog Works.

For all locomotive details see the Bord na Móna fleet list at the end of this Section.

BREFFNI HIRE PLANT LTD, RAYSTOWN BUSINESS PARK, ASHBOURNE

Gauge 5ft 3in www.breffnigroup.ie

Road/Rail plant may be present at this location

DIXON BROS. (AGRICULTURE & PLANT) LTD, CULLENTRA ROAD, RATHMORE, ENFIELD

Gauge 5ft 3in www.dixonbros.ie

XLR 8103	97 MH 6832 [R648 JFE]	4wDMR	R/R	_(Landrover (Raynesway	136483	1998	
XLR 8106	97 MH 6833 [R904 WBC]	4wDMR	R/R	_(Landrover (Raynesway	136517	1998	
RRJ03	T584 RGE 99609 976001-8	4wDMR	R/R	_(Landrover (Perm	163895	2000 2000	
	T302 VMA 99609 945017-2	4wDM	R/R	New Holland		1999	
	01 D 38049 99609 945003-2	4wDM	R/R	New Holland		2002	a

a also carries T302 VMA 99609 945017-2 in error

OLIVER DIXON (HEDGECUTTING & PLANT HIRE) LTD, ST. OLIVER'S ROAD, LONGWOOD

Gauge 5ft 3in

84 MH 585 99609 945004-0	4wDM	R/R	Ford		1984
89 MN 2449 99609 945002-4	4wDM	R/R	Ford	BC86514	1989
89 MH 3147 99609 945001-6	4wDM	R/R	Ford	BC63664	1989
90 MH 5687 99609 945005-7	4wDM	R/R	Ford		1990
90 WD 2041 99609 945013-1	4wDM	R/R	Fiat	265311	1990
01 WX 2649 99609 945011-5	4wDM	R/R	New Holland	160182B	2001
S137 TBC	4wDM	R/R	_(Landrover (Raynesway		1999

P.F. DIXON PLANT HIRE, RATHCORE, ENFIELD

Gauge 5ft 3in www.pfdixonplanthire.com

Plant Yard with Road/Rail vehicles usually present

NEW BOLIDEN GROUP, TARA MINES, KNOCKUMBER ROAD, NAVAN

Gauge 5ft 3in www.boliden.com

021	4wDM	R/R reb reb	Unilok Unilok Unilok	3028	1987 2014 2017

PRESERVATION SITES

BRIAN DARLINGTON
Gauge 2ft 0in

–	4wDM		LB	54183	1964

TAYTO ADVENTURE PARK, KILBREW, ASHBOURNE
Gauge 2ft 0in www.taytopark.ie

1954	4-4-0+4w-4wDH	s/o	SL	15982	2015

MONAGHAN

PRESERVATION SITES

BALLINASCARRA BRIDGE COMMITTEE,
former GNR(I) COOTHILL RAILWAY EMBANKMENT, near LISNALONG, COOTHILL
Gauge 3ft 0in **H 648163**

SLIABH gCUILLEAN	0-4-0DM	s/o	Dtz	57131	1960

OFFALY

INDUSTRIAL SITES

BORD NA MÓNA

Bellair **N 190329**

Seasonally operated site only. Serves West Offaly Power Station. Uses Blackwater/Boora based locomotives as required. Only service locomotives are permanently based here. Site serves Bellair Bog & Tip Head, Bellair North and Bellair South.

Blackwater System **N 002251**

In addition to the main works, locomotives are normally found out based at the following locations : West Offaly Power Station Sidings, Roscommon Bog Tea Centre, Area 4 Ballyhurt, Area 1 & 2 Bloomhill and circuit, Deryharney, Bunnahinley Bogs, Kylemore Lock, Cornavea Tea Centre, Blackwater P-Way Depot, Noggus Tea Centre, Derrybratt Tea Centre, Clongowney More Bog, Culliaghmore Bog, Clonfert Bridge, Cloonburren Bogs, Cloonburren Depot, Coolumber Bog Tea Centre, Curraghamore Crossing and Rahrabeg Tea Centre. The following areas also use locomotives on a regular basis : Drumlosh, Cornafulla, Clooniff, Cloonbeggan, Creggan, Lismanny and Kilgarven. A fuel train locomotive is also out based at the Old Clonfert Wagon Works to serve Kilmacshane and Garryhinch Bogs. Blackwater also supplies service and haulage locomotives to the Lemanaghan Group, Bellair Group and to the Pollagh Group of bogs as demand requires, these are administered from **Boora** (which see).

Boora System **N 181196**

Supplies locomotives to Derrinlough Briquette Plant and service locomotives to the following areas and stabling points : Pollagh Bog, Noggusboy, Cloonshask, Turraun Bog and Diamond, Killaranny Bog, Derrybratt, Falsk, Drinagh, Brackny, Oughter Bog Tea Centre and Galros Bog; the Lemanaghan Group, including Lemanaghan Tea Centre, West Boora complex and Clongowney More.

Croghan Works & Tip Head **N 515276**

Supplies locomotives for use at Croghan Tip Head and for bog area operations serving this tip head. Additional service locomotives are out based at Rathdrum and Ballintemple.

Derrygreenagh System **N 495382, 495384**

This is the largest single system of BNM, thus the whole area is split into four sub divisions. See also **Ballydermot Works** under Co Kildare, the remaining three sites are located within Co Offaly, see also **Edenderry EPL** and **Croghan Works and Tip Head**. Derrygreenagh Works supplies locomotives for service operations and acts as a feeder to supply additional peat, both to Croghan as required and to the EPL Power Station under Edenderry. Additional locations where locomotives are regularly reported include, Drumman, Derrygreenagh Tea Centre and Fuelling Point, Derrygreenagh West Stabling Point, Ballykeane Bog, Clonad Bog, Ballycon Track Depot, Cloncreen Quarry at Rathvilla, Esker Bog Tea Centre, Ballybeg Tea Centre, Toar Bog loading point and Tea Centre, Derryhinch Bog, Old Clonsast Works siding, Derries Bog, Clonkeen Tea Centre, Cavemount Tea Centre, Kilmurray Precast Works and siding at Ballybeg Bridge, Ballynakill Tea Centre, Gortnakeel Tea Centre, Rathvilla Area 8 and Derrylea Bog Tea Centre, Mount Lucas and Ballykeane.

Edenderry EPL & Power Station **N 613273**

Site supplies locomotives for use at the station of the same name. Peat for the power station is drawn from the Ballydermot and Derrygreenagh Group Bogs. Ash is delivered to Cloncreen Ash Tip and Cells where locomotives are out based for service duties as required.

Killaun Bog and Tip Head. Mobile Tip Head and Satellite system **N 094079**

Shares locomotive, wagon fleet and mobile tippler with Prosperous Bog in Co Kildare. Sites serves Killaun Tip Head with road transfer to West Offaly Power Station.

Lemanaghan **N 147259**

Seasonally operated system, uses Blackwater/Boora based locomotives as required. Only service locomotives are allocated to this area, under control from Boora Works (which see).

Monettia Bog **N 347165**

Serves local bog and tip head.

For all locomotive details see the Bord na Móna fleet list at the end of this Section.

PRESERVATION SITES

BORD NA MÓNA plc,
LOUGH BOORA PARKLANDS, LOUGH BOORA DISCOVERY PARK, BOORA

Gauge 3ft 0in www.loughboora.com

LM 23			4wDM		RH	244788	1946
LM 51	1		4wDM	s/o	RH	259191	1948
LM 127			4wDM		RH	379928	1954
LM 160	6		4wDM		RH	402176	1956
		rebuilt as	4wDH	s/o	BnM		1995
LM 181			0-4-0DM		Dtz	57123	1960

Sky Train Exhibit

Gauge 3ft 0in **N 182189**

LM171	Q	4wDM	RH	402983	1956	Pvd

Lough Boora Parklands Restoration Workshops,
c/o Bord na Möna, Boora Works, Boora

Gauge 3ft 0in **N 181196**

No.5	LM 38E	4wDM	RH	252246	1947	
LM 74	F 6	4wDM	RH	259760	1948	OOU
F 308		4wDM	BnM			OOU

BORD NA MÓNA plc,
MOUNT LUCAS WINDFARM VISITOR FACILITY, BALLYCON

Gauge 3ft 0in

LM 81	H	4wDM	RH	329686	1952	Pvd

BORD NA MÓNA VINTAGE MACHINERY MUSEUM,
BLACKWATER WORKS, SHANNONBRIDGE

Gauge 3ft 0in **N 002252**

LM 184		0-4-0DM	Dtz	57130	1960	
LM 254		0-4-0DM	Dtz	57835	1965	
LM 354		4wDH	DunEW			
LM 364		4wDH	DunEW		1984	
LM 367		4wDH	DunEW	LM367	1984	
LM 368		4wDH	DunEW	LM368	c1984	
LM 372		4wDH	DunEW			
LM 105	U	4wDM	RH	375344	1954	
LM 111		4wDM	RH	379079	1954	
–		4wDM	RH	422567	1958	
LM 343		4wDM	SMH	60SL746	1980	
C 55		2w-2PMR	Wkm	7680	1957	Dsm
C 80		4wPMR	BnM(BI)		1972	
F 630	341 NRI	4wDM	BnM			

TOM BRADY, WALSH ISLAND, GEASHILL

Gauge 3ft 0in

LM 197		0-4-0DM	Dtz	57139	1960
LM 78		4wDM	RH	329682	1952
LM 136	Q	4wDM	RH	382815	1955
C45		2w-2PMR	Wkm	7132	1955

JAMES DUFFY, BEEHIVE LODGE, WALSH ISLAND

Gauge 3ft 0in

LM 14	SIR 275	4wDM	RH	198290	1940

WALSH ISLAND HERITAGE COMMITTEE, BORD Na MÓNA GREEN, WALSH ISLAND
Gauge 3ft 0in **N 516197**

LM 28 G 12	4wDM	RH	249543	1947

ROSCOMMON

PRESERVATION SITES

ARIGNA MINING EXPERIENCE, DERREENAVOGGY, ARIGNA
Gauge 2ft 0in www.arignaminingexperience.ie

No.8023 ARIGNA MINES ST. ELBA	4wDM	L	8023	1936	Pvd
–	4wDM	MR	5861	1934	a
41	4wPM	Montania/OK	2563	1927	a

a not on public display

HELL'S KITCHEN PUBLIC HOUSE, MAIN STREET, CASTLEREA
Closed - open by appointment only; intending visitors should ring 00353 87 230 8152 for Mr Sean Browne
Gauge 5ft 3in www.hellskitchenmuseum.com

A55	Co-CoDE	MV	941	1956

SLIGO

PRESERVATION SITES

QUIRKY NIGHTS GLAMPING VILLAGE, ENNISCRONE
Gauge 4ft 8½in www.quirkyglamping.ie

62411 (1498)	4w-4wRER	York(BRE)		1971

TIPPERARY

INDUSTRIAL SITES

BORD NA MÓNA

Littleton (Ballydeath) **S 207535**

Locomotives are based at the main works and briquette Factory. Seasonal requirements may find some locomotives outbased on the Co Kilkenny / Laois area bogs. Additional locations are as follows : Longford Pass area, Popes Bog, Baunmore Bog, Ballybeg Bog, Derryvella Bog and Killeen Bog. Locomotives may also be found at Littleton Old Works, Lanespark, Inchirourke, Templetouchy and Carrick Hill Bogs.

For all locomotive details see the Bord na Móna fleet list at the end of this Section.

MIKE LYNCH EXCAVATIONS LTD, Junction of LIMERICK / TIPPERARY ROAD, MONARD

Gauge 5ft 3in

98 TS 4363	4wDMR	R/R	_(Landrover	1998
			(Harsco	1998
99609 943033-1	4wDHR	R/R	_(Thwaites1-96791	
			(Keltec	

MOLONEY AGRICULTURAL SERVICES, KILLBALLYBOY, CLOGHEEN, near CAHIR

Gauge 5ft 3in

05 TS 2115	99609 945010-7	4wDM	R/R	New Holland 196348B	2005
08 TS 3156	99609 945009-9	4wDM	R/R	New Holland ZOBD053922	2008
09 TS 1499	99609 945015-6	4wDM	R/R	New Holland 29BK03298	2009

PRESERVATION SITES

IRISH TRACTION GROUP, CARRICK-ON-SUIR STATION, CARRICK-ON-SUIR

Gauge 5ft 3in www.irishtractiongroup.com

B103	A1A-A1ADE	BRCW	DEL22	1956
G601	4wDH	Dtz	56118	1956
G616	4wDH	Dtz	57227	1961
226	Bo-BoDE	MV	972	1956

WATERFORD

PRESERVATION SITES

TRAMORE AMUSEMENT & LEISURE PARK, 'THE TRAIN', TRAMORE MINIATURE RAILWAY, STRAND ROAD, TRAMORE

Gauge 1ft 3in www.tramoreamusements.com

–			2-8-0PH	s/o	SL	22	1973

WATERFORD & SUIR VALLEY HERITAGE RAILWAY, KILMEADON STATION, KILMEADON

Gauge 3ft 0in www.wsvrailway.ie

LM 179			0-4-0DM	Dtz	57121	1960	
LM 256			0-4-0DM	Dtz	57837	1960	OOU
LM 259			0-4-0DM	Dtz	57840	1965	Pvd
No.3	ENTERPRISE		4wDM	MR	60S382	1969	
		rebuilt		AK		2000	
		rebuilt as	4wDH	AK		2004	
(LM 96)			4wDM	RH	375314	1954	
		rebuilt as	4wDH	HE	9904	2011	
LM 348			4wDM	SMH	60SL744	1980	

WESTMEATH

INDUSTRIAL SITES

BORD NA MÓNA

Ballivor **N 644541**

Additional stabling or loading points are located at : South Bog, North Bog, and Carranstown Bog. Locomotives might also be found on the Lisclogher Branch, at Robbersbush and also on the Bracklin Bog.

Coolnagun **N 384702**

Locomotives are kept at the main works and at the tip head on the Coole Road. Some locomotives may be outbased on the Coole Bog, Kiltareher bog and on the South Bog from time to time. Works are ongoing to lay a new 5 mile extension upon which locomotives may be found at Coolcraff Bog and Milkernagh Bog Tea Centre.

For locomotive details see the Bord na Móna fleet list at the end of this Section.

JOHN DIXON PLANT HIRE LTD, SARSFIELDTOWN, KILLUCAN
Gauge 5ft 3in www.johndixonplanthire.ie

88 G 6156 99609 945016-4	4wDM	R/R	Ford	BB86513	1988
90 MH 3006	4wDMR	R/R	Ford		1990
06 WH 4510 99609 976004-2	4wDMR	R/R	_(Landrover (Aquarius	713641	2006 2006
99 WH 4717 99609 976012-5	4wDMR	R/R	_(Landrover (Hy-Rail	162647 0307-B1	1999 1999

WEXFORD

INDUSTRIAL SITES

DOYLE AGRICULTURAL & RAILPLANT SERVICES LTD, BALLYCOURSEY, ENNISCOURTHY
Gauge 5ft 3in www.railplantservicesltd.ie

DA 082	90 WX 5496 99609 945006-5	4wDM	R/R	Ford		1990	Dsm
(X527 OVV)		2w-2DM	R/R	Benford	EY05HD203	2001	

Dealer of new or used locomotives and rail plant, with vehicles for hire, re-sale or rail conversion occasionally present.

PRESERVATION SITES

OFFICE OF PUBLIC WORKS, J.F.KENNEDY ARBORETUM, NEW ROSS
Gauge 2ft 0in

–	4wDM	RH	371538	1954	Pvd

WICKLOW

INDUSTRIAL SITES

MULLIGAN DISMANTLERS & SALVAGE LTD, SCARNAGH, INCH, GOREY
Gauge 3ft 0in www.mulliganmetal.com **T 220699**

LM 141	4wDM	RH	392142	1955	a

a currently under renovation at Christie Dunn Plant Ltd, Den Naul, Meath

BORD NA MÓNA

BORD NA MÓNA plc — IRISH TURF BOARD

The Bord operates peat bogs throughout the Irish Midlands, with locomotives kept at the locations listed below. (See also County Headings for additional sub headings within each area group or site).

Various of the larger sites (e.g. Bl, Bo, Dg & M) have constructed their own locomotives & railcars

www.bordnamona.ie

Al	Almhain North, Co.Kildare	N 801211
Bd	Ballydermot, Co.Kildare	N 659216, 656215
Be	Bellair, Co.Offaly	N 190329
Bi	Ballivor, Co.Westmeath	N 644541
Bl	Blackwater, Co.Offaly	N 002251
Bo	Boora, Co.Offaly	N 181196
Cg	Coolnagun, Co.West Meath	N 384702
Cr	Croghan, Co.Offaly	N 515276
Cm	Coolnamona, Co.Laois	S 456947
De	Derryfadda, Co.Galway	M 800432
Dg	Derrygreenagh, Co.Offaly	N 495382, 495384
Ed	Edenderry, Co.Offaly	N 613273
Gi	Gilltown Landsale, Co.Kildare	N 796338
K	Kilberry, Co.Kildare	S 664998
Kd	Kinnegad, Co.Meath	N 600441
Kn	Killaun, Co.Offaly	N 094079
Le	Lemanaghan, Co.Offaly	N 147259
Li	Littleton, Co.Tipperary	S 207535
Lu	Lullymore, Co.Kildare	N 711292, 708286, 706286
M	Mountdillon, Co.Longford	N 048688
Mo	Monettia, Co.Offaly	N 347165
Pr	Prosperous, Co.Kildare	N 831292
Te	Templetouhy, Co.Killkenny	S 228677
Ti	Timahoe, Co.Kildare **R.T.C.**	N 781284, 754335
U	Ummeras, Co.Kildare	N 623147

LOCO FLEET LIST

Gauge 3ft 0in

18			4wDM	RH	211687	1941	OOU	Lu
LM 30	G		4wDM	RH	249545	1947	OOU	Gi
(LM 34)	No.4		4wDM	RH	252239	1947		Dg
LM 35			4wDM	RH	252240	1947	Dsm	U
LM 36	E 2		4wDM	RH	252241	1947		
		rebuilt as	4wDH	BnM			OOU	M
LM 37	E		4wDM	RH	252245	1947		M
LM 39	E		4wDM	RH	252247	1947	OOU	M
LM 40	E		4wDM	RH	252251	1947		
		rebuilt as	4wDH	BnM(M)		1991		M
LM 41	E (No.1)		4wDM	RH	252252	1947		Dg
LM 46			4wDM	RH	259184	1948		
		rebuilt as	4wDH	BnM				Dg

LM 47	F			4wDM	RH	259185	1948			
			rebuilt as	4wDH	BnM					U
LM 48	F			4wDM	RH	259186	1948	OOU		U
LM 50				4wDM	RH	259190	1948	OOU		Cg
LM 52	F	15		4wDM	RH	259196	1948	OOU		U
LM 57				4wDM	RH	259203	1948			
			rebuilt as	4wDH	BnM(Dg)					Gi
LM 62				4wDM	RH	259743	1948			
			rebuilt as	4wDH	BnM(U)					Kd
LM 63	F			4wDM	RH	259744	1948	Dsm		Bi
(LM 65)				4wDM	RH	259749	1948	OOU		Li
LM 66	5			4wDM	RH	259750	1948			M
LM 67				4wDM	RH	259751	1948	Dsm		Cg
LM 69	F			4wDM	RH	259755	1948			Gi
LM 70				4wDM	RH	259756	1948	Dsm		Dg
LM 71	F			4wDM	RH	259757	1948	OOU		M
(LM 72	F)			4wDM	RH	259758	1948	Dsm		M
LM 73	8			4wDM	RH	259759	1948	OOU		M
LM 83				4wDM	RH	329690	1952			
			rebuilt as	4wDH	BnM(K)		1987	OOU		U
LM 86	J			4wDM	RH	329695	1952	OOU		M
LM 94	T			4wDM	RH	373377	1954	OOU		U
LM 97	T			4wDM	RH	375332	1954	Dsm		Lu
(LM 99)				4wDM	RH	375336	1954			
			rebuilt as	4wDH	BnM(Be)		1988			K
LM 100				4wDM	RH	375341	1954	OOU		Bd
LM 109				4wDM	RH	379064	1954	OOU		Bo
(LM 113)	U			4wDM	RH	379068	1954	OOU		Dg
LM 118				4wDM	RH	379913	1954	Dsm		Cn
LM 121				4wDM	RH	379922	1954			
			rebuilt as	4wDH	BnM					Bd
LM 123				4wDM	RH	379925	1954			
			rebuilt as	4wDH	BnM		1996			Pr
(LM 126)				4wDM	RH	379927	1954			Bo
LM 128	X			4wDM	RH	383260	1955	OOU		Dg
LM 130	P			4wDM	RH	382807	1955			
			rebuilt as	4wDH	BnM(K)		1993			K
LM 132	P			4wDM	RH	382809	1955	OOU		Kd
LM 134	Q			4wDM	RH	382811	1955	OOU		Ti
LM 135	Q			4wDM	RH	382814	1955	OOU	a	M
LM 143				4wDM	RH	392148	1956	Dsm		K
LM 144	Q			4wDM	RH	392149	1956	Dsm		Dg
(LM 145)				4wDM	RH	394023	1956	Dsm		Dg
LM 148	X			4wDM	RH	394026	1956			
			rebuilt as	4wDH	BnM					Al
LM 149	X			4wDM	RH	394028	1956			
			rebuilt as	4wDH	BnM(Dg)					Dg
LM 150	X			4wDM	RH	394027	1956			
			rebuilt as	4wDH	BnM(Dg)			OOU		Dg
LM 153	Q			4wDM	RH	394029	1956			
			rebuilt as	4wDH	BnM					Bd

LM 154	X		4wDM	RH	394030	1956			
		rebuilt as	4wDH	BnM(Bl)		2003			Bl
LM 155	X		4wDM	RH	394031	1956	OOU		M
LM 156	X		4wDM	RH	394032	1956			
		rebuilt as	4wDH	BnM(Dg)					Cr
LM 157	X		4wDM	RH	394033	1956			
		rebuilt as	4wDH	BnM(Dg)					Cr
LM 158			4wDM	RH	394034	1956	Dsm		Dg
LM 159	X		4wDM	RH	402174	1956			
		rebuilt as	4wDH	BnM		2001			Pr
LM 163	X		4wDM	RH	402178	1956			U
LM 164	Q 2		4wDM	RH	392152	1956			Dg
LM 165			4wDM	RH	402179	1956			
		rebuilt as	4wDH	BnM(Dg)					Cr
LM 166			4wDM	RH	402977	1956			
		rebuilt as	4wDH	BnM(Bi)		2014	OOU		Kd
LM 169	Q		4wDM	RH	402981	1956	Dsm		Dg
LM 173			4wDM	RH	402985	1957			Dg
LM 176			0-4-0DM	BnM		1961			
		rebuilt as	4wDH	BnM(Dg)		1999			
		rebuilt		BnM(Dg)		2016			Ed
LM 199			0-4-0DM	HE	6232	1962			
		rebuilt as	4wDH	BnM(Dg)		1995			Dg
LM 200			0-4-0DM	HE	6233	1962			Bo
LM 201			0-4-0DM	HE	6234	1962	OOU		Dg
LM 202			0-4-0DM	HE	6235	1962	OOU		Dg
LM 203			0-4-0DM	HE	6236	1962			
		rebuilt as	0-4-0DH	BnM(Bo)		2008			Bi
LM 204			0-4-0DM	HE	6237	1963			
		rebuilt as	4wDH	BnM(Dg)		1993			
		rebuilt		BnM(Dg)		2018			M
LM 205	A10		0-4-0DM	HE	6238	1963	OOU		Dg
(LM 206)	LM 335		0-4-0DM	HE	6239	1963			
		rebuilt as	0-4-0DH	BnM(Bo)		2008		b	Cr
LM 207			0-4-0DM	HE	6240	1963	OOU		Bi
LM 208			0-4-0DM	HE	6241	1963			
		rebuilt as	4wDH	BnM(Dg)		1995			Dg
LM 209			0-4-0DM	HE	6242	1963	OOU		M
LM 210			0-4-0DM	HE	6243	1963			
		rebuilt as	4wDH	BnM(Dg)		1992			Dg
LM 211			0-4-0DM	HE	6244	1963			
		rebuilt as	0-4-0DH	BnM(Bo)		2010			K
LM 212			0-4-0DM	HE	6245	1963			
		rebuilt as	0-4-0DH	BnM(Bo)		2012			Bl
LM 213			0-4-0DM	HE	6246	1963			Dg
(LM 214)			0-4-0DM	HE	6247	1963			Kd
LM 215			0-4-0DM	HE	6248	1963			Dg
LM 216			0-4-0DM	HE	6249	1963	Dsm		Gi
LM 217			0-4-0DM	HE	6250	1963	OOU		Bl
LM 218			0-4-0DM	HE	6251	1963			
		rebuilt as	4wDH	BnM(Dg)		1994			De
LM 219			0-4-0DM	HE	6252	1963			Bo

LM 220			0-4-0DM	HE	6253	1963		Bi
LM 221			0-4-0DM	HE	6254	1963		Bd
LM 222			0-4-0DM	HE	6255	1963		Cr
LM 223			0-4-0DM	HE	6256	1963		
		rebuilt as	4wDH	BnM(Dg)		1993		Cr
LM 225			0-4-0DM	HE	6304	1964		
		rebuilt as	4wDH	BnM(Dg)		1994		
		rebuilt		BnM(Dg)		2018		Dg
LM 226			0-4-0DM	HE	6305	1964		
		rebuilt as	4wDH	BnM(M)		1994		
		rebuilt		BnM(Dg)		2018		M
LM 227			0-4-0DM	HE	6306	1964		Bl
LM 228			0-4-0DM	HE	6307	1964		Mo
LM 229	(LM 294)		0-4-0DM	HE	8531	1977	OOU	Bl
LM 230			0-4-0DM	HE	6309	1964	OOU	Bl
LM 231			0-4-0DM	HE	6310	1964		Bl
LM 232			0-4-0DM	HE	6311	1964	OOU	Dg
LM 233			0-4-0DM	HE	6312	1965		Bl
LM 234			0-4-0DM	HE	6313	1965	OOU	Bl
LM 235			0-4-0DM	HE	6314	1965		Mo
LM 236			0-4-0DM	HE	6315	1965		
		rebuilt as	0-4-0DH	BnM(Bo)		2010		Bi
LM 237			0-4-0DM	HE	6316	1965		Cm
LM 238			0-4-0DM	HE	6318	1965		
		rebuilt as	4wDH	BnM(Dg)		2000		
		rebuilt		BnM(Dg)		2017		Ed
LM 239			0-4-0DM	HE	6317	1965		Al
LM 240			0-4-0DM	HE	6319	1965		
		rebuilt as	0-4-0DH	BnM(Bo)		2008		Gi
LM 242			0-4-0DM	HE	6321	1965	OOU	Dg
LM 243	A3		0-4-0DM	HE	6322	1965	Dsm	Dg
LM 244			0-4-0DM	HE	6323	1965		Bl
LM 245			0-4-0DM	HE	6324	1965		
		rebuilt as	0-4-0DH	BnM(Bo)		2012		Kn
LM 246			0-4-0DM	HE	6325	1965		Mo
LM 247			0-4-0DM	HE	6326	1965		Dg
LM 248			0-4-0DM	HE	6328	1965		
		rebuilt as	0-4-0DH	BnM(Bo)		2015		M
LM 249			0-4-0DM	HE	6327	1965		M
LM 250			0-4-0DM	HE	6329	1965	OOU	M
LM 251			0-4-0DM	HE	6330	1965	Dsm	Bi
LM 252			0-4-0DM	HE	6331	1965		
		rebuilt as	4wDH	BnM		2000		
		rebuilt		BnM(Dg)		2017		M
(LM 255)			0-4-0DM	Dtz	57838	1965	OOU	Cg
LM 261			0-4-0DM	Dtz	57842	1965	Dsm	Bl
LM 266			0-4-0DM	HE	7232	1971		Bo
LM 267			0-4-0DM	HE	7233	1971		M
LM 268			0-4-0DM	HE	7234	1971	OOU	Li
LM 269			0-4-0DM	HE	7235	1971		Bo

LM 270			0-4-0DM	HE	7237	1971		
		rebuilt as	4wDH	BnM		2000		
		rebuilt		BnM(Dg)		2017		M
LM 271			0-4-0DM	HE	7236	1971		Li
LM 272			0-4-0DM	HE	7239	1971	OOU	M
LM 273			0-4-0DM	HE	7246	1972	OOU	Bl
LM 274			0-4-0DM	HE	7238	1971		
		rebuilt as	0-4-0DH	BnM(Bo)		2012		Pr
LM 275	(LM 328)		0-4-0DM	HE	7240	1972	Dsm	Bo
(LM 276)			0-4-0DM	HE	7241	1972	Dsm	Bl
LM 277			0-4-0DM	HE	7242	1972		Be
LM 278			0-4-0DM	HE	7243	1972	OOU	Bo
LM 280			0-4-0DM	HE	7245	1972		Bl
LM 281			0-4-0DM	HE	7247	1972		Bl
LM 282			0-4-0DM	HE	7248	1972		Bl
LM 283			0-4-0DM	HE	7250	1972		Bo
LM 284			0-4-0DM	HE	7249	1972		
		rebuilt as	0-4-0DH	BnM(Bo)		2009		Al
LM 285			0-4-0DM	HE	7253	1972		Dg
LM 286			0-4-0DM	HE	7254	1972		Bl
LM 287			0-4-0DM	HE	7255	1972		Dg
LM 288			0-4-0DM	HE	7256	1972		Bl
LM 289			0-4-0DM	HE	7252	1972		U
LM 290			0-4-0DM	HE	7251	1972		Bd
LM 292			0-4-0DM	HE	8529	1977	Dsm	Bl
LM 293			0-4-0DM	HE	8530	1977		Bl
(LM 294	LM 229)		0-4-0DM	HE	6308	1964	Dsm	Bl
LM 295			0-4-0DM	HE	8532	1977		
		rebuilt as	4wDH	BnM(Bl)		1993		Bl
LM 296			0-4-0DM	HE	8534	1977	OOU	Bl
LM 297			0-4-0DM	HE	8533	1977		Kn
LM 298			0-4-0DM	HE	8538	1977		Bo
LM 299			0-4-0DM	HE	8537	1977		Bl
LM 300			0-4-0DM	HE	8535	1977	OOU	M
LM 301			0-4-0DM	HE	8536	1977		
		rebuilt as	0-4-0DH	BnM(Bo)		2009		Kd
LM 302			0-4-0DM	HE	8539	1977		
		rebuilt as	0-4-0DH	BnM(Bo)		2015		M
LM 303			0-4-0DM	HE	8540	1977		
		rebuilt as	4wDH	BnM(Dg)		2000		
		rebuilt		BnM(Dg)		2017		Ed
LM 304			0-4-0DM	HE	8543	1977		Kn
LM 305			0-4-0DM	HE	8544	1977		
		rebuilt as	0-4-0DH	BnM(Bo)		2015		Cr
LM 306			0-4-0DM	HE	8541	1977		Cm
LM 307			0-4-0DM	HE	8542	1977		
		rebuilt as	0-4-0DH	BnM(Bo)		2010		Cm
LM 308			0-4-0DM	HE	8546	1977	OOU	Ed
LM 310			0-4-0DM	HE	8547	1977	Dsm	Dg
LM 311			0-4-0DM	HE	8551	1977		Li
LM 311	(LM 332)		0-4-0DM	HE	8930	1980	OOU	De

LM 312			0-4-0DM	HE	8550	1977		
		rebuilt as	0-4-0DH	BnM(Bo)		2014		Bo
LM 313			0-4-0DM	HE	8549	1977		Bl
LM 315			0-4-0DM	HE	8922	1979		Bl
LM 318			0-4-0DM	HE	8925	1979		Cg
LM 319			0-4-0DM	HE	8926	1979		Bo
LM 320			0-4-0DM	HE	8939	1980		Bl
LM 321	TURBO		0-4-0DM	HE	8927	1977		
		rebuilt as	0-4-0DH	BnM(Bo)		2014		Bl
LM 322	PHOENIX		0-4-0DM	HE	8923	1979		Bl
LM 323			0-4-0DM	HE	8924	1979	OOU	Bl
LM 324			0-4-0DM	HE	8931	1980		Dg
LM 325			0-4-0DM	HE	8942	1981		Li
LM 326			0-4-0DM	HE	8932	1980		De
LM 327			0-4-0DM	HE	8933	1980		
		rebuilt as	0-4-0DH	BnM(Bo)		2010		Bi
LM 328	(LM 275)		0-4-0DM	HE	8935	1980		
		rebuilt as	0-4-0DH	BnM(Bo)		2014		K
LM 329			0-4-0DM	HE	8936	1980		Li
LM 330			0-4-0DM	HE	8937	1980	OOU	M
LM 331			0-4-0DM	HE	8934	1980		De
LM 333			0-4-0DM	HE	8940	1980	OOU	Li
LM 334			0-4-0DM	HE	8941	1981		De
LM 335			0-4-0DM	HE	8938	1980		Cm
LM 336			0-4-0DM	HE	8943	1981		Bo
LM 337			0-4-0DM	HE	8944	1981		Dg
LM 338			0-4-0DM	HE	8945	1981		Li
LM 339			0-4-0DM	HE	8946	1981		
		rebuilt as	0-4-0DH	BnM(Bo)		2010		Ed
LM 340			0-4-0DM	HE	8928	1980		K
LM 342			0-4-0DM	HE	8929	1980		
		rebuilt as	0-4-0DH	BnM(Bo)		2014		Bl
LM 355			4wDH	DunEW			OOU	M
LM 357			4wDH	DunEW			Dsm	Bl
LM 358			4wDH	DunEW				Bi
LM 359			4wDH	DunEW			OOU	M
LM 365			4wDH	DunEW	T4494	1984	OOU	Bl
LM 374			4wDH	HE	9239	1984		Cg
LM 375			4wDH	HE	9240	1984		M
LM 376			4wDH	HE	9241	1984		Cg
LM 377			4wDH	HE	9243	1984		Cg
LM 378			4wDH	HE	9242	1984		De
LM 379			4wDH	HE	9251	1985		Li
LM 380			4wDH	HE	9252	1985		Li
LM 381			4wDH	HE	9253	1985		Bl
LM 382			4wDH	HE	9254	1986		Dg
LM 383			4wDH	HE	9255	1986		Dg
LM 384			4wDH	HE	9256	1986		Cg
LM 385			4wDH	HE	9257	1986		M
LM 386			4wDH	HE	9258	1986		De

LM 387		4wDH	HE	9259	1986		Cm
LM 388		4wDH	HE	9272	1986	OOU	Li
LM 389		4wDH	BnM(Bl)		1994		Bl
LM 390		4wDH	BnM(Bl)		1994		De
LM 391		4wDH	BnM(Bo)		1994		Bo
LM 392		4wDH	BnM(Bo)		1995		Bl
LM 393		4wDH	BnM(Bo)		1995		Bo
LM 394		4wDH	BnM(Bo)		1996		Bo
LM 395		4wDH	BnM(Bl)		1995		
	rebuilt		BnM(Dg)		2018		Dg
LM 396		4wDH	BnM(Bl)		1995		Bo
LM 397		4wDH	BnM(Bo)		2005		Li
LM 398		4wDH	BnM(Bo)		2005		
	rebuilt		BnM(Dg)		2018		Bo
LM 399		4wDH	BnM(Bo)		2004		Bl
LM 400		4wDH	BnM(Bo)		2004		
	rebuilt		BnM(Dg)		2018		Dg
LM 401		4wDH	BnM(M)		1996		M
LM 402		4wDH	BnM(M)		1996		M
LM 403		4wDH	BnM(De)		2000		
	rebuilt		BnM(Dg)		2017		Bl
LM 404		4wDH	BnM(De)		2000		
	rebuilt		BnM(Dg)		2018		M
LM 405		4wDH	BnM(Bl)		2000		
	rebuilt		BnM(Dg)		2016		Bl
LM 406		4wDH	BnM(Bl)		2000		
	rebuilt		BnM(Dg)		2016		Dg
LM 407		4wDH	BnM(Li)		2002		Li
LM 408		4wDH	BnM(Bl)		2001		
	rebuilt		BnM(Dg)		2017		Bl
LM 409		4wDH	BnM(Bl)		2001		
	rebuilt		BnM(Dg)		2017		Bl
LM 410		4wDH	BnM(Bo)		1998		
	rebuilt		BnM(Dg)		2017		Bl
LM 411		4wDH	BnM(Bo)		1999		
	rebuilt		BnM(Dg)		2016		Bl
LM 412		4wDH	BnM(Dg)		2000		
	rebuilt		BnM(Dg)		2018		Dg
LM 413		4wDH	BnM(Dg)		2000		
	rebuilt		BnM(Dg)		2015		
	rebuilt		BnM(Dg)		2017		Ed
LM 414		4wDH	BnM(Dg)		2000		Ed
LM 415		4wDH	BnM(Dg)		2000		
	rebuilt		BnM(Dg)		2017		Ed
LM 416		4wDH	BnM(Bo)		2000		
	rebuilt		BnM(Dg)		2017		Dg
LM 417		4wDH	BnM(Bo)		2000		
	rebuilt		BnM(Dg)		2017		Ed
(LM 418)		4wDH	BnM(Bo)		2000	OOU	Ed
LM 419		4wDH	BnM(Bo)		2000		Dg
LM 420		4wDH	BnM(Bo)		2000		
	rebuilt		BnM(Dg)		2017		Ed

LM 421		4wDH	BnM(Bo)		2000		
	rebuilt		BnM(Dg)		2017		Dg
LM 422		4wDH	BnM(De)		2000		
	rebuilt		BnM(Dg)		2017		Ed
LM 423		4wDH	BnM(De)		2000		
	rebuilt		BnM(Dg)		2018		Dg
LM 424		4wDH	BnM(De)		2000		
	rebuilt		BnM(Dg)		2017		Ed
LM 425		4wDH	BnM(Bl)		2002		
	rebuilt		BnM(Dg)		2016		Bl
LM 426		4wDH	BnM(Bl)		2002		
	rebuilt		BnM(Dg)		2018		Le
LM 427		4wDH	BnM(M)		2002		
	rebuilt		BnM(Dg)		2017		M
LM 428		4wDH	BnM(M)		2001		
	rebuilt		BnM(Dg)		2017		M
LM 429		4wDH	BnM(Bo)		2004		
	rebuilt		BnM(Dg)		2018		Bl
LM 430		4wDH	BnM(Bo)		2004		
	rebuilt		BnM(Dg)		2018		Bl
LM 431		4wDH	BnM(Dg)		2004		
	rebuilt		BnM(Dg)		2017		Dg
LM 432		4wDH	BnM(Dg)		2004		
	rebuilt		BnM(Dg)		2018		Dg
LM 433		4wDH	BnM(Dg)		2004		
	rebuilt		BnM(Dg)		2018		Dg
LM 434		4wDH	BnM(Dg)		2004		Dg
LM 435		4wDH	BnM(Bo)		2005		
	rebuilt		BnM(Dg)		2017		Bl
LM 436		4wDH	BnM(Bo)		2005		
	rebuilt		BnM(Dg)		2018		Cr
LM 437		4wDH	BnM(Bo)		2006		Li
LM 438		4wDH	BnM(Bo)		2006		
	rebuilt		BnM(Dg)		2017		M
LM 439		4wDH	BnM(Bo)		2006		
	rebuilt		BnM(Dg)		2017		Bo
LM 440		4wDH	BnM(Bo)		2006		
	rebuilt		BnM(Dg)		2017		Bo
LM 441		4wDH	BnM(Bo)		2007		
	rebuilt		BnM(Dg)		2017		Dg
LM 442		4wDH	BnM(Bo)		2011		
	rebuilt		BnM(Dg)		2016		M
C 49		4wPMR	BnM	5	1958	OOU c	M
C 65		2w-2PMR	BnM		1972	DsmT	M
(C 71)		4wPMR	BnM		1972	DsmT d	M
C 77		4wPMR	BnM(Bl)		1972	OOU	Cg
F 230		4wDM	BnM			Dsm	Bl
F 836		4wDM	BnM(Bl)		2008		Bl
F 353		4wDM	BnM		1983		Bl
F 842		4wDM	BnM				Bl
F 866		4wDM	BnM		c1975		Bl

F 878	4wDM	BnM	1987			Bl
(F xxx)	4wDM	BnM		Dsm	e	Bl
RM 1	4wDH	BnM(Bl)	2000			Bl
RL 1	4wDH	BnM(M)		OOU		M
(RL 2)	4wDH	BnM(TAE/M)	2003			Bo
RM 3	4wDH	BnM(M)	2004			M
RM 4	4wDH	BnM(M)	2004			Dg
RM 5	4wDH	BnM(M)	2004			M
RM 6	4wDH	BnM(M)	2004			Bl

a carries plate 382841
b incorrectly numbered following rebuilding
c dumped at Loading Point on Derryaroge Bog (Mountdillon System) N 028718
d rebuilt as Ambulance car
e converted to a wagon for use with RM 1

SECTION 5 – OFFSHORE ISLANDS

CHANNEL ISLANDS

PRESERVATION SITES

ALDERNEY RAILWAY SOCIETY, MANNEZ QUARRY, ALDERNEY

Gauge 4ft 8½in www.alderneyrailway.com **601087**

(6	ALD 40)	4wVBT	VCG	S	6909	1927	Dsm
ALD 7	MOLLY 2	4wDM		RH	425481	1958	
D100	ELIZABETH	0-4-0DM		_(VF ¯(DC	D100 2271	1949 1949	
1044		2-2w-2w-2RER		MetCam		1960	
1045		2-2w-2w-2RER		MetCam		1960	
(PWM 3776)		2w-2PMR		Wkm	6655	1953	DsmT
PWM 3954	MARY LOU	2w-2PMR		Wkm	6939	1955	a
1	GEORGE RLC/009025	2w-2PMR		Wkm	7091	1955	
(8 9028)	SHIRLEY	2w-2PMR		Wkm	7094	1955	
7 (9022)		2w-2PMR		Wkm	8086	1958	
(9)		2w-2PMR		Wkm	9359	1963	

a currently in store c/o Nick Best, National Westminster Bank Ltd, Victoria Street, St.Annes

THE L.C. PALLOT TRUST, THE PALLOT STEAM, MOTOR & GENERAL MUSEUM, RUE DE BECHET, TRINITY, JERSEY JE3 5BE

Gauge 4ft 8½in www.pallotmuseum.co.uk **652532**

	LA MEUSE	0-6-0T	OC	LaMeuse	3442	1933	
–		0-4-0ST	OC	P	2085	1948	
–		0-4-0ST	OC	P	2129	1952	
	J.T.DALY	0-4-0ST	OC	WB	2450	1931	
(D1)		0-4-0DH		NBQ	27734	1958	

Gauge 2ft 0in

–		4wDM	s/o	MR	11143	1960	
–		4wDM		MR	60S383	1969	Dsm a

a in use as a brake van

ISLE OF MAN

INDUSTRIAL SITES

AULDYN CONSTRUCTION LTD, PLANT YARD, off PEEL ROAD, BRADDAN, ST. JOHNS (IM4 4LH)

Gauge 3ft 6in

	SHERPA	4wDM	R/R	_(Thwaites	18-93130	1997
				(Rexquote	1149	1997

Gauge 3ft 0in

LM 344	'PIGNITAS'	4wDM		SMH	60SL751	1980
			reb	BoothWKelly		2008
LM 350		4wDM		SMH	60SL748	1980
	SAINT BERNARD	4wDM	R/R	Thwaites		

Railway contractors to Isle of Man Steam Railway, Manx Electric Railway and Snaefell Mountain Railway

NATIONAL AIR TRAFFIC SERVICES LTD, LAXEY

Gauge 3ft 6in www.nats.aero **SC 432847**

–	4wDMR		Wkm	10956	1976
–	4wDHR		WkmR	11730	1991
		reb	CE	B4630	2018

TXM PLANT LTD, c/o MANX ELECTRIC RAILWAY, DHOON QUARRY, MAUGHOLD

Gauge 3ft 0in **SC 456287**

4137	V71 KCV 99709 943032-1	4wDM	R/R	_(Thwaites	1-96542	1999
				(Rexquote	1323	2000

PRESERVATION SITES

J. EDWARDS, BALLAKILLINGAN HOUSE, CHURCHTOWN, near RAMSEY

The locomotive is stored at a private site and is not available for viewing.

Gauge 3ft 0in **SC 425945**

No.14	THORNHILL	2-4-0T	OC	BP	2028	1880

GROUDLE GLEN RAILWAY CO LTD, LHEN COAN, GROUDLE GLEN, KING EDWARD ROAD, ONCHAN

Gauge 2ft 0in www.ggr.org.uk **SC 418786**

	ANNIE	0-4-2T	OC	BoothR		1997

	"BROWN BEAR"	2-4-0T	OC	GGR				a
	SEA LION	2-4-0T	OC	WB	1484	1896		
			reb	‡		1987		
	MALTBY	0-4-0DM	s/o	Bg	3232	1947		
			reb	NBRES		2017		
1	DOLPHIN	4wDM		HE	4394	1952		
2	WALRUS	4wDM		HE	4395	1952		
No.313		4wBE		WR	556801	1988		
			reb	AK	72R	2004		b

‡ rebuilt by British Nuclear Fuels Ltd, Windscale Factory, Sellafield, Cumbria
a new loco under construction
b carries worksplate BEV 313

ISLE OF MAN TRANSPORT, DEPARTMENT OF TOURISM & TRANSPORT

www.visitisleofman.com

Isle of Man Steam Railway

Locomotives are kept at : Douglas IM1 1BR SC 374754, 375755
Port Erin IM8 2WE SC 198689

Gauge 3ft 0in

No.1	SUTHERLAND	2-4-0T	OC	BP	1253	1873	Dsm	
No.4	LOCH	2-4-0T	OC	BP	1416	1874	OOU	
No.5	(MONA)	2-4-0T	OC	BP	1417	1874	OOU	a
No.8	(FENELLA)	2-4-0T	OC	BP	3610	1894		
No.9	DOUGLAS	2-4-0T	OC	BP	3815	1896	OOU	a
No.10	G.H. WOOD	2-4-0T	OC	BP	4662	1905		
No.11	MAITLAND	2-4-0T	OC	BP	4663	1905	OOU	b
No.12	HUTCHINSON	2-4-0T	OC	BP	5126	1908		
No.13	KISSACK	2-4-0T	OC	BP	5382	1910		
	CALEDONIA No.15	0-6-0T	OC	D	2178	1885		
No.17	VIKING	4wDH		Schöma	2086	1958	OOU	
18	AILSA	4wDH		_(HAB	770	1990		
				(HE	9446	1990		
			reb	HE	9342	1995		
19		0-4-0+4DMR		WkB/Dundalk		1950		
20		0-4-0+4DMR		WkB/Dundalk		1951		
21		Bo-BoDE		MPES	550/1	2013		c
–		4wDM		MR	22021	1959		d
–		4wDM		MR	40S280	1966		e
–		2w-2PMR		Wkm	8849	1961		

a in store for Isle of Man Railway & Tramway Preservation Society
b currently at Alan Keef Ltd, Lea Line, Ross-on-Wye, Herefordshire
c uses parts from an unidentified General Electric locomotive;
d based at Port Erin
e based at Douglas

Manx Electric Railway, Laxey Car Sheds, Laxey — IM4 7AZ

Gauge 3ft 0in — **SC 432846**

No.23	DR.R.PRESTON HENDRY	4w-4wWE		MER		1900		a
34		4w-4wDE		MER		2004	OOU	

No.2	2w-2PMR		Wkm	7442	1956	

a stored on un-motorised bogies, property of Isle of Man Railway & Tramway Presevation Society

Port Erin Railway Museum, Railway Station, Port Erin **IM8 2WE**
Gauge 3ft 0in **SC 198689**

No.6 PEVERIL	2-4-0T	OC	BP	1524	1875	
16 MANNIN	2-4-0T	OC	BP	6296	1926	

LAXEY & LONAN HERITAGE TRUST,
GREAT LAXEY MINE RAILWAY, off MINES ROAD, LAXEY **IM4 7NL**
Gauge 1ft7in www.laxeyminerailway.im **SC 433847**

ANT	0-4-0T	IC	GNS	20	2004	a
BEE	0-4-0T	IC	GNS	21	2004	b
WASP 263.006	4wBE		CE	B0152	1973	

a carries worksplate Lewin 684
b carries worksplate Lewin 685

MANX TRANSPORT TRUST LTD, JURBY TRANSPORT MUSEUM,
HANGAR 230, JURBY INDUSTRIAL ESTATE, JURBY **IM7 3BD**
Gauge 3ft 0in www.jtmiom.im **SC 360988**

–	4wPM	s/o	FH	2027	1937	

SECTION 6
MINISTRY OF DEFENCE

DEPOT TYPES :

DM	Defence Munitions
DSDC	Defence Storage & Distribution Centre

LOCATIONS :

ASH	DSDC Ashchurch, Gloucestershire (**R.T.C.**)	SO 932338
BIS	DSDC Bicester, Oxfordshire	SP 581203
CD	Copehill Down Battle Training Area, Salisbury Plain, Wiltshire	SU 014455
GD	DM Glen Douglas, near Arrochar, Argyll & Bute	NS 279998
KIN	DM Kineton, Warwickshire	SP 373523, 374524
LON	DM Longtown, Cumbria	NY 363682
LUD	Ludgershall Railhead, Wiltshire	SU 261507
LUL	Lulworth Ranges, East Lulworth, Dorset	SY 863822

LOCOMOTIVES

Gauge 4ft 8½in

01510 (272)		4wDH		TH	V320	1987		LON
01511 (275)		4wDH		TH	V323	1988		LON
01512 (301)	CONDUCTOR	4wDH		TH	V319	1988		
			rep	LH Group	76699	2002		BIS
01513 (302)	GREENSLEEVES	4wDH		TH	V318	1987		
			rep	LH Group	76638	2002		KIN
01514 (303)		4wDH		TH	V332	1988		
			rep	LH Group	76634	2002		LON
01521 (278)	FLACK	4wDH		TH	V333	1988		BIS
01522 (254)		4wDH		TH	272V	1977		GD
01524 (261)		4wDH		TH	301V	1982		LUD
01525 (264)	DRAPER	4wDH		TH	306V	1983		BIS
01527 (256)		4wDH		TH	274V	1977		BIS
01528 (267)		4wDH		TH	309V	1984		KIN
01548 (257)		4wDH		TH	275V	1978		LUD
01549 (258)		4wDH		TH	298V	1981		KIN
01550 (271)	STOREMAN	4wDH		TH	V324	1987		BIS
430		0-6-0DH		RH	466621	1961	OOU	CD
	ANNA	4wRE		Wkm	11547	1987		LUL
	FIONA	4wRE		Wkm	11548	1987		LUL
	BELLA	4wRE		Wkm	11549	1987		LUL
	DEBBIE	4wRE		Wkm	11550	1987		LUL
	ENID	4wRE		Wkm	11551	1987		LUL
	CLAIRE	4wRE		Wkm	11552	1987		LUL
–		4wDMR		Wkm	11621	1986		LUL
9121		4wDMR		BD	3710	1975		KIN
(9123)	1	4wDMR		BD	3712	1975	DsmT	KIN
9128	THE HORNET	4wDMR		BD	3744	1976		LON
9129		4wDMR		BD	3745	1976		BIS

761	(TNS 102)	2w-2DMR		Robel	56.27-3.AF33	1983		LON
764	(TNS 101)	2w-2DMR		Robel	56.27-10-AF32	1983		ASH
08 CP 05		4wDM	R/R	_(Ford (Wkm	 11618	1986 1986		KIN
50 KM 19	(DRC 730J KAR 536V)	4wDM	R/R	Unimog	004971	1970	OOU	KIN

SECTION 7
NETWORK RAIL, TOCS, FOCS and CONTRACTORS

All vehicles listed within this section have a gauge of 4ft 8½in unless otherwise indicated.

This section now includes internal depot based locomotives operated by Train Operating Companies (TOCS) and Freight Operating Companies (FOCS) which were previously listed in the main text of the EL Series books.

LOCOMOTIVES

MULTIPLE UNITS

Only the powered vehicles of multiple-unit sets are included in the lists.

PLANT

Vehicles listed are used primarily for material handling and/or personnel carrying. Track Maintenance Machines with these facilities are therefore not included.

The following abbreviations are used :-

CE	Civil Engineers
CTRL	Channel Tunnel Rail Link
C&W	Carriage and Wagon
EMD	Electric Maintenance Depot
EMUD	Electric Multiple Unit Depot
GWML	Great Western Main Line
IMD	Infrastructure Maintenance Depot
LMD	Light Maintenance Depot
MDU	Maintenance Delivery Unit
OHLM	Overhead Line Maintenance
OTPD	On Track Plant Depot
S&T	Signal & Telegraph
TCD	Train Care Depot
TMD	Traction Maintenance Depot
T&RSD	Traction and Rolling Stock Depot
T&RSMD	Traction and Rolling Stock Maintenance Depot

DEPOTS, STABLING POINTS AND ALLOCATIONS

AD	Ashford OTPD, Ashford, Kent	TR 021415
AFG	Arlington Fleet Group Ltd, Eastleigh Works, Eastleigh, Hampshire	SU 457185
AL	Aylesbury TMD, Buckinghamshire	SU 818134
AN	Allerton T&RSMD, Liverpool, Merseyside	SJ 414849
AS	Asfordby Test Centre, Melton Mowbray, Leicestershire	SK 726205
BA	Basingstoke MDU, Basingstoke, Hampshire	SU 640526
BC	Bedford Cauldwell LMD, Bedford, Bedfordshire	TL 044486
BH	Barrow Hill Depot, Staveley, Derbyshire	SK 414755
BM	Bournemouth EMUD, Bournemouth, Dorset	SZ 066917
BN	Bounds Green T&RSMD, Greater London	TQ 301907
CE	Crewe Electric Depot, Crewe, Cheshire	SJ 694556
CRS	Crossrail, Plumstead, Greater London	TQ 456789
CTRL	Network Rail (CTRL), Singlewell IMD, Gravesend, Kent	TQ 659702
DC	Derby MDU, Chaddesden CE Sidings, Derby, Derbyshire	SK 369359
DP	Darlington Park Lane, Darlington, Co. Durham	NZ 294138
DT	Railway Technical Centre, Derby, Derbyshire	SK 365350
DY	Derby Etches Park T&RSMD, Derby, Derbyshire	SK 368349
EA	Eastleigh Area S&T Works, Hampshire	SU 458192
EC	Edinburgh Craigentinny T&RSMD, City of Edinburgh	NT 292741
EP	Eastleigh South West Area Plant Depot, Hampshire	SU 458194
FL	Leeds Midland Road Depot, Leeds, West Yorkshire	SE 312311
GW	Glasgow Shields T&RSMD, Glasgow	NS 569638
HE	Hornsey T&RSMD, Hornsey, Greater London	TQ 313887
HB	Halkirk Ballast Tip, Highland	ND 126581
HG	Hither Green OTPD, Greater London	TQ 393743
HS	Horsham MDU, Horsham, West Sussex	TQ 179311
IS	Inverness T&RSMD, Inverness, Highland	NH 668457
KE	Kirkdale Depot, Liverpool, Merseyside	SJ 348942
KN	Kings Norton OTPD, Birmingham, West Midlands	SP 040794
LA	Laira T&RSMD, Laira, Plymouth, Devon	SX 504556
ML	Motherwell C&W Depot, North Lanarkshire	NS 749579
NC	Norwich Crown Point T&RSMD, Norwich, Norfolk	TG 246078
NG	New Cross Gate Depot, Greater London	TQ 395775, 395778
NL	Neville Hill T&RSMD, Leeds, West Yorkshire	SE 331331
OO	Old Oak Common T&RSMD, Greater London	SU 217824
PW	Paddock Wood MDU, Paddock Wood, Kent	TQ 673453
RE	Ramsgate T&RSMD, Kent	TR 372657
RF	Retford MDU, Retford, Nottinghamshire	SK 701802
RG	Reading TCD, Reading, Berkshire	SU 701741
RM	Romford MDU, Romford, Essex	TQ 499879
SG	Slade Green EMUD, Erith, Greater London	TQ 527759
SP	Slateford OTPD, Edinburgh	NT 228714
SPM	St Philips Marsh, T&RSMD, Bristol	ST 610721
SU	Selhurst TMD, Greater London	TQ 333678
SW	Swindon, GWML Electrification Depot, Wiltshire	SU 159856
TB	Three Bridges OTPD, West Sussex	TQ 287364
TM	Temple Mills Depot, Leyton, Greater London	TQ 372864, 374862
TO	Toton T&RSMD, Sandiacre, Nottinghamshire	SK 484354
TS	Tyseley T&RSMD, Birmingham, West Midlands	SP 105840
WD	Wimbledon T&RSMD, Greater London	TQ 255723
WF	Wakefield MDU, Wakefield Kirkgate, West Yorkshire	SE 340204
WG	Wigan Springs Branch OTPD, Greater Manchester	SD 593038

WK	Woking OTPD, Surrey	SU 998582
WN	Willesden EMD, Willesden, Greater London	TQ 221828
YH	York Holgate Depot, North Yorkshire	SE 585515

LOCOMOTIVES

TEST TRAIN LOCOMOTIVES

43013	MARK CARNE C.B.E.	Bo-BoDE		Crewe		1976	DT	a
43014	THE RAILWAY OBSERVER	Bo-BoDE		Crewe		1976	DT	a
43062	JOHN ARMITT	Bo-BoDE		Crewe		1977	DT	a
47714	(D1955 47511)	Co-CoDE		BT	617	1966	AS	b
73138	(E6045)	Bo-BoDE/RE		(EE	3717	1966		
				(EEV	E377	1966	DT	OOU
73951	(E6010 73104)	Bo-BoDE/RE		(EE	3572	1965		
	MALCOLM BINDED			(EEV	E342	1965		
			reb	RVEL		2014	DT	
73952	(E6019 73113 73211)	Bo-BoDE/RE		(EE	3581	1966		
	JANICE KONG			(EEV	E351	1966		
			reb	RVEL		2014	DT	
97301	(D6800 37100)	Co-CoDE		(EE	3229	1962		
				(VF	D754	1962	BH	
97302	(D6870 37170)	Co-CoDE		(EE	3348	1963		
				(EEV	D834	1963	BH	
97303	(D6878 37178)	Co-CoDE		(EE	3356	1963		
				(EEV	D842	1963	BH	
97304	(D6917 37217) JOHN TILLEY	Co-CoDE		(EE	3395	1963		
				(EEV	D861	1963	BH	

a Network Rail Measurement Train
b property of Harry Needle Railroad Co Ltd, Derbyshire

DEPOT SHUNTING LOCOMOTIVES

(D3532)	08417		0-6-0DE	Derby		1958	DT	
(D3556)	08441		0-6-0DE	Derby		1958	BN	a
(D3587)	08472		0-6-0DE	Crewe		1958	EC	b
(D3595)	08480		0-6-0DE	Hor		1958	NC	a
D3671	09007		0-6-0DE	Dar		1959	WN	
(D3685)	08523	H 061	0-6-0DE	Don		1958	IS	c
(D3747)	08580		0-6-0DE	Crewe		1959	BN	a
(D3763)	08596		0-6-0DE	Derby		1959	EC	b
(D3815)	08648		0-6-0DE	Hor		1959	IS	c
(D3837)	08670		0-6-0DE	Crewe		1960	BN	a
(D3850)	08683		0-6-0DE	Hor		1959	NC	a
(D4002)	08834		0-6-0DE	Derby		1960	AN	d
(D4122)	08892		0-6-0DE	Hor		1962	AS	d
(D4178)	08948		0-6-0DE	Dar		1962	TM	
(D4186)	08956		0-6-0DE	Dar		1962	AS	
(E6040)	73133		Bo-BoDE/RE	(EE	3712	1966		
				(EEV	E372	1966	BM	e

(01509)	LESLEY	0-6-0DH		RH	468043	1963	AL	
(01581)	TM 4150 MAGNUM "PLANT No.315"	4wDM	R/R	Trackmobile	LGN971310798	1998	YH	
–		4wBE	R/R	Zephir	2254	2009	SU	
–		4wBE		Zephir	2276	2009	HE	
–		4wBE	R/R	Zephir	2412	2012	YH	
–		4wBE	R/R	Zephir	2484	2013	BC	
–		4wBE	R/R	Zephir	2563	2015	CE	
–		4wBE	R/R	Zephir	2562	2015	TO	
DR 73912	LYNX	4w-4wDH		P&T	2884	1999	SW	f

a property of Railway Support Services Ltd, Wishaw, Warwickshire
b property of Wabtec Rail Ltd, Doncaster, South Yorkshire
c property of British American Railway Services Ltd, Stanhope, Co. Durham
d property of Harry Needle Railroad Co Ltd, Derbyshire
e property of Cambrian Transport Ltd, Barry, South Wales
f converted tamper

WHEEL LATHE & BOGIE SHUNTING LOCOMOTIVES

NP B2 WP C24		4wBE		BEMO	12162	2003	TM	OOU
	–	4wDE		_(Express (Hegenscheidt	ES402 101470	2009 2009	DY	
	–	4wBE		_(Express (Hegenscheidt	 101896	2016 2016	TO	
	–	4wBE		Express	ES407	2012	FL	
	–	4wBE		Niteq	B155	2000	SU	
	MUFFIN	4wBE		Niteq	B188	2002	SPM	
	–	4wBE		Niteq	B200	2005	OO	
	–	4wBE		Niteq	B226	2006	KE	
	–	4wBE		Niteq	B244	2007	NG	
	–	4wBE		Niteq	B245	2006	BN	
	–	4wBE		Niteq	B248	2007	HE	
	–	4wBE		Niteq	B281	2009	NG	
TM-LT-02		4wBE	R/R	Niteq	B334	2012	TM	
	–	4wBE		Niteq		2014	NG	
TM-LT-01		6wBE		Scul		2007	TM	OOU
	–	4wBE		SET		2009	SPM	
	–	4wBE		SET			HE	
	–	4wBE		Windhoff	101005675/10	2008	EC	
	–	4wBE		Windhoff	101005675/20	2008	GW	
	–	4wBE		Windhoff	101005675/30	2008	NL	
	–	4wBE		Windhoff	101005675/40	2008		
	–	4wBE		Windhoff	101005675/50	2008	LA	
	–	4wBE		Windhoff	101005675/60	2008	AN	
	–	4wBE		Windhoff	101007142	2009	TS	
	–	4wBE/DE	R/R	Zephir	2111	2007	RE	
	–	4wBE	R/R	Zephir	2455	2013	WD	
	–	4wBE	R/R	Zephir	2535	2014	RG	
	–	4wBE		Zephir	2591	2015	WD	
	–	4wBE	R/R	Zephir	2602	2016	SG	

MULTIPLE UNITS

CONVERTED ELECTRIC MULTIPLE UNIT STOCK

64664	LIWET	4w-4wRER	York(BRE)		1979	AFG	a
64707	LABEZERIN	4w-4wRER	York(BRE)		1979	AFG	a
68501	(61281)	4w-4wRER	Afd/Elh		1959	AFG	a
68504	(61286)	4w-4wRER	Afd/Elh		1959	AFG	a
68505	(61299)	4w-4wRER	Afd/Elh		1959	AFG	

a converted to EMU translator vehicle

RESEARCH AND DEVELOPMENT

These units can be seen at work throughout the rail network.

901 001	DB 999602 (S62483)	4w-4wRER	York(BRE)		1974	RE	a	
950 001	DB 999600	4w-4wDHR	York(BRE)		1987	DT		
950 001	DB 999601	4w-4wDHR	York(BRE)		1987	DT		
999 605	(S62482)	4w-4wRER	York(BRE)		1974	SU	a	
999 606	(62356)	4w-4wRER	York(BRE)		1971	DT	a	
999 800	RICHARD SPOORS	4w-4wDER	Plasser	152	2004		b	
999 801		4w-4wDER	Plasser	153	2004		b	
313 121	62549	4w-4wRER/WER	York(BRE)		1976	AFG	OOU	c
313 121	62613	4w-4wRER/WER	York(BRE)		1976	AFG	OOU	c
(1303)	62287	4w-4wRER	York(BRE)		1970	DT		

a ultrasonic test units
b Survey car.
c European Train Control System test unit.

PLANT

GENERAL PURPOSE MAINTENANCE

Depot allocations show maintenance bases for these vehicles, however many can be found away from their home base undertaking duties.

DR 76901	99709 131001-8	111 BRUNEL	4w-4wDHR	Windhoff	2013	SW	a
DR 76903	99709 131003-4	113	4w-4wDHR	Windhoff	2013	SW	a
DR 76905	99709 131005-9	115	4w-4wDHR	Windhoff	2013	SW	a
DR 76906	99709 131006-7	121	4w-4wDHR	Windhoff	2013	SW	a
DR 76910	99709 131010-9	124	4w-4wDHR	Windhoff	2013	SW	a
DR 76911	99709 131011-7	211	4w-4wDHR	Windhoff	2014	SW	a
DR 76913	99709 131013-2	213a	4w-4wDHR	Windhoff	2014	SW	a
DR 76914	99709 131014-1	221	4w-4wDHR	Windhoff	2014	SW	a
DR 76915	99709 131015-8	222	4w-4wDHR	Windhoff	2014	SW	a
DR 76918	99709 131018-2	224	4w-4wDHR	Windhoff	2014	SW	a
DR 76920	99709 131020-8	232	4w-4wDHR	Windhoff	2014	SW	a
DR 76921	99709 131021-6	233	4w-4wDHR	Windhoff	2014	SW	a
DR 76922	99709 131022-4	31	4w-4wDHR	Windhoff	2014	SW	a
DR 76923	99709 131023-2	32 GAVIN ROBERTS	4w-4wDHR	Windhoff	2014	SW	a

DR 97001	DU 94 001 URS	4w-4wDHR	Eiv de Brive	001URS	2003	CTRL
DR 97011		4w-4wDHR	Windhoff	2625	2004	CTRL
DR 97012	GEOFF BELL	4w-4wDHR	Windhoff	2626	2004	CTRL
DR 97013		4w-4wDHR	Windhoff	2627	2004	CTRL
DR 97014		4w-4wDHR	Windhoff	2628	2004	CTRL
DR 97501	99709 481001-4	4w-4wDHR	Robel	69-40-0004	2015	DP b
DR 97502	99709 481002-2	4w-4wDHR	Robel	69-40-0005	2015	PW b
DR 97503	99709 481003-0	4w-4wDHR	Robel	69-40-0006	2015	DC b
DR 97504	99709 481004-8	4w-4wDHR	Robel	69-40-0007	2015	BA b
DR 97505	99709 481005-6	4w-4wDHR	Robel	69-40-0008	2015	RF b
DR 97506	99709 481006-4	4w-4wDHR	Robel	69-40-0009	2016	RM b
DR 97507	99709 481007-2	4w-4wDHR	Robel	69-40-0010	2016	WF b
DR 97508	99709 481008-0	4w-4wDHR	Robel	69-40-0011	2016	HS b
DR 97509	99709 481009-7	4w-4wDHR	Robel	54.44.0001	2018	CRS b
DR 97510	99709 481010-5	4w-4wDHR	Robel	54.44.0002	2018	CRS b
DR 97509	99709 481011-3	4w-4wDHR	Robel	54.44.0003	2018	CRS b
DR 97509	99709 481012-1	4w-4wDHR	Robel	54.44.0004	2018	CRS b
DR 98001		4w-4wDHR	Windhoff	2516	2000	SW
DR 98002		4w-4wDHR	Windhoff	2517	2000	YH
DR 98003	ANTHONY WRIGHTON 1944-2011	4w-4wDHR	Windhoff	2506	2000	
DR 98004	PHILIP CATTRELL 1961-2011	4w-4wDHR	Windhoff	2507	2000	
DR 98005		4w-4wDHR	Windhoff	2518	2000	YH
DR 98006	JASON MCDONNELL 1970-2016	4w-4wDHR	Windhoff	2519	2000	SW
DR 98007		4w-4wDHR	Windhoff	2520	2000	SW
DR 98008		4w-4wDHR	Windhoff	2521	2000	
DR 98009	MELVYN SMITH 1953-2011	4w-4wDHR	Windhoff	2508	2000	
DR 98010	BENJAMIN GAUTREY 1992-2011	4w-4wDHR	Windhoff	2509	2000	
DR 98011		4w-4wDHR	Windhoff	2510	2000	YH
DR 98012	TERRANCE HAND 1962-2016	4w-4wDHR	Windhoff	2511	2000	YH
DR 98013	DAVID WOOD	4w-4wDHR	Windhoff	2512	2000	YH
DR 98014	WAYNE IMLACH 1955-2015	4w-4wDHR	Windhoff	2513	2000	YH
DR 98215		4wDMR	Plasser	53192A	1988	HG
DR 98216		4wDMR	Plasser	53193A	1988	WK
DR 98217		4wDMR	Plasser	53194A	1988	HG
DR 98218		4wDMR	Plasser	53195A	1988	HG
DR 98219		4wDMR	Plasser	53196A	1988	EP
DR 98220		4wDMR	Plasser	53197A	1988	EA
DR 98305		4w-4wDHR	Geismar	799	1997	AFG OOU
DR 98306		4w-4wDHR	Geismar	800	1997	AFG OOU
DR 98901		4w-4wDHR	Windhoff	2492	1998	
DR 98902		4w-4wDHR	Windhoff	2494	1998	
DR 98903		4w-4wDHR	Windhoff	2522	1999	WG
DR 98904		4w-4wDHR	Windhoff	2524	1999	WG
DR 98905		4w-4wDHR	Windhoff	2526	1999	YH
DR 98906		4w-4wDHR	Windhoff	2528	1999	SP
DR 98907		4w-4wDHR	Windhoff	2530	1999	
DR 98908		4w-4wDHR	Windhoff	2532	1999	KN
DR 98909		4w-4wDHR	Windhoff	2534	1999	KN
DR 98910		4w-4wDHR	Windhoff	2536	1999	ML

DR 98911		4w-4wDHR	Windhoff	2538	1999	ML	
DR 98912		4w-4wDHR	Windhoff	2540	2000	WG	
DR 98913		4w-4wDHR	Windhoff	2542	2000		
DR 98914	DICK PRESTON	4w-4wDHR	Windhoff	2544	2000		
DR 98915	NIGEL CUMMINS	4w-4wDHR	Windhoff	2546	2000	ML	
DR 98916		4w-4wDHR	Windhoff	2548	2000	ML	
DR 98917		4w-4wDHR	Windhoff	2550	2000		
DR 98918		4w-4wDHR	Windhoff	2552	2000	DT	
DR 98919		4w-4wDHR	Windhoff	2554	2000	EP	
DR 98920		4w-4wDHR	Windhoff	2556	2000	WK	
DR 98921		4w-4wDHR	Windhoff	2558	2000	EP	
DR 98922		4w-4wDHR	Windhoff	2560	2000	WK	
DR 98923	CHRIS LEMON	4w-4wDHR	Windhoff	2562	2000	WK	
DR 98924		4w-4wDHR	Windhoff	2564	2000	EP	
DR 98925		4w-4wDHR	Windhoff	2566	2000	TB	
DR 98926	JOHN DENYER	4w-4wDHR	Windhoff	2579	2001	EP	
DR 98976	JOHN DENYER	4w-4wDHR	Windhoff	2580	2001	EP	
DR 98927		4w-4wDHR	Windhoff	2581	2001	WK	
DR 98977		4w-4wDHR	Windhoff	2582	2001	WK	
DR 98928		4w-4wDHR	Windhoff	2583	2001	AD	
DR 98978		4w-4wDHR	Windhoff	2584	2001	AD	
DR 98929		4w-4wDHR	Windhoff	2585	2001	AD	
DR 98979		4w-4wDHR	Windhoff	2586	2001	AD	
DR 98930		4w-4wDHR	Windhoff	2587	2001	EA	
DR 98980		4w-4wDHR	Windhoff	2588	2001	EA	
DR 98931		4w-4wDHR	Windhoff	2589	2001	EA	
DR 98981		4w-4wDHR	Windhoff	2590	2001	EA	
DR 98932		4w-4wDHR	Windhoff	2591	2001	EA	
DR 98982		4w-4wDHR	Windhoff	2592	2001	EA	
APV 250	99709 231007-4	4wDH	SVI	704583	2007		c
RT 250	99709 231001-7	4wDH	SVI	704582	2007		c
–		2w-2PMR	Wkm			HB	DsmT

a Windhoff GWML Electrification factory train
b Mobile Maintenance Train power car
c ABC electrification train power car

SECTION 8
UNKNOWN LOCATIONS

The following locomotives may be found at locations which are not recorded in IRS records. Some were previously recorded as 'Unknown Owner, Unknown Location' within the main text of the EL Handbook. Where known, the last known county or country of the locomotive is listed at the end of each entry.

Gauge 4ft 8½in

No.	Name	Type		Builder	Works No.	Year	Notes / Location
42859		2-6-0	OC	Crewe	5981	1930	Dsm Lincs
	MET	0-4-0ST	OC	HL	2800	1909	South Wales
1928	ALAN GLADDEN	2-6-4T	OC	Nohab	2229	1953	Leicestershire
–		0-4-0ST	OC	WB	2702	1943	North Yorks
BARKING POWER	DUDLEY	4wDM		FH	(3294	1948?)	Suffolk
4		0-4-0DM		JF	22889	1939	Hampshire
18242	ROF CHORLEY 4	0-4-0DH		JF	4220022	1962	Warwickshire
97701		4wDE		Matisa	2655	1975	
			reb	Kilmarnock		1986	East Sussex
7	"SWANSEA JACK"	4wDM		RH	393302	1955	Mid-Wales
11348/C	W.L. No.1	4wDH		RR	10268	1967	Shropshire
(54256	14256)	2w-2-2-2wRER		BRCW		1939	Essex
A456 NWX	L84	4wDM	R/R	_(Unimog	101335	1983	
				(Zweiweg	1035	1983	North Yorkshire
(DE 320467	DB 965049)	2w-2PMR		Wkm	7564	1956	West Yorkshire
68/020		2w-2PMR		Wkm	7600	1957	DsmT N. Yorks
PSL 193	"PLIMSOLL"	4wPMR	R/R	Landrover		1957	+

\+ road vehicle with exchangeable bolt-on rail wheels

Gauge 3ft 0in

No.	Name	Type		Builder	Works No.	Year	Notes / Location
E6		4wBEF		CE	B1850A	1979	Derbyshire
–		4wDM		JF	3930044	1950	a Gloucs

a has appeared occasionally at Welland Steam Rally, Worcestershire

Gauge 900mm

No.	Name	Type		Builder	Works No.	Year	Notes / Location
–		4wDH		Schöma	6257	2008	London
–		4wDH		Schöma	6258	2008	London
–		4wDH		Schöma	6259	2008	London
–		4wDH		Schöma	6260	2008	London
–		4wDH		Schöma	6261	2008	London

Gauge 2ft 6in

No.	Name	Type		Builder	Works No.	Year	Notes / Location
695		0-6-4T	OC	KS	4408	1928	
666		4-6-2	OC	NB	17111	1906	
AK 16		2-6-2T	OC	WB	2029	1916	
No.3	CONQUEROR	0-6-2T	OC	WB	2192	1922	
9		4wDM		HE	2259	1940	Staffordshire
10		4wDM		HE	2260	1940	Staffordshire
–		4wBEF		CE	B4055	1995	
		a rebuild of		CE	B3772	1991	
L4134		2w-2PM		Wkm	3174	1942	

Gauge 750mm

–		0-6-0WT	OC	Chrz	3326	1954		North Wales

Gauge 700mm

–		0-4-4-0T	4C	OK	3770	1909		

Gauge 2ft 2in

	–	4wDH		HE	8972	1979		Derbyshire
	–	4wDH		HE	8970	1979		Derbyshire

Gauge 2ft 0in

121		2-8-2	OC	AFB	2668	1951		
	SSE 1912	4-4-0	OC	FE	265	1897		
	LISBOA	4-4-0	OC	FE	266	1897		
1		0-8-0T	OC	Hen	15540	1917		North Wales
146		2-8-2	OC	Hen	29587	1951		South Wales
–		0-6-0T	OC	Porter	6465	1920		Australia
–		0-4-0WT	OC	SLM		1944	a	Northants
762		4-6-2	OC	WB	2457	1932		
765		4-6-2	OC	WB	2460	1932		
(T2)	SALLY	4wDH		AK	No.8	1982		Cumbria
–		4wDH	s/o	AK	14	1984		Northants
–		4wDM		AK	26	1988		Cumbria
51	No.646	0-4-0PM		BgC	646	1918		Scotland
1863	166 C.P.HUNTINGTON	4w-2-4wPH	s/o	Chance	79 50166 24	1979		Gloucestershire
–		4wDM		HE	3621	1947		
	YARD No.1076	4wDM		HE	7448	1976		Shropshire
14		4wDH		HE	9081	1984		Shropshire
–		4wDM		MR	5853	1934		Leicestershire
	LIDDEL	4wDM		MR	7188	1937		Cumbria
	GELT	4wDM		MR	8696	1941		Cumbria
	TITCH	4wDM		MR	8729	1941		Devon
MPU 9	149	4wDM		MR	22119	1961	Dsm	Norfolk
87025	L201N	4wDM		MR	22238	1965		Cumbria
87032	L202N	4wDM		MR	40S412	1973	Dsm	
–		4wDM		RH	174542	1935		North Wales
	NEATH ABBEY	4wDH		RH	476106	1964		Shropshire
497760		4wDMF		RH	497760	1963		Worcestershire
X 025		4wDH		Schöma	5574	1998		Staffordshire
026		4wDH		Schöma	5575	1998		Staffordshire
–		4-2wPMR		StanhopeT		2005		West Midlands
No.9		0-6-0DM		WB	3124	1957		
	263 021	4wBE		CE	B0465	1974		
	263 076	4wBE		CE	B3766C	1991		
–		0-4-0BE		WR	P7731	1975		
	rebuilt as	2w-2BE		Ayle		1994		Lancashire
S26	25	0-4-0BE		WR	D6886	1964		Cornwall
–		0-4-0BE		WR	8079	1980		
	rebuilt as	0-4-0DM		BrownGM		1993		Durham

–		0-4-0BE		WR			Cornwall

a   either SLM 3854 or SLM 3855

Gauge 600mm

–		0-4-0WT	OC	Borsig	5913	1908	
6	LA HERRERA	0-6-0T	OC	Sabero		c1937	
RTT/767178		2w-2PM		Wkm	3170	c1943	Derbyshire

Gauge 1ft 11½in

	OGWEN	0-4-0T	OC	AE	2066	1933	Derbyshire
120	NG15 BEDDGELERT	2-8-2	OC	AFB	2667	1951	Surrey

Gauge 1ft 10¾in

	NESTA	0-4-0ST	OC	HE	704	1899	Puerto Rico

Gauge 1ft 6in

	GNR No.1	2-2-2	IC	?		c1863	Wiltshire

Gauge 1ft 3in

	AO-TE-AROA	4-4-0	OC	Herschell			Kent
–		4-6-2	OC	KraussM	8473	1929	Dsm Cumbria
3		0-4-0+0-4-0	4C	LJ		1990	Gloucestershire
–		4-4-0	OC	ThurstonTS	incomplete loco		OOU Berkshire
4	LENKA	4-4wDHR		Longleat		1984	

Gauge monorail

–		2a-2DH		AK	M001	1988	Wiltshire

SECTION 9
LOCOMOTIVE INDEX

APPENDIX 1

With the privatisation of the national railway network, a number of shunting locomotives were added to the private sector of the railway industry. These locomotives, owned by Train Operating Companies (TOC's) or Freight Operating Companies (FOC's) can often be found based at Industrial locations or hired to industrial operators. The list below comprises of such locomotives, which are not included in the 17EL index. Mainline locomotives used for shunting are excluded.

OPERATOR CODES & ALLOCATIONS

OPERATOR CODES

The following codes are used :

DBC	DB Cargo (UK) Ltd
EMT	East Midland Trains Ltd
FGW	First Great Western Ltd (part of the First Group)
FLR	Freightliner Group Ltd (part of the America Genesee & Wyoming Railway Co)
GBF	GB Railfreight Ltd (trading name of EQT Infrastructure II - part of the Hector Rail Group)
HNR	Harry Needle Railroad Company Ltd
WMT	West Midlands Trains, T/A West Midlands Railway and Northwestern Railway (an Abellio, JREast and Mitsui Joint Venture)
RSS	Railway Supply Services Ltd

ALLOCATIONS

The following codes are used :

AFG	Arlington Fleet Group Ltd, Eastleigh Works, Hampshire	SU 457185
BWM	Biffa Waste Services Ltd, Oldham Road, Manchester	SJ 859999
DAG	Dagenham Yards, Dagenham, Greater London	TQ 475831
DY	Derby Etches Park Depot, Derby, Derbyshire	SK 368349
EH	Eastleigh Depot & Yards, Eastleigh, Hampshire	SU 460200, SU 458180
FX	Felixstowe Freightliner Terminals, Felixstowe, Suffolk	TM 286325, TM 275343
IM	Immingham Depot & Yards, Immingham, Lincolnshire	(TA 198152)
LA	Laira Depot, Plymouth, Devon	SX 504556
LE	Landore Depot, Swansea	SS 658952
LHG	L.H. Group Services, Barton-under-Needwood, Staffordshire	SJ 206185
NL	Neville Hill Depot, Leeds, West Yorkshire	SE 331331
PZ	Penzance Long Rock Depot, Penzance, Cornwall	SW 492312
RG	Reading Traincare Depot, Reading, Berkshire	SU 701741
RVE	Loram UK Ltd, Derby, Derbyshire	SK 366348
SM	Southampton Maritime Depot & Terminals, Hampshire	SU 380129, SU 396127
SO	Soho Depot, Birmingham, West Midlands	SP 039882
SPM	St. Philips Marsh Depot, Bristol	ST 610721
TPK	Trafford Park Terminals, Manchester, Greater Manchester	SJ 800961
TS	Tyseley Depot, Tyseley, Birmingham, West Midlands	SP 105840
WM	Whitemoor Yard, March, Cambridgeshire	TL 415984
WSP	Springs Branch, Wigan, Lancashire	SD 593038

SHUNTING LOCOMOTIVES

(D3520)	08405		0-6-0DE	Derby	1958	RSS	NL	a
(D3525)	08410		0-6-0DE	Derby	1958	FGW	PZ	
(D3598)	08483		0-6-0DE	Hor	1958	FGW	LA	
(D3666)	09002		0-6-0DE	Dar	1959	GBF	WM	
(D3687)	08525	DUNCAN BEDFORD	0-6-0DE	Dar	1959	EMT	NL	
(D3689)	08527		0-6-0DE	Dar	1959	HNR	IM	b
(D3692)	08530		0-6-0DE	Dar	1959	FLR	SM	
(D3693)	08531		0-6-0DE	Dar	1959	FLR	FX	
(D3720)	09009		0-6-0DE	Dar	1959	GBF	BWM	
(D3752)	08585	VICKY	0-6-0DE	Crewe	1959	FLR	SM	
(D3783)	(08616)	TYSELEY 100	0-6-0DE	Derby	1959	WMT	TS	
(D3791)	08624	PAUL RAMSEY - RAMBO	0-6-0DE	Derby	1959	FLR	FX	
(D3799)	08632		0-6-0DE	Derby	1959	RSS	NL	a
(D3808)	08641	FRED	0-6-0DE	Hor	1959	FGW	LA	
(D3811)	08644	LAIRA DIESEL DEPOT 50 YEARS 1962 - 2012	0-6-0DE	Hor	1959	FGW	LA	
(D3812)	08645	MIKE BAGGOTT	0-6-0DE	Hor	1959	FGW	LE	
(D3830)	08663	ST. SILAS	0-6-0DE	Hor	1959	FGW	SPM	
(D3857)	08690	DAVID THIRKILL	0-6-0DE	Hor	1959	EMT	NL	OOU
(D3858)	08691	TERRI	0-6-0DE	Hor	1959	FLR	LHG	
(D3870)	08703		0-6-0DE	Hor	1960	DBC	WSP	
(D3903)	08735		0-6-0DE	Crewe	1960	DBC	EH	OOU
(D3953)	08785		0-6-0DE	Derby	1960	FLR	TPK	
(D3963)	08795		0-6-0DE	Derby	1960	FGW	LE	
(D3973)	08805		0-6-0DE	Derby	1960	WMT	SO	
(D3990)	08822	DAVE MILLS	0-6-0DE	Derby	1960	FGW	SPM	
(D4004)	08836		0-6-0DE	Derby	1960	FGW	RG	
(D4129)	08899	MIDLAND COUNTIES RAILWAY 175 1839-2014	0-6-0DE	Hor	1962	EMT	DY	
(D4138)	08908	IVAN STEPHENSON	0-6-0DE	Hor	1962	EMT	NL	OOU
(D4155)	08925		0-6-0DE	Hor	1962	GBF	WM	
D4164	08934		0-6-0DE	Dar	1962	GBF	DAG	
(D4180)	08950	DAVID LIGHTFOOT	0-6-0DE	Dar	1962	EMT	NL	

a on hire to East Midlands Trains Ltd
b on hire to GB Railfreight Ltd

INDUSTRIAL RAILWAY SOCIETY

This book has been produced by the INDUSTRIAL RAILWAY SOCIETY
which was founded in 1949 as the
Birmingham Locomotive Club, Industrial Locomotive Information Section

The Society caters for those interested in privately owned locomotives and railways. Members receive the INDUSTRIAL RAILWAY RECORD in addition to regular bulletins containing topical news and update amendments to Society-produced Handbooks. Access is available to a well stocked library and visits are arranged to industrial railways, etc.

If you are interested in industrial railways why not join the Society now.

Further details can be obtained by sending two first class stamps to
David Kitching, 30 Carleton Road, Poynton, Cheshire, SK12 1TL (admin@IRSociety.co.uk)

or visit the Society website at **www.irsociety.co.uk**

INDUSTRIAL RAILWAY RECORD

This profusely illustrated magazine contains articles of lasting interest concerning a wide variety of industrial locomotives and railways of all gauges, at home and abroad. Accurate line drawings of locomotives and rolling stock, together with carefully prepared maps, are a regular feature.

The INDUSTRIAL RAILWAY RECORD is available to non-members by direct subscription.
Enquiries regarding subscriptions and back numbers should be sent with two first class stamps to :
David Kitching, 30 Carleton Road, Poynton, Cheshire, SK12 1TL (admin@IRSociety.co.uk)

PUBLICATIONS by the IRS

Apart from "Handbook EL", the Society produces a range of volumes titled "Industrial Railways & Locomotives" which record all known past and present locomotives in particular areas or countries, with descriptions of the railways and the industries served. These Handbooks include a good range of maps, and a selection of photographs. A selection of titles currently available (hardback binding) are as follows:-

Cumberland - 464 pages 212 photographs (colour and b & w) and 47 maps - £35.00.
Kent - 458 pages with 149 photographs (colour and b& w) and 50 maps - £35.00.
Sussex & Surrey - 286 pages with 97 photographs (colour and b & w) and 38 maps - £27.95.
South Western England (Cornwall, Devon & Somerset) - 383 pages with 164 photographs (colour and b & w) and 59 Maps - £29.95.
Private Railways of County Durham - 429 pages, 314 photographs, 86 maps - £29.95.
Spain - 518 pages, 168 photographs (colour and b & w), and 54 maps - £28.00.
India and South Asia - 556 pages, 185 photos (colour and b & w) and 44 maps - £29.95

Dyfed & Powys, Nottinghamshire, North Staffordshire, South Staffordshire, Essex, Lincolnshire and Rutland, Hertfordshire and Middlesex, Buckinghamshire, Bedfordshire and Northamptonshire are also available, please enquire for details, availability and prices.

Titles available covering industrial railways history, locomotive builders etc :

Kerr, Stuart's Internal Combustion Locomotives by A.C. Baker.
- 145 pages with 130 photographs and drawings. Hardbound. £30.00.

The Railway Foundry, Leeds, Hudswell Clarke & Co Ltd – The Diesel Era by R.N. Redman.
- 204 pages, 210 photographs. Hardbbound £29.95.

The Railways Products of Baguley Drewry Ltd of Burton on Trent, Civil & Etherington.
- This 372 page book is the definitive history of this important locomotive builder. 441 photographs. Hardbound with dust jacket and printed on high quality paper - £29.99.

Skinningrove Iron and Steel Works (2018 reprint) C. Shepherd – 100 pages, 205 photographs
- (colour and b & w), 23 maps.Hardbound - £24.95.

Orders to IRS Sales, 24 Dulverton Road, Melton Mowbray, LE13 0SF
Add 20% for postage and packing

HUNSLET LOCOMOTIVES

for industrial use

Our standard range of diesel locomotives incorporating Hunslet patent hydraulic transmission, includes a type ideally suited to your application. The 153 h.p. 0-4-0 type 2-speed hydraulic locomotive is just one example of what we can offer in the medium power range. It is a QUALITY LOCOMOTIVE—the class of locomotive which is designed and built to give a life-time of trouble-free service.

THE HUNSLET ENGINE CO. LTD. LEEDS 10 LONDON OFFICE: LOCOMOTIVE HOUSE, BUCKINGHAM GATE, WESTMINSTER, S.W.1

[1] BON ACCORD (AB 807 of 1897) at Beamish Museum, Co. Durham on 25th March 2017, visiting from the Royal Deeside Railway, Milton of Crathes, Aberdeenshire (Michael Denholm)

[2] Ed Murray & Sons Ltd, hire locomotive EEV 4003 of 1971, under test at the Chasewater Railway, Brownhills West, Staffordshire on 12th July 2018 (Phil Civil)

[3] PWR AO.296V.03 of 1992 is seen inside Honister Slate Mine, Borrowdale, Cumbria during an IRS visit on 17th May 2018 (Alex Betteney)

[4] No.2 (HL 2859 of 1911) at Tanfield Railway, Marley Hill, Co. Durham on 7th September 2013 (Michael Denholm)

[5] SAMSON (Beamish BM2 of 2014) new build replica of a Lewin locomotive of 1874 at Beamish Museum, Co. Durham on 7th April 2016 (Michael Denholm)

[6] ELIZABETH (TH 276V of 1977) crosses Forty Foot Road, Middlesbrough, heading onto A.V. Dawson Ltd's Ayrton Store site on 21st July 2018 with an enthusiast railtour (Cliff Shepherd)

[7] 10222 (RR 10222 of 1965) at the Llanelli & Mynydd Mawr Railway, Cynheidre, Carmarthenshire on 27th August 2018 (Derek Horton)

[8] 1 (CE B4618.1 of 2016) at Crossrail, Plumstead, Greater London during construction work on 11th June 2017 (IRS Collection)

[9] Niteq B200 of 2005 at Old Oak Common Depot, Greater London on 2nd September 2017 (Ed Copcutt)

[10] No.20 (HC DM1120 of 1958) at the Lancashire Mining Museum, Astley Green, Greater Manchester on 19th May 2018 (Chris Weeks)

[11] SIR PETER GADSDEN (AK 84 of 2008) at Ironbridge Gorge, Blists Hill, Telford, Shropshire on 15th July 2017 (Chris Weeks)

[12] Ed Murray & Sons Ltd, Hire locomotives 261 (GECT 5430 of 1977) and 268 (GECT 5465 of 1977) at ICL UK (Cleveland) Ltd, Tees Dock, Teesside on 2nd August 2018 (Jim Stancliffe)

[13] SPONDON (Spondon of 1928) at the Derbyshire Dales Narrow Gauge Railway, Rowsley South, Derbyshire on 12th May 2018 (Brian Cuttell)

[14] No.9 RICHBORO (HC 1243 of 1917) at the Aln Valley Railway, Alnwick, Northumberland on 6th May 2018 (Dennis Graham)

[15] 2758 CEFN COED (WB 2758 of 1944) at Cefn Coed Colliery Museum, Neath on 10th June 2018 (Dennis Graham)

[16] CHUQUIANTA (Coulliet 810/Decauville 36 of 1885) at the private Richmond Light Railway, Kent on the open day of 13th August 2016 (Dennis Graham)

[17] Schöma 6844 of 2015 (a rebuild of narrow gauge Schöma 6321 of 2008) at Crossrail, Plumstead, Greater London on 11th June 2017 (Dennis Graham)

[18] D2112 (03112 Sdn of 1960) at the Rother Valley Railway, Robertsbridge, East Sussex, on 26th July 2018 (Dennis Graham)

[19] Muir-Hill 110 of 1925 at the private Silverleaf Poplar Railway, Lincolnshire on 14th October 2017 (Dennis Graham)

[20] 8.708 (MaK 160.0008 of 1996) and 8.716 (MaK 160.0016 of 1996) at the former SSI UK, Redcar Blast Furnace, Teesside on 10th July 2015. These locos are now at British Steel Scunthorpe. (Jim Stancliffe)

[21] RH 210479 of 1942 at the Somerset & Dorset Railway Museum Trust, Washford, Somerset on 10th June 2016 (Peter Nicholson)

[22] Metroplitan Railway 1 (Neasden 3 of 1898), visiting from the Buckinghamshire Railway Centre, with L92 (5786 Sdn of 1930, visiting from the South Devon Railway, at the Avon Valley Railway, Bitton, Gloucestershire on 18th October 2015 (Peter Nicholson)

[23] HENBURY (P 1940 of 1937) and visiting locomotive TEDDY (P 2012) at Bristol Harbour Railway, on 27th June 2015 (Peter Nicholson)

[24] 4 (Unimog 240933 of 2016 with King/Zagro rail conversion 4345 of 2016) at Crossrail, Plumstead, Greater London, during construction work on 11th June 2017

(IRS Collection)

[25] MYFANWY (RSHD 8366 / WB 3211 of 1962) at the Chasewater Railway, Brownhills West, Staffordshire on 28th May 2018 (Alex Betteney)

[26] VIW 4441 of 1943 with fictitious NCB No.35 number visiting Tanfield Railway, Marley Hill, Co. Durham, from Tenterden Railway, Kent, on 10th June 2016 (Michael Denholm)

[27] SAO DOMINGOS (OK 11784 of 1928) at The Great Bush Railway, near Uckfield, East Sussex on 20th August 2017 (Dennis Graham)

[28] Zephir 2631 of 2016 shunts an Bi-modal IEP unit onto the test track at Hitachi, Newton Aycliffe, Co. Durham on 13th May 2016 (Alex Betteney)

[29] LM 99 (RH 375336 of 1954, rebuilt BnM(Be) in 1988) at Bord na Móna, Kilberry Works, Co. Kildare on 6th June 2018 (Andrew Waldron)

[30] LM 163 (RH 402178 of 1956) at Bord na Móna, Ummeras Works, Co. Kildare on 7th June 2018. (Andrew Waldron)

[31] LM 409 (BnM(Bl) of 2001, rebuilt BnM(Dg) of 2017) passes Bord na Móna, Blackwater Works, Co. Offaly with a train of peat for West Offaly Power Station on 25th April 2018 (Dennis Graham)

[32] Zephir 2734 of 2017 awaits its next duty at LUAS Broombridge Depot, Dublin on 30th May 2018 (Andrew Waldron)

SPECIAL OFFER TO PURCHASERS OF THIS BOOK

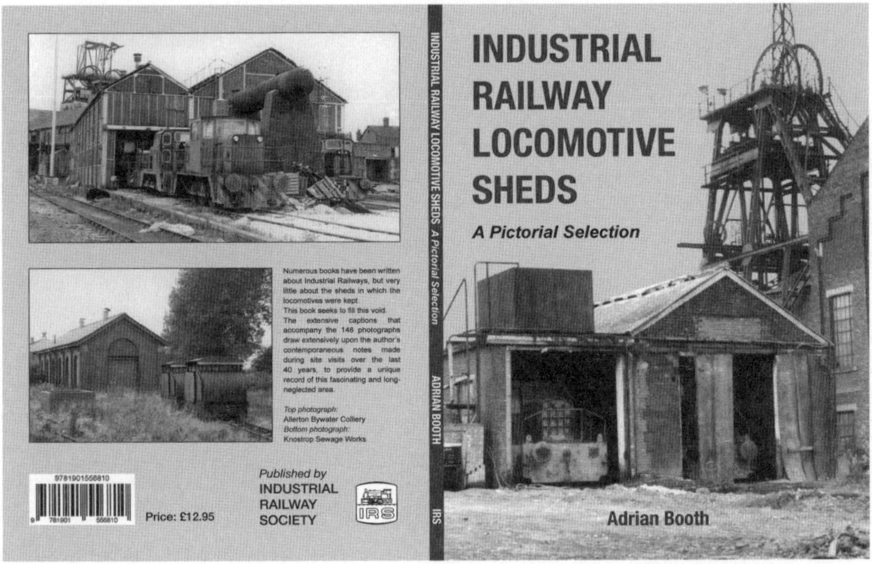

The Society has published three volumes of books showing photographs and extended captions about the sheds where industrial locomotives were kept.

The normal public price for these books is **£12.95** but we are offering purchasers of this book the chance to buy all three for just **£29.90** which is a **saving** of **£8.95** (plus postage and packing).

This offer is available (while stocks last), until **31/12/2019**, through our online shop which accepts Visa, Mastercard, and American Express cards as well as Paypal.

To take advantage of this offer please go to:

www.irsshop.co.uk

And add the books to the basket.

When you checkout the goods enter Voucher Code No. **KU8OG8V1**

and the offer will automatically be applied.